AND DIGITAL

DESIGN

S Edition

Other titles in the EDN Series for Design Engineers

Electromagnetics Explained: A Handbook for Wireless, RF, EMC, and High-Speed Electronics by Ron Schmitt, 0-7506-7403-2, Hardcover, 359 pgs., $34.99

Practical RF Handbook, Third Edition by Ian Hickman
0-7506-5639-8, Paperback, 304 pgs., $39.99

Power Supply Cookbook, Second Edition by Marty Brown
0-7506-7329-X, Paperback, 336 pgs., $39.99

Radio Frequency Transistors, Norman Dye and Helge Granberg
0-7506-7281-1, Paperback, 320 pgs., $49.99

Troubleshooting Analog Circuits by Robert A. Pease
0-7506-9499-8, Paperback, 217 pgs., $34.99

The Art and Science of Analog Circuit Design, edited by Jim Williams
0-7506-7062-2, Paperback, 416 pgs., $34.99

ANALOG AND DIGITAL FILTER DESIGN

Second Edition

STEVE WINDER

 Newnes

An imprint of Butterworth-Heinemann

Amsterdam Boston London New York Oxford Paris San Diego San Francisco
Singapore Sydney Tokyo

Newnes is an imprint of Elsevier Science.

Copyright © 2002, Elsevier Science (USA). All rights reserved.

 Recognizing the importance of preserving what has been written, Elsevier Science prints its books on acid-free paper whenever possible.

 Elsevier Science supports the efforts of American Forests and the Global ReLeaf program in its campaign for the betterment of trees, forests, and our environment.

Library of Congress Cataloging-in-Publication Data
Winder, Steve.
  Analog and digital filter design / Steve Winder.—2nd ed.
    p. cm.
  Rev. ed. of: Filter design. c1997.
  Includes bibliographical references.
  ISBN 0-7506-7547-0 (pbk. : alk.paper)
  1. Electric filters—Design and construction. I. Winder, Steve. Filter design. II. Title.

TK7872.F5 W568 2002
621.3815′324—dc21

                                              2002071430

British Library Cataloguing-in-Publication Data
A catalogue record for this book is available from the British Library.

The publisher offers special discounts on bulk orders of this book.
For information, please contact:

Manager of Special Sales
Elsevier Science
225 Wildwood Avenue
Woburn, MA 01801-2041
Tel: 781-904-2500
Fax: 781-904-2620

For information on all Newnes publications available, contact our World Wide Web home page at: http://www.newnespress.com

10 9 8 7 6 5 4 3 2 1

Printed in the United States of America

CONTENTS

CHAPTER 3   Poles and Zeroes   83

CHAPTER 4   Analog Lowpass Filters   125

CHAPTER 5   Highpass Filters   **147**

CHAPTER 6   Bandpass Filters   **173**

PREFACE

This book is about **analog** and **digital filter design**. The analog sections include both **passive** and **active filter designs**, a subject that has fascinated me for several years. Included in the analog section are filter designs specifically aimed at radio frequency engineers, such as impedance matching networks and quadrature phase all-pass networks. The digital sections include **infinite impulse response (IIR)** and **finite impulse response (FIR) filter design**, which are now quite commonly used with digital signal processors. Infinite impulse response filters are based on analog filter designs.

Detailed circuit theory and mathematical derivations are not included, because this book is intended to be an aid in practical filter design by engineers. The circuit theory and mathematical material that is included is of an introductory nature only. Those who are more academically minded will find much of the information useful as an introduction. A more in-depth study of filter theory can be found in academic books referred to in the bibliography. Equations and supplementary material are included in the Appendix.

Designing filters requires the use of mathematics. Fortunately, it is possible to successfully design filters with very little theoretical and mathematical knowledge. In fact, for passive analog filter design the mathematics can be limited to simple multiplication and division by the use of look-up tables. The design of active analog filters is slightly more difficult, requiring both arithmetic and algebra combined with look-up tables. The equations behind many of the look-up tables are included in the Appendix.

Digital FIR filters perform their function by first passing a digitized signal through a series of discrete delay elements and then multiplying the output of each delay element by a number (or coefficient). The values produced from all the multiplication functions at each clock period are then added together to give an output. Hence digital filter designs do not produce component values. Instead, they produce a series of numbers (coefficients) that are used by the multiplication functions. There are no design tables; the series of coefficients is produced by an algebraic equation, so the designer must be familiar with arithmetic and algebra in order to produce these coefficients.

The principles behind digital filters are based on the relationship between the time and frequency domains. Although digital filters can be designed without knowledge of this relationship, a basic awareness makes the process far more understandable. The relationship between the time and frequency domains can be grasped by performing a practical test: apply a range of signals to both the input of an oscilloscope and the input of a spectrum analyzer, and then compare the instrument displays. More formally, Fourier and Laplace transforms are used to convert between the time and frequency domains. A brief introduction to these is given in chapter 3. Whole books are devoted to the Fourier and Laplace transforms; references are given in the Bibliography.

All the designs described in this book have been either built by myself or simulated using circuit analysis software on a personal computer. As is the case in all filter design books, not every possible design topology is included. However, I have included useful material that is hard to find in other filter design books, such as Inverse Chebyshev filters and filter noise bandwidth. I have researched many filter design books and papers in search of simple design methods to reduce the amount of mathematics required.

Chapters have been arranged in what I think is a logical order. A summary of the chapters in this book follows.

Chapter 1 gives examples of filter applications, to explain why filter design is such an important topic. A description of the limitations for a number of filter types is given; this will help the designer to decide whether to use an active, passive, or digital filter. Basic filter terminology and an overview of the design process are also discussed.

Chapter 2 describes the frequency response characteristics of filters, both ideal and practical. Ideally, filters should not attenuate wanted signals but give infinite attenuation to unwanted signals. This response is known as a brick wall filter: it does not exist, but approximations to it are possible. The four basic responses are described (i.e., flat or rippled passband and smooth or rippled stopband) and show how standard Bessel, Butterworth, Chebyshev, Cauer, and Inverse Chebyshev approximations have one of these responses. Graphs describe the shape of each frequency response.

A very important topic of this chapter is the use of normalized lowpass filters with a 1 rad/s cutoff frequency. Normalized lowpass filters can be used as a basis for any filter design. For example, a normalized lowpass filter can be scaled to design a lowpass filter with any cutoff frequency. Also, with only slightly more difficulty, the normalized design can be translated into highpass, bandpass, and bandstop designs. Tables of component values for some normalized approximations are given. Formulae for deriving these tables are also provided, where applicable.

The subject of Inverse Chebyshev filters are covered in some detail, because information on this topic has been difficult to find. Natural application of Inverse Chebyshev design techniques leads to a stopband beginning at $\omega = 1$. This may be academically correct, but I describe how to obtain a more practical 3dB cutoff point. I also give explicit formulae for finding third-order passive filters, and show a method of finding component values for higher orders.

Chapter 3 provides the foundation for filter design theory. This leads from transfer function equations to pole and zero locations in the s plane. The s plane and its underlying Laplace transform theory are described. This should give the reader a feel for how the filter behaves if it has a certain pole-zero pattern or a certain transfer function. Pole and zero placing formulae and the tables derived from them are given for normalized lowpass filter responses.

Pole and zero locations are important in active filter design. With only knowledge of the normalized lowpass pole and zero locations for a certain transfer function, an active filter can be designed. Pole and zero locations can be scaled or converted for highpass, bandpass, or bandstop designs.

Chapters 4 to 7 describes how to design active or passive lowpass, highpass, bandpass, and bandstop filters to meet most desired specifications. Separate chapters describe each type because the reader is usually interested only in a particular type, for a given application, and will not want to search the book to find the information. Formulae are given for the denormalization of the component values or pole-zero locations that were given in earlier chapters.

Chapter 8 describes the diplexer and its application and performance. Diplexers are passive filters and are used in RF design to split signals from different frequency bands in either a highpass/lowpass or a bandpass/bandstop combination. One of the most common applications is in terminating mixer ports in radio frequency system designs.

Chapter 9 describes the use of phase-shift networks, with examples for flattening the group delay response of Butterworth filters. One application is the Weaver single sideband modulator, which uses a phase-shift network to cancel out the unwanted sideband of an AM radio transmission. A description of the Weaver single sideband modulator are given, both in mathematical terms and with practical applications. This chapter also provides details of how to go about the design of passive and active phase-shift networks.

Chapter 10 is very practical in orientation, describing how different materials and component types can affect the performance of filters. Capacitor dielectric and component lead lengths can be critical for a good filter performance. Details on the construction of inductors using ferrite cores are given, and transformer construction using similar techniques is included. Active filter components are also described (amplifier parameters can have a significant effect), as are measurement techniques.

Chapter 11 describes current software availability, including integrated circuit–specific software. The actual filter design process can be considerably automated. Indeed, I have written a program with Number One Systems Ltd. called *FILTECH*, which designs and simulates filter circuits. I outline how *FILTECH* operates at a systems level. There are also other programs on the market. Some of these only design active filters; they are offered free because they enable users to design filters using certain manufacturers' integrated circuits.

Executable PC programs, capable of designing useful filters, are supplied at www.bh.com/companions/0750675470. This chapter basically serves as a user guide, describing their operation. These programs are far simpler than *FILTECH* and give a netlist compatible with SPICE-like analysis programs.

Chapter 12 describes how transmission lines can be used to filter signals. Quarter-wave lines of either short or open circuit termination can be used to pass or stop certain frequencies. One application of this is to allow a radio carrier signal into a receiver from an antenna while preventing internal radio signals from radiating back to the antenna.

Printed circuit board (PCB) filters are also described. Tracks on a PCB can be transmission lines when the signal frequency is high. The width of a track on a printed circuit board defines its impedance; sections of wider or narrower track become inductive or capacitive. Concatenation of narrow and wide track sections can therefore form an LC (inductor capacitor) filter.

Phase-locked loop filters are usually quite simple, but poor design can cause instability of the loop. Many people avoid designing phase-locked loops for this reason. Chapter 13 provides some examples that may help remove some of this fear.

Chapter 14 provides an introduction to switched capacitor filters. Commercial filter ICs (integrated circuits) are described and plots of some practical examples are given. Problems with this type of filter are described, as are some of the benefits such as being able to make the filter cutoff programmable or adjustable.

Chapter 15 outlines the process of digital filtering. In this chapter I cover the data sampling operation (under-sampling, over-sampling, interpolation, and decimation) and the advantages or problems of each. A brief outline of digital filtering techniques provides some understanding of digital signal processing. Digital signal processors (DSPs) are described, along with the mathematical methods by which they handle data during signal processing.

Chapters 16 and 17 cover digital filtering in a little more depth. Chapter 16 covers Finite Impulse Response (FIR) filters and Chapter 17 covers Infinite Impulse Response (IIR) filters. Equations needed to find multiplier coefficients are included with worked examples.

CHAPTER 1

INTRODUCTION

This chapter gives an introduction to filters and signals, and the terminology used in relation to filters. Experienced engineers may wish to skip this chapter.

Fundamentals

Why Use Filters?

Why are you so interested in filters? This was a question put to me when I was planning this book. It is a very good question. I have been involved with electronic system design for a number of years and have found that the performance of an electronic filter can determine whether the system is successful. Detection of a wanted signal may be impossible if unwanted signals and noise are not removed sufficiently by filtering. Electronic filters allow some signals to pass, but stop others. To be more precise, filters allow some signal frequencies applied at their input terminals to pass through to their output terminals with little or no reduction in signal level.

Analog electronic filters are present in just about every piece of electronic equipment. There are the obvious types of equipment, such as radios, televisions, and stereo systems. Test equipment such as spectrum analyzers and signal generators also need filters. Even where signals are converted into a digital form, using analog-to-digital converters, analog filters are usually needed to prevent aliasing. Computers use filters: to reduce EMI (electro-magnetic interference) emissions from their power lead; to smooth the output of the switched-mode power supply; to limit the video bandwidth of signals going to the display.

What Are Signals?

Before describing filters in detail, it is important to understand the characteristics of signals. A signal can be described in the time domain or in the frequency domain. What does this mean?

The time domain is where an event, such as a change in amplitude, is measured over time. All alternating current (AC) signals vary in amplitude over a certain time period. Some signals are periodic, which means that the same pattern of variation is repeated again and again. Signals are measured and displayed in time domain by an oscilloscope. A line is drawn horizontally across the screen at a steady rate, and the signal amplitude is used to change the vertical position of the line. An increasingly positive going signal forces the line to rise toward the top of the screen, and an increasingly negative going signal forces the line toward the bottom of the screen.

The frequency domain is where the amplitude of a signal is measured relative to its frequency. A spectrum analyzer is used to display the amplitude across a range of frequencies (the spectrum). The simplest type of signal is a pure sinusoid, which is periodic in the time domain and has energy at only one frequency in the frequency spectrum. The frequency is determined by the number of cycles per second and is given the name Hertz (Hz). The frequency can be found by measuring the period of one complete cycle (in seconds) and taking the inverse: frequency = 1/period. Other signals, such as such as human speech, a square wave, or impulsive signals, contain energy at many frequencies. Figure 1.1 shows the relationship between time and frequency domains for a simple sinusoidal signal.

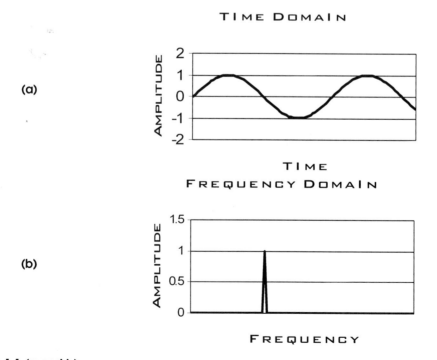

Figure 1.1 (a and b)

Time and Frequency Relationship

Decibels

The amplitude of a signal is measured in volts. The r.m.s. (root means square) voltage of AC signals is used, rather than the peak voltage, because this gives the same power as a DC signal having that voltage. However, because the signal level has to be multiplied by the gain or loss of components (such as filters) in the signal path, decibels are used. This make the mathematics simpler, because once the voltage is expressed in decibel notation, gains can be added and losses can be subtracted.

The number of decibels relative to one volt is expressed as dBV, and is given by the expression $20\log(V)$. That is, measure the voltage (V), take the logarithm of it, and multiply the result by 20. If the voltage level is 0.5 volts, this is expressed as -6 dBV. If this signal is amplified by an amplifier having a gain of 10 ($+20$ dB), the output signal will be $-6 + 20 = +14$ dBV.

Signal power can be expressed in decibels too. The most common unit of power is the milliwatt, and the number of decibels relative to one milliwatt is expressed as dBm. The formula for expressing power (P) in decibels is $10\log(P)$, hence a milliwatt equals 0 dBm. However, the signal is measured in terms of volts and converted to power using $P = V^2/R$, where R is the load resistance. In filter designs the half-power signal level (-3 dB) is often used as a reference point for the filter's passband.

The Transfer Function

Both analog and digital filters can be considered a "black box." Signals are input on one side of the black box and output on the other side. The amplitude of the output signal voltage (or its equivalent digital representation) depends on the filter design and the frequency of the applied input signal. The output voltage can be found mathematically by multiplying the input voltage by the **transfer function**, which is a frequency-dependent equation relating the input and output voltages. The transfer function is illustrated in Figure 1.2.

Figure 1.2

Transfer Function

$F(w) = Vout / Vin$

The relationship between input and output will be a function of frequency w (omega), given in terms of radians per second. Radians/sec are used as the unit of frequency measure because in an analog filter this gives a value for reactive impedance that is directly proportional to the frequency. An inductor that has a value of one Henry has an impedance of $1\,\Omega$ at 1 rad/s.

The transfer function, $F(\omega)$, is frequency dependent. For example, suppose that at $w = 0.5$, $F(\omega)$ is equal to 1 and hence $V_{out} = V_{in}$. Now suppose that at $\omega = 2$, $F(\omega)$ is equal to 0.01, hence V_{out} is $V_{in} \div 100$. In decibels, the gain is $-40\,dB$, since it is $20\log(V_{out}/V_{in})$; since the gain is negative, this can be referred to as a (positive) attenuation, or signal loss, of $40\,dB$. The function $F(\omega)$ is flawed because it assumes that the source and load impedance has no effect.

For the most common filter types, the transfer function is often presented in graphical form. The graph has a number of curves showing signal gain (loss) versus frequency. As the filter design grows more complex, the steepness of the curve increases. This means that a design engineer can determine the simplest filter for a given performance, by comparing one curve with another.

An imaginary "brick wall" lowpass filter, illustrated in Figure 1.3, is ideal in that it has an infinitely steep change in its frequency response at a certain cutoff frequency. It passes all signals below the cutoff frequency with a gain of 1. That is, signals below the cutoff frequency have their amplitude multiplied by 1 (i.e., they are unchanged) as they pass through the filter. Above the cutoff frequency, the filter has a gain of 0. Signals above the cutoff frequency have their amplitude multiplied by 0 (i.e., they are completely blocked) and there is no output. The "brick wall" filter is impossible for reasons that will be described later.

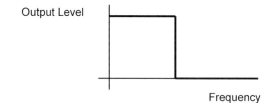

Output Level

Frequency

Figure 1.3

The Ideal "Brick Wall" Filter

Filter Terminology

The range of signal frequencies that are allowed to pass through a filter, with little or no change to the signal level, is called the **passband**. The passband **cutoff frequency** (or cutoff point) is the passband edge where there is a 3 dB reduction in signal amplitude (the half-power point). The range of signal frequencies that are reduced in amplitude by an amount specified in the design, and effectively prevented from passing, is called the **stopband**. In between the passband and the stopband is a range of frequencies called the **skirt** response, where the reduction in signal amplitude (also known as the **attentuation**) changes rapidly. These features are illustrated in Figure 1.4, which gives the frequency response of a lowpass filter.

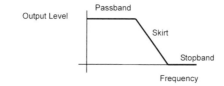

Figure 1.4

Frequency Domain Features of
a Lowpass Filter

Frequency Response

There are four possible frequency domain responses: lowpass, highpass, band-pass, and bandstop. Simplistic graphical representations are given below in Figure 1.5

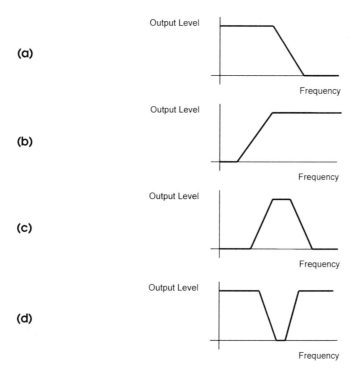

Figure 1.5 (a–d)

Frequency Domain Responses
(a) **Lowpass filters** pass low frequencies. That is, they allow frequencies from DC up to what is known as the cutoff frequency with minimal loss of amplitude.
(b) **Highpass filters** pass high frequencies. They have the opposite function to that of lowpass filters, in that they allow frequencies above the cutoff to pass with minimal loss. They do not pass DC.
(c) **Bandpass filters** pass a band of frequencies between the lower and upper cutoff points. The upper cutoff determines the maximum frequency passed (with minimal loss). The lower cutoff decides the minimum frequency to be passed; DC is blocked.
(d) **Bandstop filters** stop a band of frequencies between the lower and upper cutoff points. They are the opposite of bandpass filters and allow two frequency bands to pass. One band that is passed goes from DC to the lower cutoff frequency. The other band passed covers all frequencies above the upper cutoff point.

The designer must determine the cutoff frequencies, the stopband attenuation, and whether a lowpass, highpass, bandpass, or bandstop filter is required. Sometimes this specification will be supplied by the system designer, but this may be left to the filter designer to decide for him or herself.

Phase Response

Radians/sec are used as the unit of frequency measure because in an analog filter this gives a value for reactive impedance, which is directly proportional to the frequency. Ohms law states that current can be expressed as the ratio of voltage to load resistance. This is true for DC measurements with a purely resistive load. For AC measurements with loads that include reactive elements like capacitors and inductors, the current can be expressed as the ratio of voltage to load impedance. If there is some reactance in the load, the current through the load is not in phase with the voltage across it.

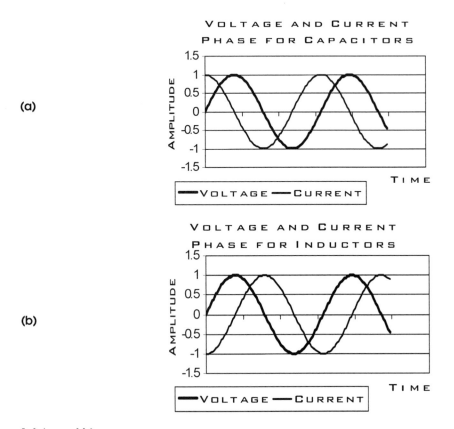

Figure 1.6 (a and b)

Voltage and Current Phase Relationship for Capacitors and Inductors

The power dissipated at a resistive load is the product of voltage and current averaged over one sine wave cycle. This is the r.m.s. voltage times the r.m.s. current. No power is dissipated in a purely reactive load because over one complete sine wave cycle the product of voltage and current is zero. Instead, energy is stored in capacitors and inductors, which is the reason for the phase difference between voltage and current at a reactive load.

Inductors have an impedance given by the expression $X_L = jwL$. Capacitors have an impedance given by the expression $X_C = 1/jwC$, which is equivalent to $X_C = -j/wC$. The symbol "j" indicates a phase shift of 90° (or −90° for the "$-j$" term). This means that if a sinusoidal voltage is applied across a pure inductor, the peak current flow occurs ¼ cycle after the peak voltage is applied. The $-j$ term describing the capacitor's impedance means that the peak current flow through a capacitor occurs ¼ cycle before the peak voltage is applied. Because the voltage and current are not in phase, the impedance is described as reactance rather than as resistance.

Analog Filters

Missing from the simple black box diagram in Figure 1.2 are the source and load impedance. The resistance of these is crucial to analog filter design. Quite often the source and load are equal in value, typically 50 Ω for radio frequency applications, 75 Ω for television applications, and 600 ω for telephony applications. However, some applications require unequal source and load resistance, and some require values different from the ones listed. A modified black box diagram is given in Figure 1.7.

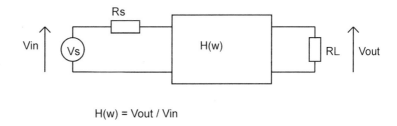

Figure 1.7

Transfer Function with Source and Load

H(w) = Vout / Vin

The output voltage is always measured at the filter's output, but the input voltage is not measured at the filter's input. The input voltage is measured at the voltage source (i.e., the electro-motive force [e.m.f.]) because the source impedance, Rs, is part of the filter design, even though it is not physically part of the filter. The practicalities of measuring the source voltage are described in Chapter 10. When the filter is designed for zero source impedance, the filter's input voltage and the source voltage are identical, so the voltage at the filter input is measured.

Analog filters can be passive or active. Passive filters use only resistors, capacitors, and inductors, as shown in Figure 1.8. Passive designs tend to be used where there is a requirement to pass significant direct current (above about 1 mA) through lowpass or bandstop filters. They are also used more in specialized applications, such as in high-frequency filters or where a large dynamic range is needed. (Dynamic range is the difference between the background noise floor and the maximum signal level.) Also, passive filters do not consume any power, which is an advantage in some low-power systems.

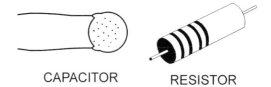

Figure 1.8

Components of a Passive Filter   CAPACITOR   RESISTOR

The main disadvantage of using passive filters containing inductors is that they tend to be bulky. This is particularly true when they are designed to pass high currents, because large diameter wire has to be used for the windings and the core has to have sufficient volume to cope with the magnetic flux.

Very simple analog lowpass or highpass filters can be constructed from resistor and capacitor (RC) networks. In the lowpass case, a potential divider is formed from a series resistor followed by a shunt capacitor, as illustrated in Figure 1.9. The filter input is at one end of the resistor and the output is at the point where the resistor and capacitor join. The RC filter works because the capacitor reactance reduces as the frequency increases. It should be remembered that the reactance is 90° out of phase with resistance.

At low frequencies the reactance of the capacitor is very high and the output voltage is almost equal to the input, with virtually no phase difference. At the cutoff frequency, the resistance and the capacitive reactance are equal and the filter's output is $1/\sqrt{2}$ of the input voltage, or −3 dB. At this frequency the output will not be in phase with the input: it will lag by 45° due to the influence of the capacitive reactance. At frequencies above the 3 dB attenuation point, the output voltage will reduce further. The rate of attenuation will be 6 dB per doubling of frequency (per octave). As the frequency rises, the capacitive reactance falls and the phase shift lag approaches 90°.

Although this is a description of a lowpass filter, a highpass response can be obtained by swapping the components. Placing a capacitor in series with the input, followed by a shunt resistor, gives a highpass filter with the same 3 dB frequency, but with a 45° phase lead. However, as the frequency rises, the attenuation and phase shift decrease. Lowpass and highpass RC networks are illustrated in Figure 1.9.

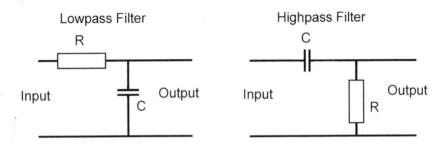

Figure 1.9

Lowpass and Highpass RC Networks

Now that you have an understanding of simple filters, I shall consider more complex passive filters. If the series resistor in the lowpass filter is now replaced by a series inductor, to form an LC network, the frequency response changes. The reactance of the series element is increasing while that of the shunt element is reducing, so the rate of increase in attenuation is doubled compared to a simple resistor-capacitor (RC) or resistor-inductor (RL) filter. At frequencies significantly above the passband, the rate of increase in attenuation with frequency is 12 dB/octave. Also the phase shift is doubled; it is 90° at the cutoff frequency and rises to a maximum of 180° at very high frequencies.

Note that the simple LC network is actually a series tuned circuit. If there were no series source or shunt load resistances present, there would be a magnification of the applied voltage by the inductor's Q factor. The Q of an inductor is given by the ratio of inductive reactance divided by its series resistance. Series source resistance or shunt load resistance is needed to limit the Q and to give a smooth passband response. Another effect of high Q values is that they would produce ringing at the output if an impulse were applied at the input.

As more reactive elements are connected in a ladder of series inductors and shunt capacitors, so the rate of attenuation beyond the passband increases in proportion. The rate of attenuation will be $n \times 6$ dB/octave, where n is the number of reactive components in the ladder and is known as the **filter order**. The filter order is also equal to the number of poles in the frequency response. Poles will be described in Chapter 3.

Active analog filters use operational amplifiers (op-amps) as the "active" element; these can be housed in a number of package types as illustrated in Figure 1.10. Op-amps are combined with resistors and capacitors to produce a filter with the appropriate frequency response. Thus they avoid the use of inductors. Because there are gain and bandwidth limitations for all op-amps, the performance of the filter can be restricted. Active filter designs were once restricted to frequencies below 100 kHz, but wide bandwidth op-amps (particularly current feedback types) are now allowing filter designs up to a few megahertz (MHz). This makes them suitable for video signal filtering.

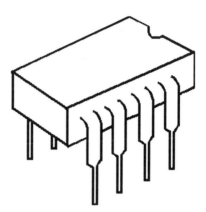

Figure 1.10

An Operational Amplifier
(op-amp)

Active filters have the advantage of being smaller than passive types, and integrated circuit designs allow them to be miniaturized further. Unfortunately active filters do have disadvantages: op-amps add noise to the signals; the signal's amplitude is limited by the op-amp's output slew rate and the power supply voltage; and harmonic distortion can also be introduced, particularly at the output stage.

Active filters are more suited to designs that are not very demanding, where rapid changes in amplitude occur as the frequency of the signal is changed. Even in a nondemanding filter design the signals within a filter circuit can be many times the applied voltage. For example, a signal may have an amplitude of, say, one volt, and this may be multiplied typically to perhaps ten volts within the filter. Devices within the filter must therefore be able to handle signals with large amplitudes at frequencies well beyond the passband required.

Integrated circuit (IC) filters are now quite common because they can be much smaller than active filters using op-amps and very much smaller than passive filters. Their small size supports the general trend to miniaturize equipment. The IC filters fall into two categories: continuous time and switched capacitor.

Continuous time filters use a number of op-amp circuits within the IC, and often integrating resistors and capacitors too. The filter response is selected by the addition of further resistors or capacitors around the IC. Continuous time filters tend to have a limited frequency range because of the integrated component values that have been provided.

Switched capacitor IC filters use the principle of rapidly charging and discharging a capacitor to replace a resistor, as shown in Figure 1.11. The effective resistor value depends on the rate of switching of the charge and discharge cycle. As the switching speed is changed, the effective resistance of the circuit also changes. The filter can thus be tuned by changing the switch clocking frequency. This type of filter generates signals at the switching frequency, and they tend to be generally noisy. Most switched capacitor filters are lowpass types and are limited in their frequency range to below 100 kHz.

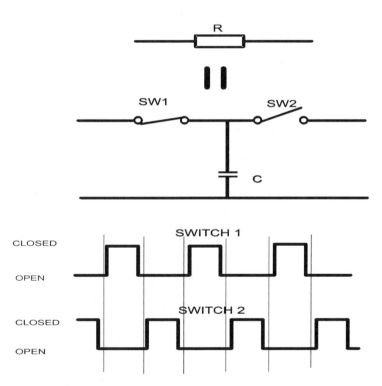

Figure 1.11

Switched Capacitor "Resistor Equivalent"

The Path to Analog Filter Design

At this point it would be helpful to know the overall process to design an analog filter. These processes will be described fully in later chapters, but a description now will help put it all into perspective.

All analog filters are designed from a normalized lowpass model. This model is a set of component values that are normalized for a $\omega = 1$ rad/s at the passband edge. Passive filter models have component values that are normalized for a $1\,\Omega$ load. Normalization allows the use of a table or set of component values, in conjunction with a single graph, to determine any filter design. This is a very powerful method, but transforming and scaling are necessary for each filter design undertaken.

Component values are scaled to produce an analog lowpass filter with a more practical passband and, in the case of passive filters, a more practical load resistance. The scaling process requires simple arithmetic to multiply and divide by certain factors. The result of scaling is that the cutoff point is changed from 1 rad/s to the required frequency and the load impedance is changed from $1\,\Omega$ to the required value.

Highpass filters can be produced from a lowpass model. The frequency response is the reciprocal of the lowpass response; so the attenuation of a lowpass filter model at $\omega = 2$ is the same as the equivalent highpass at $\omega = 0.5$. Passive highpass filter components are the reciprocal of the normalized lowpass filter. This means that where there are capacitors in the lowpass model, they are replaced by inductors in the highpass model. Similarly, where there are inductors in the lowpass model, they are replaced by capacitors in the highpass model.

Bandpass and bandstop filters are more complex but can still be derived from a normalized lowpass model. As an illustration, I will consider a bandpass filter and describe how to find out whether the filter specification is demanding, and hence I will be able to determine the filter order required to achieve it. First, I need to find out the bandwidth of the passband. Second, I need to find out the stopband attenuation and the width of the passband skirt.

If the desired passband (between the points where the filter provides less than 3 dB attenuation) extends from 20 kHz to 24 kHz, the passband bandwidth is 4 kHz. Suppose a 40 dB stopband attenuation is required at frequencies below 10 kHz and above 40 kHz. The width of the passband skirt is thus 30 kHz, being the difference between the two. The ratio of skirt width to bandwidth is $30 \div 4 = 7.5$. In terms of the lowpass model, the passband width is 1 rad/s, and hence the skirt response at 7.5 rad/s must provide the desired 40 dB attenuation. This is not very demanding, so a simple filter will do.

Bandstop filters have the inverse response of the bandpass filters described above. The normalized frequency of attenuation is given by the 3 dB bandwidth divided by the width of the stopband.

Active filters do not use normalized component value tables. Instead, they use something called pole and zero locations. (Do not worry too much about this now; it will be described in more depth in later chapters.) The pole and zero locations can be used in calculations to produce normalized component values for any given active filter circuit. As with passive filters, the frequency is normalized to 1 rad/s, hence the values have to be scaled to give a particular frequency response. Highpass, bandpass, and bandstop filters can be produced by transforming the equations before frequency scaling.

The ratios used in frequency transformation and scaling are summarized in Table 1.1. In all of these ratios, the resultant frequency is always greater than one.

Filter Type	Normalized Frequency
Lowpass	F_{atten}/F_{3dB}
Highpass	F_{3dB}/F_{atten}
Bandpass	BW_{atten}/BW_{3dB}
Bandstop	BW_{3dB}/BW_{atten}

Table 1.1

Filter Scaling Factors

Digital Filters

Signal Processing for the Digital World

An important relationship between the time domain and the frequency domain occurs when two signals are multiplied together. This relationship is important in both digital filter design and radio systems. Consider signals "cosA" multiplied by "cosB," where "A" and "B" are proportional to frequency. Trigometric identities are used to give the relationship $\cos A.\cos B = 0.5\cos(A + B) + 0.5\cos(A - B)$.

In the time domain, when one sinusoidal signal is **modulated** by the other having a different frequency there are two effects: (1) the peak amplitude of the resultant signal is greater than either of the source signals; (2) the waveform is no longer sinusoidal and the rate of change of the waveform varies over time, being alternately faster then slower compared to that of the highest frequency source signal. The highest frequency source signal is usually referred to as a **carrier signal** and the lowest frequency source signal is usually referred to as a **modulating signal**. The product of the two is an **amplitude modulated carrier**, as shown in Figure 1.12.

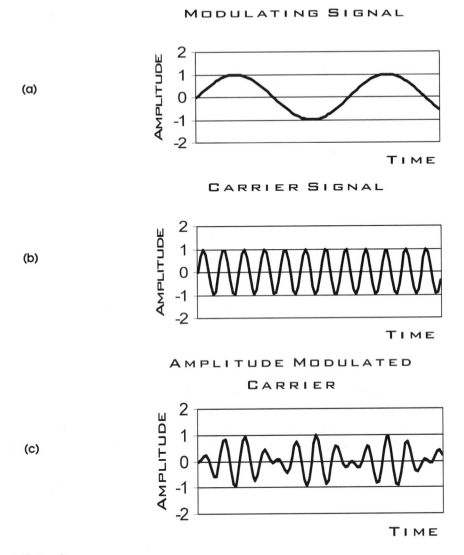

Figure 1.12 (a–c)

Multiplying Signals in the Time Domain

In the frequency domain, multiplying one signal by another (known as **mixing** in radio frequency design terms) causes frequency shifting. Suppose the two signals cos A and cos B, described above, are $\cos(\omega_1 t)$ and $\cos(\omega_2 t)$. Each of these signals has energy and produce lines on the spectrum analyzer display at a single frequency, ω_1 and ω_2. When mixed together there are two new signals produced with energy at new frequencies, which are the sum and difference frequencies given by $\omega_1 + \omega_2$ and $\omega_1 - \omega_2$. An example of this is shown in Figure 1.13.

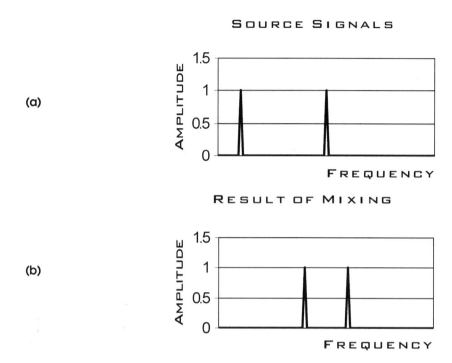

Figure 1.13 (a and b)

Multiplying Signals in the Frequency Domain

The relationship between time and frequency domains for multiplied signals is important for digital filter designers. When analog signals are sampled, they are effectively multiplied by an impulsive sampling signal. An periodic sampling pulse that is very short has spectral energy at multiples (**harmonics**) of the sampling frequency. The energy of every harmonic is equal to that of the lowest (**fundamental**) frequency, Fs. This means that the analog signal "A" is multiplied by the fundamental and every harmonic of the sampling signal. Thus spectral spreading occurs with energy appearing at Fs ± A, 2Fs ± A, 3Fs ± A, 4Fs ± A, and so on. When converting the sampled signal back into analog form, a further sampling operation reverses the frequency spreading process and results in all the spectral energy being concentrated at frequency "A."

The analog signal must be frequency limited prior to sampling, to less than half the sampling frequency. Otherwise the resultant spectral energy from mixing products will overlap in the frequency domain (which is known as **aliasing** and illustrated in Figure 1.14). If this happens, when signals are converted back into analog form, they have the wrong frequency. Filters are therefore placed before the sampling device to prevent aliasing, and these are known as **anti-alias** filters.

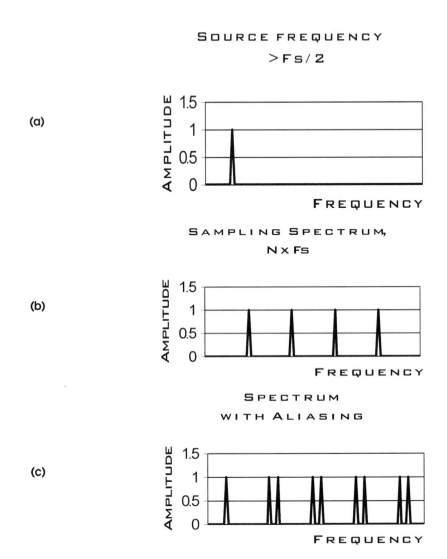

Figure 1.14 (a–c)

Alias Signal Generation due to Sampling

The "Brick Wall" Filter

A further relationship between the time and frequency domains can be used to explain why the "brick wall" filter cannot exist. More importantly, it can be used to explain how digital finite impulse response (FIR) filters work. This relationship is the impulse response of a "brick wall" filter, which has a sin(x)/x envelope in the time domain, as shown in Figure 1.15.

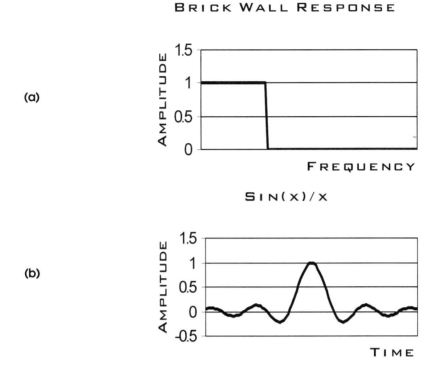

Figure 1.15 (a and b)

Time and Frequency Domain Response of "Brick Wall" Filter

I have shown how a sin(x)/x envelope produces a "brick wall" frequency response. Another relationship that is very useful for our analysis is that a very short impulse contains equal energy at all frequencies. If such an impulse is applied to the input of a filter, the frequency spectral energy at the output will be the same as the filter's frequency response. This is because the spectrum at the output of a filter is the input spectrum multiplied by the frequency response. The impulse response measured in the time domain at the filter's output will therefore have a shape that can be related to the frequency response measured in the frequency domain.

For any function, including filtering, there is an inverse relationship between the impulse response in the time domain and the frequency response in the frequency domain. A short impulse response means that the output pulse width is similar to the input pulse width. This occurs when the "function" performs little or no processing on the signal passing through. A long impulse response means that an output signal is present for some time after the input impulse signal has ended. This occurs if the function performs a high level of processing on the signal passing through such that it causes a sudden reduction in the output signal level relative to the input signal level as the frequency is changed.

The reason why the "brick wall" filter cannot be built is because of the relationship between the time and frequency domains. Just as a voltage step function (a sudden change in the time domain) has frequency components that extend across a wide band, a step function in the frequency domain has voltage components that extend across a wide period of time. The frequency domain can be considered to cover both positive and negative frequencies, so a 1 kHz sine wave can be represented by a pair of spectral lines at +1 kHz and −1 kHz. The step frequency response will, by reciprocity, have time domain components at positive and negative time, relative to the event. Since a response cannot occur before an event has taken place (i.e., negative time), the step frequency response cannot exist.

Digital FIR filters make use of the impulse-response relationship by taking samples of the analog input signal and passing these through a multistep delay line. At each step in the delay line the signal is used as the input to a multiplier: the other input to the multiplier is a fixed value. The fixed values for each multiplier are arranged so that the array overall has the equivalent of a sampled $\sin(x)/x$ envelope. The output of every multiplier is then summed to produce the filter's output. A single input pulse will produce a $\sin(x)/x$ envelope at the output. A single pulse has energy at all frequencies, and the $\sin(x)/x$ envelope has the spectral energy of the filter's frequency response. Thus a sampled analog signal fed into the FIR filter will be filtered in the frequency domain response due to pulse shaping in the time domain.

The impulse response can be shortened (**truncated**) by making extreme values equal to zero, symmetrically on either side of the response peak. The frequency response is degraded by truncating the impulse response, particularly due to the sudden change to zero values. However, modifying the values to give a smoother response by shaping using a **window** results in a frequency response that is closer to the desired "brick wall." Windowing a truncated sin(x)/x envelope is illustrated in Figure 1.16.

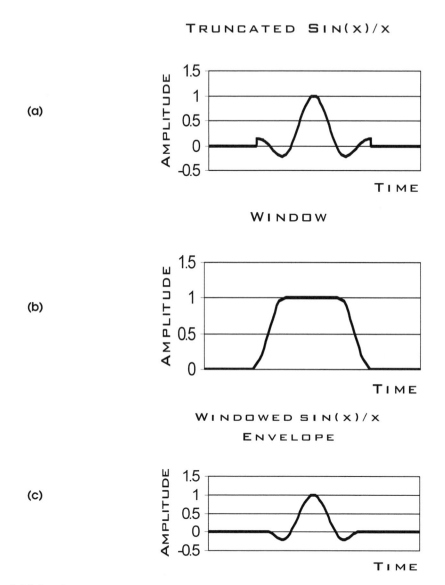

Figure 1.16 (a–c)

Windowing a Truncated Time Domain Response

To make a practical filter, the peak in the sin(x)/x envelope cannot occur at time equals zero, since producing an output signal before the input pulse has occurred is impossible. Instead, the peak in the envelope is moved to midway along the delay line. The first nonzero value in the sin(x)/x envelope is taken from the input to the delay line, and the last nonzero value is taken from the end of the delay line.

Digital Filter Types

Digital filters are becoming more widespread in use and are replacing analog filters in many systems. Digital filters process signals in the time domain. Analog signals have first to be sampled and digitized at discrete (clock) intervals using an analog-to-digital converter.

Because the analog signal is sampled, care has to be taken to prevent errors such as aliasing. Aliasing, which was described earlier, occurs when the analog signal has spectral energy at frequencies above half the sampling frequency. The analog and sampling signals mix in such a way that it is impossible to recover the original signal when it is converted back to analog. To prevent aliasing, the highest frequency of the input signal must be filtered. In telecommunications, the upper voice frequency is limited to 3.4 kHz with a very steep skirt (i.e., a sharp **roll-off**), so that there is no discernable energy at 4 kHz or higher. The voice frequency is then sampled at 8 kHz.

What would happen if an analog signal at, say, 5 kHz is passed then sampled at 8 kHz? Mixing between the 5 kHz signal and the 8 kHz signal would cause signals to be generated at the sum and difference frequencies. Thus signals at 3 kHz and 13 kHz would be produced. When converted back to analog, the 13 kHz signal would be outside the passband of the output filter, but the 3 kHz signal would be inside the passband and thus appear at the output as an alias.

Once digitized, the signals are digitally filtered by either a dedicated IC or a digital signal processor (DSP) using a filtering software. Within every digital filter there are delay elements, multiplying functions and adders, which process the digitized signal. There are two types of digital filter: Finite Impulse Response (FIR) filters, which are also described as nonrecursive filters; and Infinite Impulse Response (IIR) filters, which are recursive because part of the output signal is fed back to the input.

The recursive approach in digital filtering processes uses negative feedback in order to obtain a sharp roll-off using the minimum of delay, summing, and multiplying elements. The feedback comprises a small fraction of the output signal. Because of the delays, any sudden change in the input signal affects the output for some time (possibly forever, if there is any instability in the design). Recursive filters are said to have an Infinite Impulse Response (IIR). Some designs are sensitive to the filter coefficients used as multiplying factors, and truncating the coefficient (limiting the number of decimal places) can result in positive feedback and hence oscillation.

Nonrecursive, or moving average, filters take several successive samples then sum them (perhaps with a scaling factor at each tapping point) to produce the average of several samples. As time goes on the samples ripple through the filter, so any sudden change at the input only affects the output for the duration of its passage through the delay line. A nonrecursive filter is said to have a Finite Impulse Response (FIR). The FIR filter is inherently stable, but truncating the coefficients used as multiplying factors can lead to a nonideal frequency response, such as a passband that is not flat (i.e., one that has high levels of **ripple**).

The Path to Digital Filter Design

To design a Finite Impulse Response (FIR) filter, the required number of delay elements (or filter taps) must be calculated. This is determined by a number of factors: the window function to be used, the sampling clock frequency, and the ratio of passband frequency to stopband frequency. Once the number of taps is known, the multiplying coefficients are found using the $\sin(x)/x$ envelope. Each coefficient is then multiplied by the chosen window function. If a high-pass, bandpass, or bandstop filter is required, frequency scaling equations are also used to convert the response from the lowpass prototype.

The design of Infinite Impulse Response (IIR) filters are based on analog designs. The signal paths are arranged so that the output depends on both the input signal and the output signal. Input signals are fed into a delay line, and multiplying factors are used in the same way as in FIR filters to provide a feed-forward source. In addition, output signals are fed into a second delay line, and multiplying factors are used to provide a feedback source. These two sources are then combined to produce the output. The required analog frequency response is transformed using simple equations to give the feed-forward and feedback filter coefficients.

Exercises

1.1   The signal power into a filter at a particular frequency is 6 mW and the output power is 0.3 mW. What is the attenuation of the filter at this frequency? If the input voltage is 2 V, what will be the output voltage?

1.2   A second-order filter with a cutoff frequency of 1 MHz gives a signal attenuation of 12 dB at 2 MHz. What will be the attenuation at 4 MHz?

1.3   If the filter described in Exercise 1.2 has an input signal level of 10 mW, what will be the output level at 2 MHz and 4 MHz?

1.4   A simple RC lowpass filter has an input voltage of 10 V. What will be the voltage across (a) the resistor and (b) the capacitor at the −3 dB point?

CHAPTER 2

TIME AND FREQUENCY RESPONSE

This chapter describes filter frequency and time domain responses for a number of filter response types (e.g., Butterworth) and filter orders. This information on the frequency and time domain responses will be of use for all filter designs, whether passive, active, or digital.

Normalized frequency response graphs are used, with the passband edge usually being at a frequency of 1 rad/s (for the reasons discussed in Chapter 1). The frequency domain is described in terms of attenuation relative to this normalized frequency. Hence, the attenuation at, say, 10 times the cutoff frequency will be the value given on the graph where the curve crosses the frequency axis at 10 rad/s. On the frequency response graphs there is one curve for each filter order, thus 10 curves allow the relative performance of different filter orders to be compared. Higher filter orders give greater stopband attenuation but require more components.

Tables of normalized component values are given in this chapter for analog passive lowpass filters. Formulae used to derive many of these component values are given in the Appendix. The use of these tables to produce lowpass, highpass, bandpass, and bandstop filters will be given in Chapters 4, 5, 6, and 7 respectively. Tables for the design of analog active filters and digital IIR filters will be given in Chapter 3.

Filter Requirements

Filters are intended to pass some signal frequencies but stop others. Before a design can commence, the designer needs to consider the signals that need to be processed in this way. Does the filter have to pass DC? Are the signals impulsive? Which frequencies must pass and which must be stopped? How much attenuation (i.e., reduction in signal amplitude) is required? Once this information is known and the type of filter has been selected, the appropriate normalized frequency response curve can be used to find the filter order required. The lowest filter order to achieve the desired stopband attenuation is usually chosen because the filter will be simpler and lower cost.

For example, if the design must pass DC and signals up to a frequency F1, but attenuate signals above frequency F2, a lowpass filter is required. The ratio of F2 to F1 will be, by necessity, greater than one. Let's say the frequency ratio is 2.0: the passband must be flat and the attenuation required is 40 dB. Graphs showing the attenuation versus frequency for a number of filter orders are given in this chapter; use the Butterworth response graph shown in Figure 2.10 to find the required filter order. Since the plot is normalized, the frequency axis is equal to the stopband to passband ratio in the final (denormalized) filter. For 40 dB attenuation at $\omega = 2$ (the frequency ratio is equal to two), the required filter order is seven (n = 7).

In this example a Butterworth response was used because it has a smooth passband. Other responses also have a smooth passband: these are the Bessel and the Inverse Chebyshev. The Bessel response does not have sufficient attenuation at a frequency ratio of two, no matter how high the filter order. The Inverse Chebyshev response would require a fifth-order filter, to give 40 dB attenuation, but practically would be more difficult to make. This chapter will provide the designer with the information so that he or she can make the correct approach.

The most popular responses will be described in more detail later. In Chapter 1, the passband and stopband of a filter were described. The frequencies that are intended to pass through the filter with very little loss determine the passband. Those frequencies where a certain level of attenuation is required determine the stopband. There are four basic responses that can be made from the combination of flat or rippled passband and smooth or rippled stopband. I will show how standard Bessel, Butterworth, Chebyshev, Cauer, and Inverse Chebyshev approximations have one of these responses. Graphs will be used to describe the shape of each frequency response.

The subject of Inverse Chebyshev filters will be covered in some detail. In particular I will show how to obtain a more practical 3 dB cutoff point, rather than have the filter normalized at the stopband. I will also give tables for third- and fifth-order passive filters.

Practical filters are characterized by passband, stopband, and skirt response, as shown in Figure 2.1. The passband is the region where the loss is less than at the cutoff point. If the cutoff point is at, say, 1 dB, then all frequencies at which the loss is lower than 1 dB are in the passband. The stopband is the region of

high loss; it is the frequency band where the loss is greater than the desired attenuation. Clearly the stopband can be anywhere; it will depend on the desired
attenuation and the filter design. The skirt is the transition frequency response
and is between the passband and the stopband. The steepness of the skirt can
be important; Bessel filters have a gentle slope in the skirt response, while Cauer
filters have very steep skirts. As a rough guide, the steeper the skirt, the poorer
the impulse response.

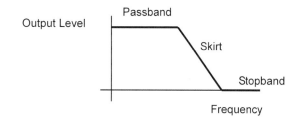

Figure 2.1

Practical Filter Response

A filter can have a smooth passband, or one with ripple. The stopband can either
have a smooth decay or a series of ripples peaking at a certain stopband attenuation. Thus four combinations of passband and stopband responses are possible. Bessel and Butterworth filters have a smooth passband and a smooth decay
in the stopband. Chebyshev filters have ripple in the passband but have a smooth
decay in the stopband. Inverse Chebyshev filters have a smooth passband with
ripples in the stopband. Cauer (or elliptic) filters have ripple in the passband
and in the stopband. All four variants are shown in Figure 2.2.

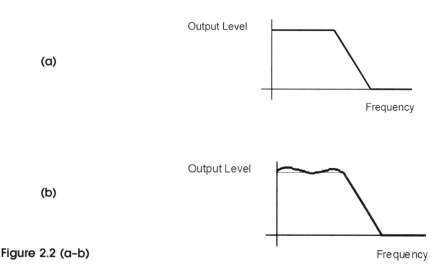

(a)

(b)

Figure 2.2 (a–b)

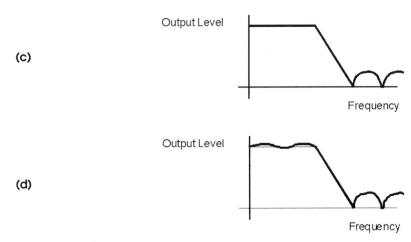

Figure 2.2 (c–d)

Passband and Stopband Response

(a) Smooth Passband. This can be approximated by Bessel and Butterworth responses. The Bessel response has a very slow change of attenuation beyond the passband, but it has excellent impulse performance. The Butterworth response is generally used to provide a smooth passband filter.

(b) Passband Ripple. Chebyshev filters have ripple in the passband; this allows the initial rate of attenuation to increase more rapidly with frequency than a Butterworth filter of equal order. The steepness of the skirt depends on the ripple allowed. Ripple can be below 0.01 dB, or as high as 3 dB, although ripple values beyond 1 dB are not normally used.

(c) Stopband Ripple. The Inverse Chebyshev response has stopband ripple. The nulls in output level within the stopband allow the skirt to have a very steep rate of attenuation increase. The advantage over the Chebyshev filter is that it has a smooth passband, which gives low variation in group delay. The disadvantage is that more components are needed in the circuit design.

(d) Passband and Stopband Ripple. This response can be satisfied using the Cauer response. The Cauer response is sometimes known as the elliptic response. Cauer filters have the same degree of complexity as Inverse Chebyshev filters, but ripple in the passband as well as the stopband allows the steepest of skirts.

The Time Domain

As signals pass through a filter they are delayed. Bessel filters are special in that they introduce an almost constant delay to all frequencies within the passband. This means that relative to the input, the phase of output signals changes in proportion to the applied frequency. Other types of filter (Butterworth, Chebyshev, Inverse Chebyshev, and Cauer) introduce a phase change in the output signal that is not proportional to the frequency. The rate of change in phase with frequency is known as the **group delay**. The group delay increases with the number of filter stages, so a fourth-order filter will produce a greater delay than a third-order filter.

The group delay is the delay seen by all signal frequencies as they pass through the filter. For example, a signal of 1 kHz may see a phase shift of 36°, which is a delay of 0.1 ms (the period of a 1 kHz signal is 1 ms and 36° is 0.1 cycle). If the phase change is proportional to frequency, a 2 kHz signal will see a phase shift of 72°, which is also a delay of 0.1 ms (the period of a 2 kHz signal is 0.5 ms and 72° is 0.2 cycle). This represents a constant group delay because both signals are delayed by the same amount.

An example of a nonconstant group delay filter would be one where the 1 kHz signal is delayed by 0.1 ms, as before. Now suppose that the 2 kHz signal sees a phase shift of 90°, which is a delay of 0.125 ms. The timing relationship between the two signals has changed because the 2 kHz signal is delayed by 0.025 ms more than the 1 kHz signal.

The consequence of a nonconstant group delay can be seen when pulses are applied to the filter input. Pulses contain signal harmonics several times the fundamental frequency of the pulse. As these harmonics propagate through the filter, they each experience different delays. Summing the delayed fundamental and harmonic signals results in slowly rising and falling pulse edges and causes ripple on top of the pulse. This distortion can produce errors when subsequent circuits process the pulse.

Butterworth, Chebyshev, Inverse Chebyshev, and Cauer filters have a group delay that increases with frequency and reaches a peak value close to the cutoff point. Beyond the cutoff point, the group delay gradually reduces to a constant value. An example of group delay for a third-order Butterworth filter is given in Figure 2.3. The delay shown is for a filter having a 1 rad/s cutoff frequency, which is about 2 seconds at low frequencies. The delay is inversely proportional to the cutoff frequency, so a filter having a 1 kHz (6283 rad/s) cutoff frequency will have a delay of 2/6283, or 318 μs, at low frequencies.

Further information on group delay is provided in Chapter 9, which describes all-pass filters. All-pass filters have a phase response that can be used to correct group delay variations of band-shaping filters. All-pass filters can also be used to create Hilbert transform filters. These have two filter branches whose outputs have the same signal amplitude but with a 90° phase difference. Hilbert filters are used to create single-sideband modulation of a radio carrier signal.

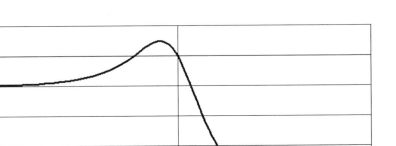

Figure 2.3

Group Delay of Butterworth Filter

Analog Filter Normalization

A normalized filter is one in which the passband cutoff point is at $\omega = 1$ radian per second. This is $1/2\pi$ Hz or about 0.159 Hz. Some may think that normalization to 1 Hz would be a good idea, but at 1 rad/s the impedance of reactive components is simply, $X_L = L$ and $X_C = 1/C$, which makes calculations simpler. Normalizing to 1 Hz would introduce 2π factors into the equations: $X_L = 2\pi L$ and $X_C = 1/2\pi C$. Passive filters are normalized for a 1Ω load impedance. The reason for normalization is to make the calculation of values simple, which in turn makes the filter design simple.

Passive analog filters can be designed using the tables of normalized component values given in this chapter. One set of normalized component values can be used to design passive lowpass, highpass, bandpass, and bandstop filters with any load impedance. The procedure is to first select the type of response required and then determine the filter order using the frequency response graphs. The tables are then used to provide a set of normalized component values. Using the selected set of values, scale for the frequency, impedance, and frequency response (lowpass, highpass, etc.) as required.

Active filters are designed using S-plane pole and zero locations, which are described in more detail in Chapter 3. Basically, a table of pole and zero locations are used in conjunction with simple equations to find component values. They can be used in a similar way to tables of normalized component values, like those used in passive filter design, by scaling for the frequency and response required. The source or load impedance does not affect pole and zero locations.

The process of denormalizing pole and zero locations, or passive component values, is explained in Chapters 4, 5, 6, and 7. These describe lowpass, highpass, bandpass, and bandstop responses, respectively.

Normalized Lowpass Responses

I will now describe Bessel, Butterworth, Chebyshev, Inverse Chebyshev, and Cauer responses, describing the frequency and time domain characteristics of each response in turn. Tables of normalized passive filter component values and the processes to denormalize them will be given in this chapter. Pole and zero locations, and circuit designs relevant to active filters will be given in the next chapter.

Bessel Response

The Bessel response is smooth in the passband, and attenuation rises smoothly in the stopband. The stopband attenuation increases very slowly until the signal frequency is several times higher than the cutoff point. Far away from the cutoff point the attenuation rises at $n \times 6\,dB$/octave, where n is the filter order and an octave is the doubling of frequency. For example, a third-order filter will give an $18\,dB$/octave rise in attenuation. The slow rise in attenuation near the cutoff frequency gives it an excellent time domain response, but this is not very useful in removing unwanted signals just outside the passband.

The natural cutoff frequency for the Bessel response is that which gives a one-second delay. This is not a constant value, but depends on the filter order. To make the design process simpler, the Bessel response can be scaled to give a $3\,dB$ cutoff frequency at $\omega = 1$ for all filter orders. To do this the frequency components of the transfer function have to be scaled. This has been done for the fifth-order Bessel filter, and the response curve is shown in Figure 2.4. In this graph, the attenuation versus frequency plot is given with the frequency axis normalized. Also in Figure 2.4 is the attenuation versus frequency plot of a natural cutoff Bessel response for comparison. The frequency at which 3 dB attenuation occurs in the natural Bessel response is 2.42 times that of the modified response, which is designed to give a 3 dB cutoff point at 1 rad/s.

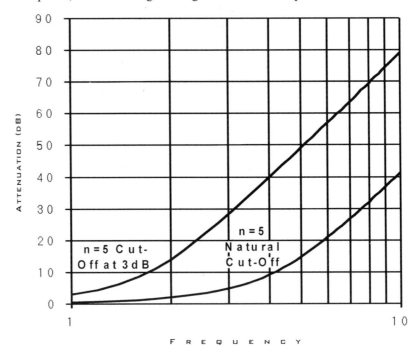

Figure 2.4

Bessel Response with Natural and 3 dB Cutoff Points

The constant factors for all filter orders up to ten are given in Table 2.1. The values given were found by practical experiment and are approximately $\sqrt{(2n-1)\cdot \ln 2}$, where n is the filter order.

Order, n	Normalizing Factor
1	1
2	1.36
3	1.75
4	2.13
5	2.42
6	2.7
7	2.95
8	3.17
9	3.39
10	3.58

Table 2.1

Bessel Normalizing Factors

The Bessel transfer function has been used to produce plots of attenuation versus frequency, for filter orders from one to ten. These plots can be used to find the filter order needed for any given attenuation and are given in the graph of Figure 2.6. Note that for low frequency ratios the attenuation is almost independent of the filter order. This means that, for example, at twice the cutoff frequency, the attenuation will be about 15 dB regardless of the filter order. At higher frequencies, an increase in the filter order does produce more attenuation.

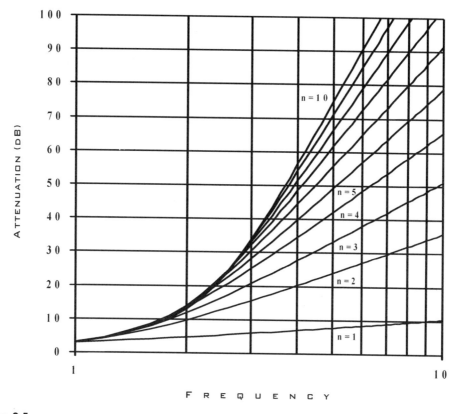

Figure 2.5

Bessel Attenuation Values Filter Order (N)

Bessel Normalized Lowpass Filter Component Values

Passive filter values have been tabulated by Weinberg for a number of source impedance values and for a normalized 1Ω load. These tables are normalized for a one-second delay, but, using the scaling factors described previously, I have recalculated them for a 3 dB cutoff point at 1 rad/s. Components have also been reordered and scaled so that for Rs equal to or greater than 1Ω the network is C1, L2, C3, L4, and so on. If Rs is equal to or is less than 1Ω, then the first component is a series inductor, and the ladder network is then L1', C2', L3', C4', and onward. Fifth- and sixth-order filter example circuits are given in Figures 2.6, 2.7, 2.8, and 2.9.

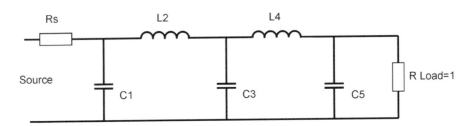

Figure 2.6

Fifth-Order Lowpass Rs ≥ 1

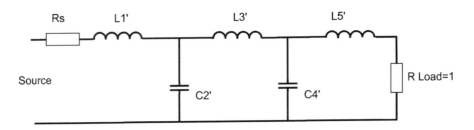

Figure 2.7

Fifth-Order Lowpass Rs ≤ 1

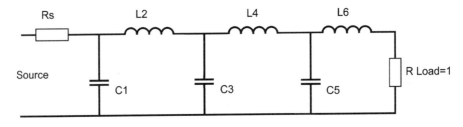

Figure 2.8

Sixth-Order Lowpass Rs ≥ 1

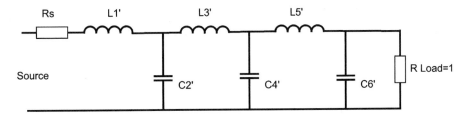

Figure 2.9

Sixth-Order Lowpass Rs ≤ 1

Those of you who have been paying attention will have noticed that a ladder network beginning with either a capacitor or an inductor can be used if the source and load impedance values are equal. Minimum inductor circuits are preferred because capacitors are easier to obtain and are generally cheaper and smaller.

Bessel passive filter networks cannot be calculated using formulae. Component values are found using the transfer function and continued fractional division. The reason for including a number of component tables here is that continued fractional division is very time-consuming and only undertaken by heroes! This technique is given in texts on circuit theory. Those of you with mathematical interests may like to try it out with some low-order Bessel designs. The answers are in the tables! Tables 2.2 to 2.7 have been adapted from Weinberg.[1]

Order	C1	L2	C3	L4	C5	L6	C7	L8	C9	L10
1	1.0000									
2	1.36	0.4539								
3	1.4631	0.8427	0.2926							
4	1.5012	0.9781	0.6127	0.2114						
5	1.5125	1.0232	0.7531	0.4729	0.1618					
6	1.5124	1.0329	0.8125	0.6072	0.3785	0.1287				
7	1.5087	1.0293	0.8345	0.6752	0.5031	0.3113	0.1054			
8	1.5044	1.0214	0.8392	0.7081	0.5743	0.4253	0.2616	0.0883		
9	1.5006	1.0127	0.8361	0.722	0.6142	0.4963	0.3654	0.2238	0.0754	
10	1.4973	1.0045	0.8297	0.7258	0.6355	0.5401	0.4342	0.3182	0.1942	0.0653
Rs = 0	L1′	C2′	L3′	C4′	L5′	C6′	L7′	C8′	L9′	C10′

Table 2.2

Bessel LC Values Rs = ∞ or Rs = 0

Order	C1	L2	C3	L4	C5	L6	C7	L8	C9	L10
1	2.000									
2	0.576	2.148								
3	0.3374	0.9705	2.2034							
4	0.2334	0.6725	1.0815	2.2404						
5	0.1743	0.5072	0.804	1.111	2.2582					
6	0.1365	0.4002	0.6392	0.8538	1.1126	2.2645				
7	0.1106	0.3259	0.5249	0.702	0.869	1.1052	2.2659			
8	0.0919	0.2719	0.4409	0.5936	0.7303	0.8695	1.0956	2.2656		
9	0.078	0.2313	0.377	0.5108	0.6306	0.7407	0.8639	1.0863	2.2649	
10	0.0672	0.1998	0.327	0.4454	0.5528	0.6493	0.742	0.8561	1.0781	2.2641
Rs = 1	L1′	C2′	L3′	C4′	L5′	C6′	L7′	C8′	L9′	C10′

Table 2.3

Bessel LC Values Rs = 1

Order	C1	L2	C3	L4	C5	L6	C7	L8	C9	L10
1	1.5000									
2	0.2601	3.5649								
3	1.8572	0.9174	0.3176							
4	0.112	1.2952	0.5202	3.7824						
5	1.90385	1.0764	0.7836	0.493	0.169					
6	0.0666	0.7824	0.3131	1.6752	0.5405	3.8122				
7	1.9045	1.0748	0.8555	0.6914	0.51605	0.3198	0.1084			
8	0.0452	0.5354	0.2173	1.1718	0.3608	1.7153	0.5329	3.8041		
9	1.8996	1.0566	0.8530	0.7334	0.62415	0.505	0.37225	0.2282	0.0769	
10	0.0332	0.3951	0.1617	0.8818	0.2739	1.2879	0.3678	1.6913	0.5242	3.7953
Rs = ½	L1′	C2′	L3′	C4′	L5′	C6′	L7′	C8′	L9′	C10′

Table 2.4

Bessel LC Values Rs = 2

Order	C1	L2	C3	L4	C5	L6	C7	L8	C9	L10
1	1.3333									
2	0.16633	4.9409								
3	1.7267	0.8925	0.30905							
4	0.07391	1.9268	0.34357	5.341						
5	1.7731	1.0578	0.7725	0.4857	0.16642					
6	0.04401	1.161	0.20655	2.4886	0.3553	5.3325				
7	1.7763	1.0611	0.8492	0.6868	0.5120	0.3168	0.1075			
8	0.02986	0.79535	0.1436	1.74255	0.2387	2.5499	0.34997	5.3066		
9	1.7704	1.0434	0.8478	0.7302	0.621	0.5025	0.3701	0.2268	0.0764	
10	0.02198	0.5878	0.107	1.3131	0.1811	1.9189	0.2434	2.5156	0.3444	5.2902
Rs = $\frac{1}{3}$	L1′	C2′	L3′	C4′	L5′	C6′	L7′	C8′	L9′	C10′

Table 2.5

Bessel LC Values Rs = 3

Order	C1	L2	C3	L4	C5	L6	C7	L8	C9	L10
1	1.25									
2	0.1221	6.3116								
3	1.6622	0.8804	0.3051							
4	0.05495	2.5466	0.2552	6.8631						
5	1.70895	1.04932	0.76775	0.482	0.16522					
6	0.03267	1.5398	0.1542	3.3024	0.2646	6.8512				
7	1.7112	1.0536	0.8459	0.6844	0.510	0.31624	0.107			
8	0.02232	1.05656	0.1071	2.316	0.1785	3.3884	0.2609	6.8142		
9	1.7046	1.036	0.8452	0.728	0.6195	0.5004	0.369	0.2264	0.0762	
10	0.01643	0.7819	0.08019	1.7463	0.1357	2.5532	0.1822	3.3444	0.2563	6.7902
Rs = $\frac{1}{4}$	L1′	C2′	L3′	C4′	L5′	C6′	L7′	C8′	L9′	C10′

Table 2.6

Bessel LC Values Rs = 4

Order	C1	L2	C3	L4	C5	L6	C7	L8	C9	L10
1	1.125									
2	0.05889	11.768								
3	1.5624	0.8608	0.2986							
4	0.02705	5.0202	0.12546	12.929						
5	1.6102	1.036	0.7598	0.4762	0.1634					
6	0.01623	3.05343	0.07641	6.5521	0.1307	12.9068				
7	1.6111	1.04312	0.84037	0.6797	0.5066	0.31384	0.10624			
8	0.0111	2.1008	0.05326	4.6079	0.08876	6.7381	0.12902	12.8264		
9	1.6038	1.0251	0.8408	0.7241	0.6169	0.49904	0.36722	0.22512	0.07576	
10	0.00816	1.5566	0.03938	3.479	0.06766	5.08897	0.09093	6.6556	0.1267	12.774
$Rs = \frac{1}{8}$	L1′	C2′	L3′	C4′	L5′	C6′	L7′	C8′	L9′	C10′

Table 2.7

Bessel LC Values Rs = 8

Butterworth Response

The Butterworth response has a smooth passband and a smooth increase in stopband attenuation. It differs from the Bessel response in that the attenuation in the stopband rises by $n \times 6$ dB/octave almost immediately outside the passband. Figure 2.10 gives a graph showing how the attenuation rises with frequency and filter order. A curve is given in the graph for each filter order up to ten. Using the graph, it is possible to determine the filter order required to give a certain level of attenuation at some multiple of the cutoff frequency.

For example, suppose the desired specification of a filter is that it has 60 dB attenuation at three times the cutoff frequency. Using Figure 2.10, the $\omega = 3$ axis and 60 dB attenuation axis cross at a point midway between the curves of $n = 6$ and $n = 7$. A seventh-order filter will meet the specification.

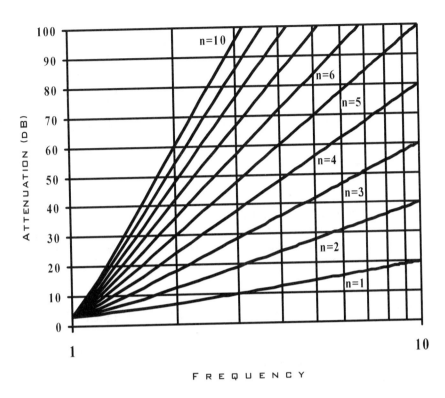

Figure 2.10

Attenuation of the Butterworth Response Values Filter Order (n)

Butterworth Normalized Lowpass Component Values

Butterworth passive lowpass filters have a ladder network of series inductors with shunt capacitors at their connection nodes. The first component in this ladder can be either a series inductor or a shunt capacitor; the following components then alternate: for example, series L, shunt C, series L, shunt C, and so on. This is shown in Figure 2.11.

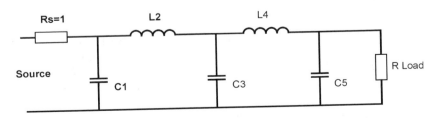

Figure 2.11

First Component Is Shunt C

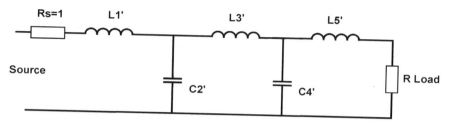

Figure 2.12

First Component Is Series L

Recursive formulae exist for the element values of passive Butterworth filters. The equations are for a filter with 3 dB attenuation at the passband edge and 1Ω source. These equations are related to those for Chebyshev filters and can be determined for any nominal source impedance.

Normalized Component Values for RL >> RS or RL << RS

For a filter having a load impedance value much greater than that of the source impedance (more than 10 times Rs) the load is considered to be of infinite impedance (open circuit) and the last component must be a shunt capacitor. This makes sense because if the load were open circuit, a series inductor would have no effect. On the other hand, a shunt capacitor provides a load for the filter to drive into, reducing the output impedance of the filter closer to that of the source. Odd-order filters therefore begin with a shunt C and even-order filters begin with a series L.

Conversely, for a load impedance much less than the source (less than Rs/10) the load is considered to be zero ohms, and the last component must be a series inductor. If the load were zero (taking the extreme to prove the point) a shunt capacitor would have no effect because the load is bypassing it. Series impedance is needed to raise the output impedance of the remaining network. Odd-order filters therefore begin with a series L and even-order filters begin with a shunt C.

Table 2.8 gives element values for passive Butterworth response filters with zero or infinite source impedance. This table has been produced using the results produced by the formulae given. However, the order of the components has been reversed to normalize on a 1Ω load, rather than a 1Ω source.

A simple diplexer uses lowpass and highpass filter sections that have equal cutoff frequencies. This is used to split a signal path into separate low- and high-frequency paths, without the losses associated with conventional power-splitter circuits. Zero source impedance filter designs are needed to obtain the correct diplexer response.

Order	C1	L2	C3	L4	C5	L6	C7	L8	C9	L10
1	1.00000									
2	1.41422	0.70711								
3	1.50000	1.33333	0.50000							
4	1.53074	1.57716	1.08239	0.38268						
5	1.54509	1.69443	1.38196	0.89443	0.30902					
6	1.55292	1.75931	1.55291	1.20163	0.75787	0.25882				
7	1.55765	1.79883	1.65883	1.39717	1.05496	0.65597	0.22521			
8	1.56073	1.82464	1.72874	1.52832	1.25882	0.93705	0.57755	0.19509		
9	1.56284	1.84241	1.77719	1.62019	1.40373	1.14076	0.84136	0.51555	0.17365	
10	1.56435	1.85516	1.81211	1.68689	1.51000	1.29209	1.04062	0.76263	0.46538	0.15643
Rs = 0	L1'	C2'	L3'	C4'	L5'	C6'	L7'	C8'	L9'	C10'

Table 2.8

Normalized Butterworth Element Values, Rs = ∞ or Rs = 0

Normalized Component Values for Source and Load Impedances within a Factor of Ten

If the load impedance value is close to the source impedance (say within a factor of 0.1 or 10 times), either shunt C or series L can be used as the first component. The last component will depend on whether the filter has an odd or even order.

Practically, most passive filters have equal source and load impedance. Table 2.9 gives element values for equal source and load impedance filters, normalized for one ohm. Various transformations are then used to convert them into any lowpass, highpass, bandpass, or bandstop designs. Details of how to do this for each specific design will be given in Chapters 4, 5, 6, and 7, respectively.

The format of Table 2.9 is to use the first set of component labels if the ladder begins with a shunt capacitor: C1, L2, C3, L4, and so on. If the first component is a series inductor, then use the lower set of component labels: L1', C2', L3', C4', and so forth. Notice the symmetry in the table; the reason behind this is that the component values are derived from equations that contain sine and cosine functions. These are natural functions that contain circular symmetry.

Order	C1	L2	C3	L4	C5	L6	C7	L8	C9	L10
1	2.0000									
2	1.41421	1.41421								
3	1.00000	2.00000	1.00000							
4	0.76537	1.84776	1.84776	0.76537						
5	0.61803	1.61803	2.00000	1.61803	0.61803					
6	0.51764	1.41421	1.93185	1.93185	1.41421	0.51764				
7	0.44504	1.24698	1.80194	2.00000	1.80194	1.24698	0.44504			
8	0.39018	1.11114	1.66294	1.96157	1.96157	1.66294	1.11114	0.39018		
9	0.34730	1.00000	1.53209	1.87938	2.00000	1.87938	1.53209	1.00000	0.34730	
10	0.31287	0.90798	1.41421	1.78201	1.97538	1.97538	1.78201	1.41421	0.90798	0.31287
	L1′	C2′	L3′	C4′	L5′	C6′	L7′	C8′	L9′	C10′

Table 2.9

Normalized Butterworth Element Values, Rs = Rl = 1

The most common filter designs have equal source and load impedance. For these, the component values given in Table 2.9 should be used. Less common but still popular are filter designs where the source and load are different by a factor of 10 or more (when the load is 10 times or one-tenth of the source impedance). For these filter designs, Table 2.8 should be used. There are obviously an infinite number of less common loads that could be applied. Under these circumstances the reader should make use of the equations given in the Appendix to calculate the element values needed.

Chebyshev Response

The Chebyshev response has ripples in the passband but a smooth increase in stopband attenuation. By allowing the passband response to have ripples, the stopband attenuation rises sharply just beyond the cutoff frequency. Further beyond the cutoff frequency, the attenuation rises by $n \times 6$ dB/octave, which is the same as the Butterworth. However, for a filter of equal order measured at the same frequency, a Chebyshev response will produce more stopband attenuation. This is because of the sudden rise in attenuation immediately beyond the cutoff point.

The Chebyshev response has a disadvantage in the time domain; its group delay has a greater peak level near the passband edge than the Butterworth response. Also, there are ripples in the group delay that make equalization with all-pass filters more difficult than in the Butterworth case.

Five graphs displaying attenuation versus frequency curves for filter orders up to ten are given in Figures 2.13 to 2.17. The graphs are for Chebyshev filters with a passband ripple of 0.01 dB, 0.1 dB, 0.25 dB, 0.5 dB, and 1 dB, respectively. In each case a 3 dB cutoff point is used.

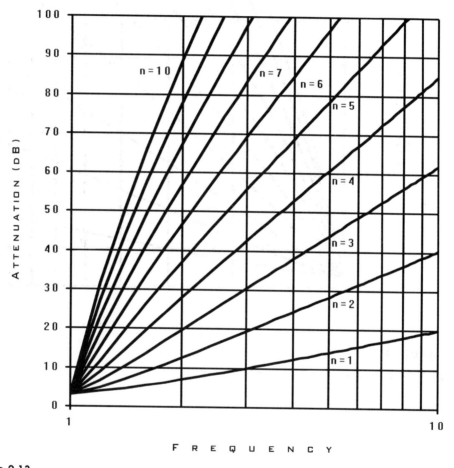

Figure 2.13

0.01 dB Chebyshev Attenuation Values Filter Order

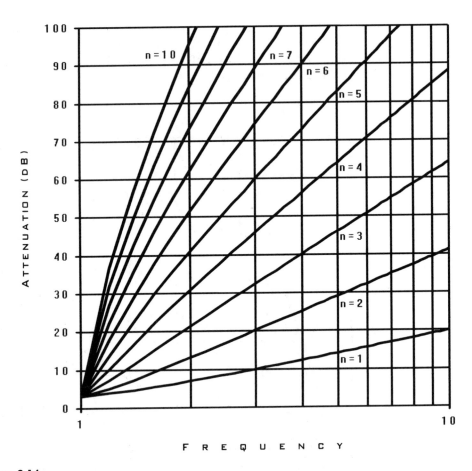

Figure 2.14

0.1 dB Chebyshev Attenuation Values Filter Order

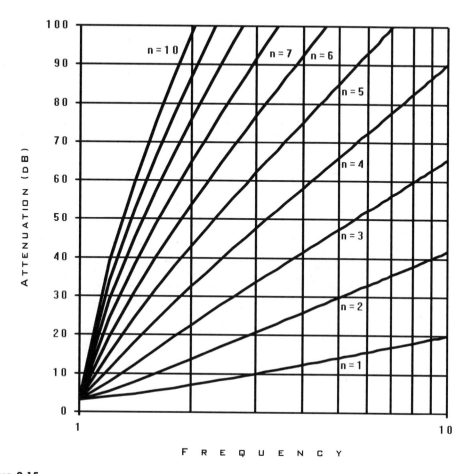

Figure 2.15

0.25 dB Chebyshev Attenuation Values Filter Order

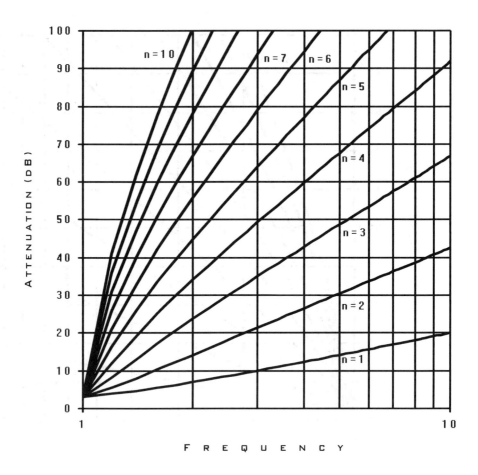

Figure 2.16

0.5 dB Chebyshev Attenuation Values Filter Order

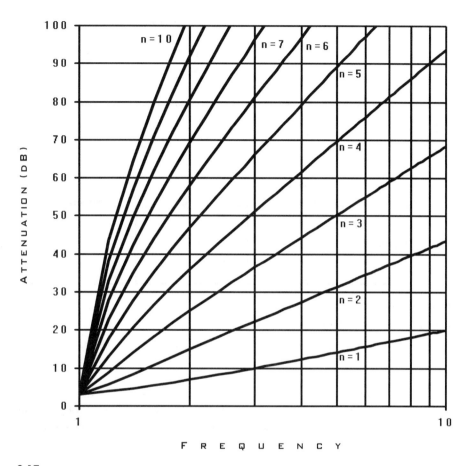

Figure 2.17

1 dB Chebyshev Attenuation Values Filter Order

Normalized Component Values

Chebyshev passive lowpass filters are like Butterworth and have a ladder network of series inductors with shunt capacitors at their connection nodes. The first component in this ladder can be either a series inductor or a shunt capacitor, the following components then alternate: that is, series L, shunt C, series L, shunt C, and so forth. This is shown in Figures 2.18 and 2.19.

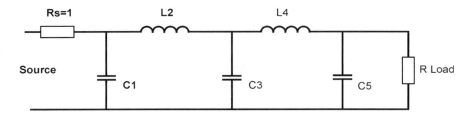

Figure 2.18

First Component Is Shunt C

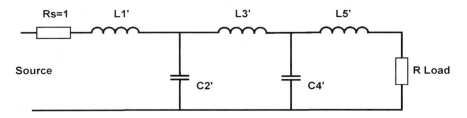

Figure 2.19

First Component Is Series L

Recursive formulae exist for the element values of passive Chebyshev filters and are given in the Appendix. These equations are for a filter with passband attenuation equal to the ripple and 1Ω source. An equation to modify these element values to give a passband attenuation of 3 dB is also given. These equations are related to those for Butterworth filters and can be determined for any nominal source impedance.

As in the Butterworth case, for a filter having a load impedance much greater than the source impedance (more than 10 times Rs), the source is considered to be of zero impedance (short circuit) and the first component must be a series inductor. This series inductor is part of the filter and provides source impedance for the following circuits. Even-order filters begin with a series L and end with a shunt C. Odd-order filters begin and end with a series L.

Conversely, for load impedance much less than the source (less than Rs/10), the source is considered to be infinite impedance and the first component must be a shunt capacitor. A series inductor would have no effect if the source were infinite impedance, so shunt impedance is needed to lower the input impedance for the remaining network. Even-order filters therefore begin with a shunt C and end with a series L. Odd-order filters begin and end with a shunt C.

Equal Load Normalized Component Value Tables

Since most passive filters have equal source and load impedance, the normalized values for these are given in Tables 2.10 to 2.14. Another useful filter is designed for infinite or zero source impedance, and element values for these are given in Tables 2.16 to 2.20. For other loads the reader should make use of the equations given in the Appendix to calculate the required element values. Transformation to any lowpass, highpass, bandpass, or bandstop designs is then possible. Details will be given in Chapters 4, 5, 6, and 7, respectively.

The format of Tables 2.10 to 2.14 is to use the first set of component labels if the ladder begins with a shunt capacitor: C1, L2, C3, L4, and so on. If the first component is a series inductor, then use the lower set of component labels: L1′, C2′, L3′, C4′, and so forth. Notice that there is symmetry in the tables because the component values are derived from sine and cosine functions that possess circular symmetry.

Order	C1	L2	C3	L4	C5	L6	C7	L8	C9
3	1.18111	1.82142	1.18111						
5	0.97660	1.68494	2.03666	1.68494	0.97660				
7	0.91273	1.59470	2.00209	1.87037	2.00209	1.59470	0.91273		
9	0.88538	1.55131	1.96146	1.86164	2.07173	1.86164	1.96146	1.55131	0.88538
	L1′	C2′	L3′	C4′	L5′	C6′	L7′	C8′	L9′

Table 2.10

Normalized Chebyshev Element Values, 0.01dB Ripple

Order	C1	L2	C3	L4	C5	L6	C7	L8	C9
3	1.43286	1.59373	1.43286						
5	1.30134	1.55594	2.24110	1.55594	1.30134				
7	1.26152	1.51955	2.23927	1.68038	2.23927	1.51955	1.26152		
9	1.24466	1.50168	2.22199	1.68293	2.29571	1.68293	2.22199	1.50168	1.24466
	L1′	C2′	L3′	C4′	L5′	C6′	L7′	C8′	L9′

Table 2.11

Normalized Chebyshev Element Values, 0.1dB Ripple

Order	C1	L2	C3	L4	C5	L6	C7	L8	C9
3	1.63306	1.43616	1.63306						
5	1.53996	1.43493	2.44027	1.43493	1.53996				
7	1.51189	1.41692	2.45311	1.53492	2.45311	1.41692	1.51189		
9	1.50000	1.40755	2.44460	1.54062	2.50767	1.54062	2.44460	1.40755	1.50000
	L1′	C2′	L3′	C4′	L5′	C6′	L7′	C8′	L9′

Table 2.12

Normalized Chebyshev Element Values, 0.25 dB Ripple

Order	C1	L2	C3	L4	C5	L6	C7	L8	C9
3	1.86369	1.28036	1.86369						
5	1.80691	1.30248	2.69145	1.30248	1.80691				
7	1.78962	1.29608	2.71773	1.38476	2.71773	1.29608	1.78962		
9	1.78229	1.29208	2.71630	1.39214	2.77344	1.39214	2.71630	1.29208	1.78229
	L1′	C2′	L3′	C4′	L5′	C6′	L7′	C8′	L9′

Table 2.13

Normalized Chebyshev Element Values, 0.5 dB Ripple

Order	C1	L2	C3	L4	C5	L6	C7	L8	C9
3	2.21565	1.08839	2.21565						
5	2.20715	1.12798	3.10248	1.12798	2.20715				
7	2.20391	1.13061	3.14695	1.19368	3.14695	1.13061	2.20391		
9	2.20246	1.13079	3.15397	1.20201	3.20772	1.20201	3.15397	1.13079	2.20246
	L1′	C2′	L3′	C4′	L5′	C6′	L7′	C8′	L9′

Table 2.14

Normalized Chebyshev Element Values, 1 dB Ripple

Only odd-order values are given in Tables 2.10 to 2.14. This is because even-order Chebyshev filters cannot be used if the source and load are equal. In fact, the even-order passive Chebyshev filter must have a normalized load resistance of greater than unity if the first component is a series inductor (the last component is therefore a shunt capacitor across the load). Conversely, if the first component is a shunt capacitor, the last component will be a series inductor feeding the load, and the normalized resistance of the load must be less than unity. The maximum and minimum load impedance limits for a number of Chebyshev filter passband ripple values are given in Table 2.15.

Ripple	Minimum Load, with Parallel Shunt Capacitor	Maximum Load, Fed by Series Inductor
0.01 dB	1.100746883	0.90847407
0.1 dB	1.355361345	0.73781062
0.25 dB	1.619565248	0.61744965
0.5 dB	1.984055712	0.5040181
1 dB	2.659722586	0.37597906

Table 2.15

Load Impedance Limits for Even-Order Chebyshev Filters

Normalized Element Values for Filters with RS = 0 or RS = ∞

Zero source impedance filters are used in the design of diplexers, and these are discussed further in Chapter 8. Normalized component values for zero or infinite **source impedance** filters are given in Tables 2.16 to 2.20. If a filter is required with a zero or infinite **load impedance**, instead of the source, the order of components given in these tables is simply reversed, so that the first reactive component is connected to the load.

Order	C1	L2	C3	L4	C5	L6	C7	L8	C9	L10
1	1.00000									
2	1.41336	0.74228								
3	1.50124	1.43296	0.59054							
4	1.52930	1.69459	1.31270	0.52307						
5	1.54664	1.79501	1.64491	1.23650	0.48829					
6	1.55130	1.84753	1.79009	1.59789	1.19066	0.46868				
7	1.55932	1.86709	1.86566	1.76514	1.56334	1.16096	0.45636			
8	1.55903	1.88502	1.89902	1.85578	1.74349	1.53932	1.14133	0.44834		
9	1.56456	1.88838	1.92421	1.89768	1.84251	1.72607	1.52167	1.12734	0.44269	
10	1.56262	1.89792	1.93251	1.92894	1.89081	1.83103	1.71295	1.50890	1.11738	0.43868
Rs = 0	L1′	C2′	L3′	C4′	L5′	C6′	L7′	C8′	L9′	C10′

Table 2.16

Normalized 0.01dB Chebyshev Element Values, Rs = ∞ or 0

Order	C1	L2	C3	L4	C5	L6	C7	L8	C9	L10
1	1.00000									
2	1.40488	0.82725								
3	1.51328	1.50900	0.71642							
4	1.51567	1.77396	1.45978	0.67474						
5	1.56126	1.80689	1.76588	1.41728	0.65065					
6	1.53633	1.88669	1.83342	1.75125	1.39590	0.63933				
7	1.57477	1.85775	1.92103	1.82699	1.73396	1.37856	0.63075			
8	1.54355	1.91231	1.90251	1.92697	1.82167	1.72463	1.36955	0.62633		
9	1.58037	1.87275	1.95841	1.90942	1.92294	1.81361	1.71504	1.36113	0.62232	
10	1.54689	1.92121	1.92274	1.97115	1.91128	1.92054	1.80936	1.71000	1.35669	0.62020
Rs = 0	L1′	C2′	L3′	C4′	L5′	C6′	L7′	C8′	L9′	C10′

Table 2.17

Normalized 0.1dB Chebyshev Element Values, Rs = ∞ or 0

Order	C1	L2	C3	L4	C5	L6	C7	L8	C9	L10
1	1.00000									
2	1.38934	0.90986								
3	1.53459	1.52828	0.81651							
4	1.49240	1.83405	1.51681	0.79093						
5	1.58646	1.78565	1.83856	1.48558	0.76996					
6	1.51127	1.92783	1.82548	1.83907	1.47670	0.76375				
7	1.60115	1.82834	1.96618	1.82342	1.82607	1.46285	0.75593			
8	1.51783	1.94883	1.87811	1.97925	1.82513	1.82364	1.45940	0.75382		
9	1.60726	1.84134	1.99659	1.88630	1.97701	1.81910	1.81573	1.45231	0.75000	
10	1.52086	1.95624	1.89342	2.01317	1.89184	1.97865	1.81878	1.81444	1.45086	0.74915
Rs = 0	L1′	C2′	L3′	C4′	L5′	C6′	L7′	C8′	L9′	C10′

Table 2.18

Normalized 0.25dB Chebyshev Element Values, Rs = ∞ or 0

Order	C1	L2	C3	L4	C5	L6	C7	L8	C9	L10
1	1.00000									
2	1.36144	1.01565								
3	1.57200	1.51790	0.93182							
4	1.45345	1.91162	1.53954	0.92395						
5	1.62994	1.73996	1.92168	1.51377	0.90343					
6	1.46994	1.99084	1.79019	1.93593	1.51606	0.90305				
7	1.64643	1.77716	2.03065	1.78918	1.92388	1.50337	0.89478			
8	1.47565	2.00848	1.83056	2.05041	1.79671	1.92786	1.50504	0.89433		
9	1.65329	1.78899	2.05701	1.83833	2.04815	1.79101	1.91988	1.49810	0.89112	
10	1.47828	2.01478	1.84229	2.07746	1.84692	2.05357	1.79404	1.92217	1.49949	0.89185
Rs = 0	L1′	C2′	L3′	C4′	L5′	C6′	L7′	C8′	L9′	C10′

Table 2.19

Normalized 0.5 dB Chebyshev Element Values, Rs = ∞ or 0

Order	C1	L2	C3	L4	C5	L6	C7	L8	C9	L10
1	1.00000									
2	1.30223	1.19145								
3	1.65199	1.45972	1.10778							
4	1.37686	2.05105	1.51740	1.12742						
5	1.72155	1.64455	2.06119	1.49297	1.10354					
6	1.38984	2.11627	1.70474	2.09336	1.50789	1.11259				
7	1.74142	1.67712	2.15585	1.70229	2.07901	1.49453	1.10192			
8	1.39431	2.13071	1.73338	2.18479	1.71600	2.09151	1.50218	1.10717		
9	1.74970	1.68810	2.17984	1.73916	2.18069	1.70937	2.08153	1.49435	1.10119	
10	1.39636	2.13592	1.74170	2.20597	1.75099	2.19124	1.71609	2.08873	1.49916	1.10462
Rs = 0	L1′	C2′	L3′	C4′	L5′	C6′	L7′	C8′	L9′	C10′

Table 2.20

Normalized 1 dB Chebyshev Element Values, Rs = ∞ or 0

Inverse Chebyshev Response

This response has a smooth passband and nulls in the stopband. This combination is a compromise that gives a reasonably sharp roll-off in the frequency response and a reasonably low overshoot in its impulse response. For any given frequency response, the filter order required for an Inverse Chebyshev will be the same as required for a Chebyshev filter. The advantage of using the

Inverse Chebyshev design is that the Q factor of its components is lower than in the Chebyshev design and therefore easier to achieve. The disadvantage is that Inverse Chebyshev designs are more complex and require more components.

The underlying method used to find the component values, which will be described in the next chapter, is pole positions derived from Chebyshev designs. The disadvantage of this is that the frequency response stopband is normalized to $\omega = 1$, instead of the usual 3 dB attenuation frequency. This description is not very helpful to practicing engineers because the 3 dB point will vary, depending upon the stopband attenuation and the filter order. Fortunately, it is possible to correct this and produce pole and zero positions based on a 3 dB cutoff. Passive filter component values can also be corrected to give a 3 dB cutoff frequency.

Inverse Chebyshev filters have a smooth passband with a gentle roll-off, a steep skirt, and ripples in the stopband. Poles and zeroes will be explained in the next chapter, but you may like to know that the "inverse" in Inverse Chebyshev filters comes from the filter pole positions, which are the inverse of those for Chebyshev filters. Pole and zero positions can be obtained using formulae, and these can be used directly in the design of active filters. Formulae to find the zero positions are given in the Appendix.

Inverse Chebyshev filters can achieve the same performance as Chebyshev filters of the same order, however they are more complex. The smooth passband with a gentle roll-off in the frequency domain transforms into the time domain as a group delay that is flatter than Chebyshev designs. The other advantage is that circuit elements require a lower Q factor; this makes them easier to produce.

These filters have not been popular because there are no simple algorithms to find passive filter component values. The exception to this is equations for third-order filters, which were derived by John Rhodes, Professor at the University of Leeds in the U.K., and these are presented in the Appendix. Rhodes's book, *Theory of Electrical Filters* (Wiley, 1976) is difficult to read, and for Inverse Chebyshev filters Rhodes assumes a highpass prototype. Some conversion is needed for a lowpass prototype and to give 3 dB attenuation at the passband edge at a frequency of $\omega = 1\,rad/s$, but the results are given in this chapter.

Table 2.21 lists the zero locations for filter orders up to ten. These values are for filters having a stopband beginning at $\omega = 1$.

Order	Zero 1	Zero 2	Zero 3	Zero 4	Zero 5
2	1.41421				
3	1.15470				
4	1.08239	2.61313			
5	1.05146	1.70130			
6	1.03528	1.41421	3.86370		
7	1.02572	1.27905	2.30477		
8	1.01959	1.20269	1.79995	5.12583	
9	1.01543	1.15470	1.55572	2.92380	
10	1.01247	1.12233	1.41421	2.20269	6.39245

Table 2.21

Zero Locations for Inverse Chebyshev Filters

Filters can be normalized to the 3 dB cutoff frequency, instead of the start of the stopband. If the zero locations relative to this 3 dB point are required, the values given in Table 2.21 must be divided by the frequency where the 3 dB point occurs. The 3 dB cutoff frequency is less than $\omega = 1$ rad/s.

Component Values Normalized for 1 Rad/s Stopband

Normalized component values for some passive Inverse Chebyshev filters have been published in Huelsman.[2] These component values are for a filter having a stopband beginning at $\omega = 1$ rad/s. These values are not reproduced here; please refer to Huelsman's book for further details.

I have used Rhodes' equation to produce normalized component values for third-order Inverse Chebyshev filters (see Table 2.22). In addition, I have used the "impedance synthesis" method, described in Huelsman, combined with circuit analysis software to produce normalized component values for fifth-order filters (see Table 2.23). Tables 2.22 and 2.23 are normalized with respect to a 1 rad/s stopband frequency.

Atten	C1	L2	C2	C3
20	1.171717	2.343437	0.320043	1.171717
25	1.49178	2.983563	0.251377	1.49178
30	1.866437	3.732877	0.200918	1.866437
35	2.309844	4.619692	0.162349	2.309844
40	2.838492	5.676988	0.132112	2.838492
45	3.471945	6.943896	0.108009	3.471945
50	4.233615	8.467236	0.088577	4.233615
55	5.151636	10.30328	0.072792	5.151636
60	6.259915	12.51984	0.059905	6.259915
65	7.599384	15.19878	0.049346	7.599384
70	9.219512	18.43904	0.040675	9.219512
75	11.18013	22.36028	0.033542	11.18013
80	13.55366	27.10734	0.027668	13.55366
85	16.42774	32.85551	0.022827	16.42774
90	19.90854	39.8171	0.018836	19.90854
95	24.12459	48.24921	0.015544	24.12459
100	29.23161	58.46326	0.012829	29.23161

Table 2.22

Passive Third-Order Inverse Chebyshev (1 Rad/s Stopband)

Atten	C1	L2	C2	C3	L4	C4	C5
25	0.034826	0.976387	0.926384	1.997745	1.53603	0.224925	0.479829
30	0.200989	1.250222	0.723478	2.227904	1.752054	0.197192	0.573384
35	0.357922	1.542176	0.586514	2.499294	1.989518	0.173656	0.674494
40	0.512368	1.855072	0.487587	2.810959	2.25281	0.15336	0.784536
45	0.669038	2.192623	0.412524	3.164524	2.545923	0.135704	0.904982
50	0.831614	2.559145	0.353442	3.563026	2.872931	0.120257	1.037435
55	1.003257	2.959464	0.305633	4.010791	3.238221	0.106692	1.183653
60	1.18689	3.398898	0.266118	4.513096	3.64663	0.094743	1.345577
65	1.385372	3.883307	0.232922	5.076139	4.103559	0.084193	1.525356
70	1.601619	4.419142	0.20468	5.707146	4.615038	0.074862	1.725375
75	1.838693	5.013537	0.180413	6.414394	5.187834	0.066596	1.948289
80	2.099869	5.674402	0.159402	7.207034	5.829529	0.059266	2.197056
85	2.388703	6.410517	0.141098	8.095517	6.548621	0.052758	2.474977
90	2.709089	7.231668	0.125076	9.091458	7.354638	0.046976	2.785739
95	3.065322	8.148752	0.111	10.20834	8.258267	0.041836	3.133466
100	3.462159	9.173953	0.098595	11.46061	9.271494	0.037264	3.522772

Table 2.23

Passive Fifth-Order Inverse Chebyshev (1 Rad/s Stopband)

The component values given in Huelsman and in Tables 2.22 and 2.23 can be modified to give a filter with a 3 dB passband cutoff frequency at 1 rad/s. This is possible by using the normalization frequency correction formula, given in the Appendix. The frequency correction values for some filters are given in Table 2.24. Series inductors and shunt capacitors are reduced in value by multiplying them by the factor ω_{3dB}. To produce useful design tables, I have carried out this frequency correction; the results are presented later in Tables 2.25 to 2.29.

Consider a fifth-order lowpass filter with a stopband attenuation of 40 dB. The 3 dB attenuation point occurs at a frequency of 0.61882 rad/s and there are two zeroes beyond the stopband: at 1.05146 rad/s and 1.7013 rad/s. Normalizing the design to give a 3 dB point at 1 rad/s by scaling the component values, the stopband becomes 1/0.61882, which is about 1.616 rad/s. The zeroes are then scaled in a similar way to become 1.05146/0.61882 and 1.7013/0.61882, which are 1.699 rad/s and 2.749 rad/s, respectively.

The normalized Inverse Chebyshev tables published by Huelsman relate to the minimum inductor circuit designs given in Figures 2.20 to 2.23.

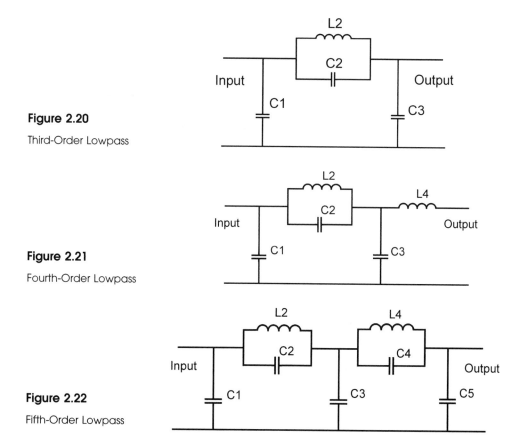

Figure 2.20

Third-Order Lowpass

Figure 2.21

Fourth-Order Lowpass

Figure 2.22

Fifth-Order Lowpass

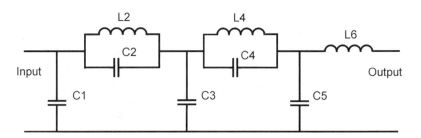

Figure 2.23

Sixth-Order Lowpass

The pattern of circuit design as the filter order increases can be seen from the examples. A seventh-order filter will have an extra two capacitors; C6 will be connected in parallel with L6, and C7 will be between the output and the common rail. Odd-order filters are symmetrical, but even-order filters have a single series inductor to the load.

Minimum capacitor designs are also possible, although less likely to be used since inductors are much harder to produce than capacitors. In this design, inductors replace capacitors and capacitors replace inductors; Figure 2.24 gives a circuit schematic example. In this design, series inductors replace shunt capacitors, and a series resonant shunt arm replaces the parallel resonant series arm. The value of the shunt capacitor is equal to the series inductor in the minimum inductance design. Similarly, the value of the shunt inductor is equal to that of the series capacitor in the minimum inductance design.

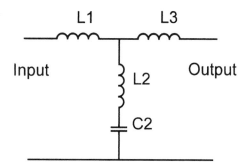

Figure 2.24

Minimum Capacitor Lowpass Filter

A highpass Inverse Chebyshev design can be produced by replacing lowpass prototype components with their complement; that is, replace capacitors with inductors and replace inductors with capacitors. This is shown in Figure 2.25.

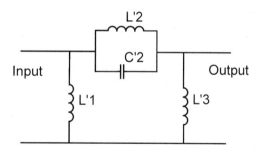

Figure 2.25

Third-Order Highpass

Normalized 3 dB Cutoff Frequencies and Passive Component Values

Using the formula to normalize the lowpass filter design to give a 3 dB cutoff frequency, recalculated component values are given in Tables 2.25 to 2.27. Table 2.24 gives normalizing ω_{3dB} frequencies that were used in the conversion; denormalization can be achieved by dividing component values by these frequencies.

Filter Order	20 dB Stopband	30 dB Stopband	40 dB Stopband
3	0.65	0.47233	0.33229
4	0.77384	0.63008	0.49672
5	*	0.73314	0.61882
6	0.88763	0.80101	0.70627
8	*	0.87924	0.81470
10	*	*	0.87438

Table 2.24

3 dB Cutoff Frequencies for Some Filters

Note: * indicates that Huelsman did not publish component values for these designs. These cutoff frequency values can be calculated easily using the formula given, but they were not required to produce the following tables of normalized component values.

Order	C1	L2	C2	C3	L4	C4	C5	L6
3	0.76162	1.52324	0.20803	0.76162				
4	0.39649	1.21732	0.40753	1.53955	0.71872			
6	0.059471	0.75761	0.48265	1.27903	0.78414	0.93651	0.63648	0.43328
	L'1	C'2	L'2	L'3	C'4	L'4	L'5	C'6

Table 2.25

Passive 20 dB Stopband Inverse Chebyshev Component Values

Order	C1	L2	C2	C3	L4	C4	C5	L6
3	0.88157	1.76315	0.09490	0.88157				
4	0.57065	1.5230	0.21595	1.71479	0.76245			
5	0.42037	1.2845	0.14457	1.63337	0.91659	0.53040	0.014736	
6	0.19029	0.93256	0.31931	1.43027	1.10124	0.54080	0.87389	0.46067
8	0.042388	0.62053	0.35045	0.92416	0.87687	0.84673	1.16705	0.76148
	L'1	C'2	L'2	L'3	C'4	L'4	L'5	C'6

Order	C6	C7	L8
8	0.68946	0.45019	0.32491
	L'6	L'7	C'8

Table 2.26

Passive 30 dB Stopband Inverse Chebyshev Component Values

Order	C1	L2	C2	C3	L4	C4	C5	L6
3	0.94320	1.88641	0.04390	0.94320				
4	0.67795	1.71867	0.11893	1.83035	0.78962			
5	0.31706	1.14795	0.30173	1.73949	1.39408	0.094902	0.48549	
6	0.03875	0.78742	0.58800	1.6670	1.6366	0.14145	1.1995	0.48188
8	0.11801	0.71751	0.26022	1.08035	1.10767	0.57551	1.30591	0.94266
10	0.0343	0.51572	0.27602	0.74526	0.84735	0.71164	1.10288	0.91953
	L'1	C'2	L'2	L'3	C'4	L'4	L'5	C'6

Order	C6	C7	L8	C8	C9	L10
8	0.47819	0.60325	0.33988			
10	0.81058	0.97738	0.67178	0.55477	0.3546	0.25988
	L'6	L'7	C'8	L'8	L'9	C'10

Table 2.27

Passive 40 dB Stopband Inverse Chebyshev Component Values

Normalized component values for the third-order and fifth-order Inverse Chebyshev, which I have calculated using Rhodes' equation and the "impedance synthesis" method, have been modified to give a 3 dB cutoff frequency of 1 rad/s. These values are given in Tables 2.28 and 2.29.

Atten	C1	L2	C2	C3	
20	0.761617	1.523236	0.208028	0.761617	1.538459
25	0.831004	1.662009	0.140031	0.831004	1.795155
30	0.881585	1.763171	0.094901	0.881585	2.117136
35	0.917735	1.83547	0.064503	0.917735	2.516901
40	0.943194	1.88639	0.043899	0.943194	3.009447
45	0.96094	1.921881	0.029894	0.96094	3.613069
50	0.97322	1.946442	0.020362	0.97322	4.350115
55	0.981676	1.963354	0.013871	0.981676	5.247801
60	0.987479	1.97496	0.00945	0.987479	6.339305
65	0.991452	1.982906	0.006438	0.991452	7.664891
70	0.994168	1.988338	0.004386	0.994168	9.273599
75	0.996023	1.992047	0.002988	0.996023	11.22473
80	0.997288	1.994579	0.002036	0.997288	13.59046
85	0.998152	1.996305	0.001387	0.998152	16.4582
90	0.99874	1.997482	0.000945	0.99874	19.93382
95	0.999141	1.998284	0.000644	0.999141	24.14526
100	0.999415	1.998831	0.000439	0.999415	29.24832

Table 2.28

Passive Third-Order Inverse Chebyshev (3 dB at 1 Rad/s Cutoff)

Atten	C1	L2	C2	C3	L4	C4	C5
25	0.027514	0.771385	0.731881	1.5783	1.213526	0.1777	0.379085
30	0.147353	0.91659	0.530412	1.63337	1.284504	0.14457	0.420372
35	0.241809	1.04188	0.396244	1.6885	1.3441	0.11732	0.455682
40	0.317066	1.147964	0.301731	1.73949	1.394094	0.094903	0.48549
45	0.377312	1.236555	0.232648	1.78467	1.435803	0.076532	0.510375
50	0.425619	1.309766	0.180891	1.82355	1.470361	0.061548	0.530957
55	0.464354	1.369777	0.141461	1.85638	1.498799	0.049382	0.547849
60	0.495392	1.418654	0.111074	1.883705	1.522054	0.039544	0.561626
65	0.520236	1.458262	0.087467	1.906195	1.540971	0.031616	0.572803
70	0.5401	1.490228	0.069022	1.92457	1.556288	0.025245	0.581833
75	0.555966	1.515943	0.054552	1.93952	1.568645	0.020137	0.589104
80	0.568626	1.536577	0.043165	1.9516	1.578584	0.016049	0.594943
85	0.57872	1.553099	0.034184	1.96133	1.586558	0.012782	0.599622
90	0.586762	1.566308	0.02709	1.96912	1.592942	0.010175	0.603364
95	0.593166	1.576853	0.021479	1.9754	1.598045	0.008096	0.606352
100	0.598263	1.585264	0.017037	1.9804	1.602119	0.006439	0.608737

Table 2.29

Passive Fifth-Order Inverse Chebyshev (3 dB at 1 Rad/s Cutoff)

Cauer Response

The Cauer response has ripple in the passband and in the stopband. Cauer filters are used where it is necessary to have a sharp transition between the passband and stopband, that is, a very steep skirt response. The drawback is that the filter circuit is more complex: passive filters require series or parallel tuned sections; active filters require three or four operational amplifiers per section. A further drawback is that, because of the sharp transition between the passband and stopband, the phase of the output signal changes rapidly close to the cutoff frequency, which results in a large group delay variation. This type of filter will not be suitable for handling pulsed signals if one of the harmonics frequencies coincides with a peak in the group delay.

Cauer filters are named after a German scientist, W. Cauer, but are commonly called elliptic filters because elliptic integrals are used in the calculation of their transfer function. Tables of normalized component values and pole-zero positions have been published by Zverev,[3] and their use is reasonably simple. His tables show the attenuation that can be expected for a given set of values. For the average user these tables are entirely adequate for the design of both active and passive filters.

Amstutz[4] has published computer programs that calculate pole, zero, and component values for both symmetrical (odd-order) and nonsymmetrical (even-order) filters. These programs will not be described here; readers interested in pursuing this subject further are recommended to read Amstutz's article and an explanation given by Cuthbert[5] (who has published a BASIC version of Amstutz's FORTRAN program).

Passive Cauer Filters

A passive Cauer filter has the same circuit configuration as an Inverse Chebyshev filter; there are mid-element tuned circuits that produce zeroes in the frequency response. These are shown in Figures 2.26 to 2.29.

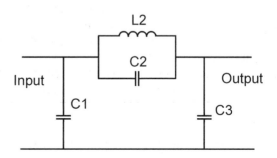

Figure 2.26

Third-Order Lowpass

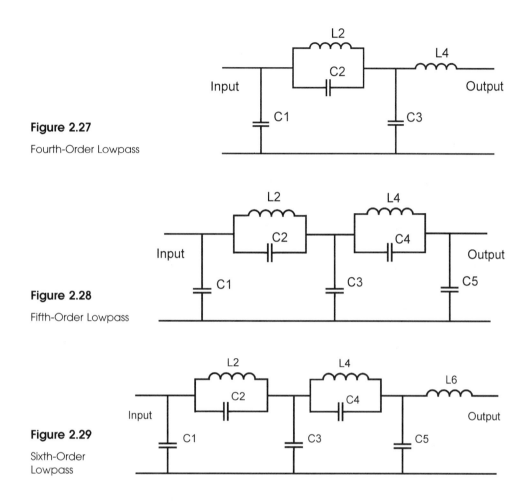

Figure 2.27

Fourth-Order Lowpass

Figure 2.28

Fifth-Order Lowpass

Figure 2.29

Sixth-Order
Lowpass

The pattern of circuit design as the filter order increases can be seen from the examples. A seventh-order filter will have an extra two capacitors; C6 will be connected in parallel with L6, and C7 will be between the output and the common rail. Odd-order filters are symmetrical, but even-order filters have a single series inductor to the load.

Minimum capacitor designs are also possible, although less likely to be used since inductors are much harder to produce than capacitors. In this design, inductors replace capacitors and capacitors replace inductors. Figure 2.30 gives a circuit schematic for this filter topology. In this design series inductors replace shunt capacitors, and a series resonant shunt arm replaces the parallel resonant series arm. The value of the shunt capacitor is equal to the series inductor in the minimum inductance design. Similarly, the value of the shunt inductor is equal to that of the series capacitor in the minimum inductance design.

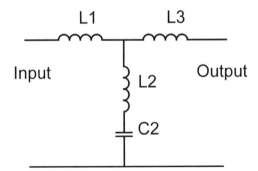

Figure 2.30

Minimum Capacitor Lowpass
Filter

Normalized Cauer Component Values

For Cauer filters, there are many combinations of passband ripple, stopband attenuation, and stopband frequency. This can result in many tables that present component values. Extensive tables of passive filter component values have been published [2, 3]. A small number of example values are given in Table 2.30. These are adapted from an abstract of a Ph.D. dissertation by Baez-Lopez.[6] The first column gives the minimum stopband attenuation (loss), in dBs, that can be expected.

Loss (dB)	Stopband	Order	C1	L2	C2	C3	L4	C4	C5	L6	C6	C7
30	2.5	3	0.9472	1.0173	0.1205	0.9472						
30	2	4	0.7755	1.1765	0.1796	1.3347	0.9338					
40	2.5	4	0.8347	1.2744	0.1053	1.3722	0.9325					
40	1.5	5	1.0279	1.2152	0.1513	1.6318	0.9353	0.4408	0.8155			
50	2	5	1.0876	1.2932	0.07317	1.7938	1.1433	0.20038	0.9772			
50	1.5	6	0.8659	1.2740	0.1855	1.4311	1.2723	0.33007	1.2825	1.0332		
50	1.2	7	1.0503	1.2487	0.16123	1.4838	0.8287	0.81542	1.2872	0.8743	0.58918	0.7539
			L1′	C2′	L2′	L3′	C4′	L4′	L5′	C6′	L6′	L7′

Table 2.30

Cauer Filter Component Values

The second column gives the normalized stopband frequency. The normalized passband frequency is unity, so a stopband value of 2.5 means that the stopband attenuation (in the lowpass prototype) begins at 2.5 times the cutoff frequency. When denormalized, a passband of 1 kHz will result in a stopband beginning at 2.5 kHz.

The third column gives the order of a Cauer filter that will meet the specification. In some cases the specification will be exceeded.

The Cutoff Frequency

There are two schools of thought concerning the cutoff frequency. The purists would say that it depends upon the filter design. The Butterworth response has a natural cutoff frequency at the point where signal loss through the filter is 3 dB. However, Chebyshev and Cauer filters have a natural cutoff at the point where the attenuation is equal to the passband ripple. Bessel filters are designed from their group delay characteristics, so their natural cutoff point depends on the filter order. Inverse Chebyshev filters have a natural frequency at the edge of the stopband, because their response is derived from inverting Chebyshev pole positions.

My view is that a 3 dB cutoff point should be used for all filter responses. All filters can be normalized to have a 3 dB cutoff frequency by suitable scaling. This gives some consistency and allows a direct comparison of performance to be made. It also makes sense from an engineering perspective because the transmitted power is halved at this point. An example where this is used is in a diplexer. A diplexer comprises two filters that are connected together, and each one is required to have a 3 dB cutoff frequency in order for the overall response to be correct. Diplexers are described further in Chapter 8 and are used for frequency band separation of signals.

References

1. Weinberg, L. "Additional Tables for Design of Optimum Ladder Networks." *Journal of the Franklin Institute*, August 1957: 127–138.

2. Huelsman, L. P. *Active and Passive Analog Filter Design*. New York: McGraw-Hill, 1993.

3. Zverev, A. I. *Handbook of Filter Synthesis*. New York: John Wiley & Sons, 1967.

4. Amstutz, P. "Elliptic Approximation and Elliptic Filter Design on Small Computers." *IEEE Circuits and Systems*, vol. CAS-25, no. 12, December 1978: 1001–1011.

5. Cuthbert, Thomas R. *Circuit Design Using Personal Computers*. New York: John Wiley & Sons, 1983.

6. Baez-Lopez, David. "Synthesis and Sensitivity Analysis of Elliptical Networks." Ph.D. dissertation, University of Arizona, 1979.

Exercises

2.1 Which filter responses allow DC to flow?

2.2 What are the passband, stopband, and skirt?

2.3 Which filter types have ripple in the passband?

2.4 Which types have ripple in the stopband?

2.5 Why do engineers use filters with ripple in both passband and stopband?

2.6 Which filter type has a constant delay that is not frequency dependent?

2.7 Why are component values "normalized"?

CHAPTER 3

POLES AND ZEROES

This chapter contains material that is essential for those involved with analog active filter design. Active filters can be designed from a set of numbers known as the pole and zero locations. The pole and zero locations are obtained from a filter's transfer function, and typical pole and zero location patterns will be illustrated. This should give the reader a feel for how a filter will behave if it has a certain pole-zero pattern, or a certain transfer function.

Tables of pole and zero locations are given in this chapter and can be used in formulae (given in the next four chapters) to find resistor and capacitor values. These normalized lowpass pole and zero values can be used to design lowpass, highpass, bandpass, or bandstop filters. Scaling either the poles and zeroes, or scaling the component values obtained from them, allows the frequency response to be changed from the normalized 1 rad/s cutoff frequency. Pole and zero placing formulae are given in the Appendix and include "natural" and 3 dB attenuation limited passbands. These formulae allow tables of pole and zero values to be produced.

Knowledge of the origin or theory of poles and zeroes is not essential. However, this information is provided for those readers who would like to understand the ideas behind them. Poles and zeroes are located on a two-dimensional plane, known as the S-plane. In the S-plane, one axis is "real" and is related to signal decay. The other axis is "imaginary" and is related to frequency. The S-plane will be explained further in a later section, as will an introduction to the Laplace Transform.

This book is about filters, not poles and zeroes, so the theoretical coverage here will be kept to a minimum. However, the excellent book by Robin Maddock called *Poles and Zeros in Electrical and Control Engineering* would be useful reading for anyone wishing to pursue this subject further.[1] As Maddock's title indicates, he prefers the plural "zeros" rather than "zeroes," which is my preference; it's just personal choice.

Frequency and Time Domain Relationship

In the frequency domain there are three parameters to consider: frequency, amplitude, and phase. The amplitude and phase versus frequency give the transfer function of a network in the frequency domain. The transfer function can be measured if a pure sine wave is applied at the network's input, and the amplitude and phase of the output signal is recorded for each frequency. Through analysis of the network, an equation for the transfer function can also be found. Transfer functions are normally calculated, then used for comparison with the actual circuit implementation.

The time domain parameters include delay, rise time, overshoot, and ringing. Delay can be measured by applying a step-input voltage to a network, where the time for the output to reach 50% of the final value is measured. Rise time uses the same step input and the time difference between the output reaching 10% and 90% of the final output level. Overshoot and ringing are related, and they also use a step input. Overshoot is where the output rises above the steady state final value; the maximum output is recorded in terms of percent above the nominal output. Ringing is where there is insufficient damping and the output has an exponentially decaying sinusoidal waveform superimposed upon it.

Frequency and time domain transfer functions were described in Chapter 2. It was made clear that a relationship between the frequency response and the time domain response exists. If a filter's frequency response has a gentle transition between the passband and the stopband, it also has constant group delay (the Bessel response). If the frequency response has a steep roll-off outside the passband, it's group delay (in the time domain) peaks where the change in attenuation is greatest.

The time domain response can be converted into the frequency domain using the Fourier Transform. Unfortunately, this transform can only be applied to continuous periodic signals, so a variant of this, the Laplace Transform, is used instead. The Laplace Transform is used to analyze transient signals; that is, signals that appear at time, $t = 0$. When the Laplace Transform is applied to a signal that is a function of time, $f(t)$, it produces a response as a function of complex frequency, $F(s)$. The frequency response $F(s)$ is complex, where $S = \sigma \pm j\omega$. This leads us nicely to the S-plane.

The S-Plane

The S-plane can be used to describe both time and frequency domain responses. It is just a graphical representation of mathematical ideas. However, these visual aids are very powerful in helping us to understand filters and signals.

Frequency Response and the S-Plane

The transfer function $F(\omega)$ is frequency dependent. However, this response is complex because both the amplitude and the phase depend on frequency. For this reason, the frequency response is described in terms of s, rather than frequency, ω. Hence, the transfer function becomes $F(s)$, where $S = \sigma \pm j\omega$.

The transfer function becomes an infinite value and is referred to as a "pole" when the denominator becomes zero. The value of s that makes the denominator zero could be a real value or a complex value, depending on the transfer function. Thus poles occur at certain values of $s = \sigma \pm j\omega$. In some cases (when $j\omega = 0$), the value of $s = \sigma$ alone. Invariably, the value of σ is negative. If the value of $j\omega$ is not zero, there are a pair of values for s and these are $s = \sigma + j\omega$ and $s = \sigma - j\omega$.

The transfer function becomes zero when the numerator becomes zero. The value of s needed to make the numerator equal zero is referred to as the "zero" location. Responses like Bessel, Butterworth, and Chebyshev have a numerator value of 1, hence there are no zeroes. These filters are referred to as "All-Pole" filters. Responses like Inverse Chebyshev and Cauer have numerators that depend on powers of s, and hence have zeroes. Invariably, the numerator zero locations occur at values of $s = 0 + j\omega$. That is, the zero is on the (imaginary) frequency axis.

Before looking at specific responses, whether they are Butterworth, Chebyshev, Inverse Chebyshev, or others, I will give a brief outline of how poles, zeroes, and the transfer function are related. This section assumes knowledge of complex numbers; the S-plane (showing pole-zero diagrams) is introduced without explanation but is described in more detail later.

Let's start by considering the transfer function of a filter. Say, for example, a simple second-order filter is formed from a series inductor followed by a shunt capacitor. The transfer function of this filter can be expressed algebraically as:

$$T(s) = \frac{k}{as^2 + bs + c} \quad \text{where} \quad S = \sigma \pm j\omega.$$

The values of k, a, and b determine the shape of the transfer function. If a Butterworth response is required, $k = 1$, $a = 1$, $c = 1$, and $b = \sqrt{2}$. If $\omega = 1$, $s = j\omega = j1$. Considering the other terms in the equation, $s^2 = j^2 = -1$ and $bs = j\sqrt{2}$. So the transfer function becomes:

$$T(\omega) = \frac{1}{-1 + j\sqrt{2} + 1} = \frac{1}{j\sqrt{2}} = 0.7071 \angle -90°$$

The output is 0.7071, or −3 dB, and the output phase is 90° behind the input.

A pole is said to exist where the transfer function would have a value of infinity. This is when the denominator of the equation is equal to zero. So poles exist at location x when $ax^2 + bx + c = 0$.

The well-known root-finding equation: $x = \dfrac{-b \pm \sqrt{b^2 - 4ac}}{2a}$ can be used to find the pole locations that, in this case, are at $-0.7071 + j0.7071$ and $-0.7071 - j0.7071$. On a pole-zero diagram this looks like Figure 3.1. Note that the poles both lie on the circumference of a circle with a radius of 1. They have a negative real part with equal and opposite imaginary parts (i.e., they are symmetrically placed above and below the real axis).

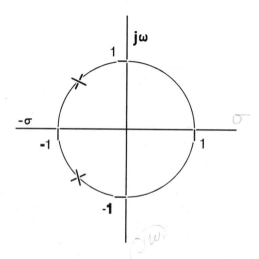

Figure 3.1

Second-Order Pole-Zero Diagram

Now let's consider a more complicated filter. The transfer function is given by general expression:

$$T(\omega) = \frac{ds^2 + e}{as^2 + bs + c}.$$

Not only is this equation for a third-order filter having three poles, it also has two zeroes. The roots of the denominator expression give the position of the poles. The position of the zeroes is given by the numerator expression. In the case of an Inverse Chebyshev filter, having a stopband starting at $\omega = 1$, the constants in the numerator are $d = 1$ and $e = 1.3333$. This gives the required zeroes at $\pm j1.15470$. Note that these are on the imaginary "j" axis and have no "real" part. The constants in the denominator depend on the required stopband attenuation.

If a 30 dB stopband attenuation is required the constants are $a = 1$, $b = 0.44086$, and $c = 0.23621$. The resultant poles are at -0.53578 and $-0.22043 \pm j0.43315$.

The pole-zero diagram for such an Inverse Chebyshev filter is shown in Figure 3.2. The 3 dB cutoff point for a filter with these pole positions is 0.47233 rad/s.

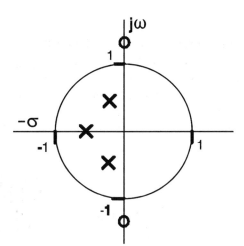

Figure 3.2

Pole Zero Diagram for Inverse
Chebyshev Filter

One of the most obvious differences is that the poles are nowhere near the unit circle. This is because of the way in which their positions are calculated, to give a stopband rather than the 3 dB attenuation point, starting at $\omega = 1$. When corrected to give a 3 dB point at $\omega = 1$, the pole and zero positions change. The zeroes move to $\pm j2.8228$, the poles become -1.1342 and $-0.4666 \pm j0.917$. These are plotted as shown in Figure 3.3.

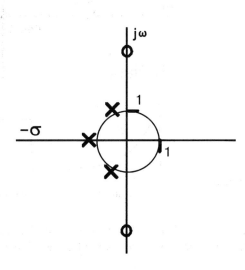

Figure 3.3

Normalized Inverse Chebyshev
Pole Zero Diagram

If the filter is third-order or higher, finding the pole positions are more difficult. Fortunately, the pole (and zero) positions of many filter designs have been published. There are also equations available for many designs to allow the pole and zero positions to be calculated.

Impulse Response and the S-Plane

The S-plane is a surface that has real and imaginary axes. In other words, $S = \sigma \pm j\omega$, with σ representing the real axis and $j\omega$ representing the imaginary axis. Because the Laplace Transform converts transient time domain signals into the frequency domain, positions on the S-plane describe signals that are transient in the frequency domain. A diagram best describes this; see Figure 3.4.

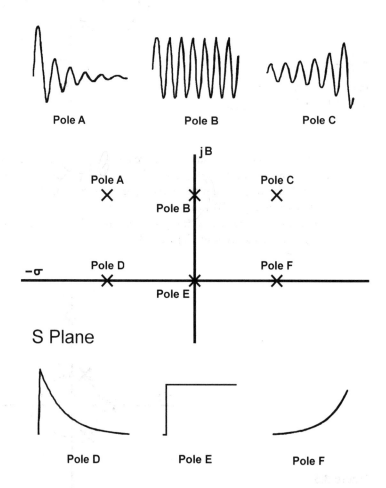

Figure 3.4

Transient Signals in the S-Plane

The real σ axis defines the decay when subject to an impulse. If $\sigma = 0$, the signal level rises immediately to its final value; that is, a step function (as shown by pole E). If α is negative the signal will decay with time. The more negative σ becomes, moving left and away from the imaginary ω axis, the faster the signal decays. If σ is to the right of the ω axis, the amplitude of the signal rises by the initial step value then grows exponentially.

The imaginary $j\omega$ axis describes the oscillatory nature of the signal and is often called the frequency axis. Moving the pole away from the S-plane origin causes the oscillation frequency to increase. If a point on the imaginary axis represents a signal, the amplitude response has a step increase to a level that is then maintained forever (as shown by pole B). Actually this signal would be represented by two points, both with the same value of ω, one above the real axis and one below. A sine wave has both a positive and a negative frequency, which is an interesting concept.

Complex signals, or responses, can comprise two or more points in the S-plane. For example a signal that combined a decaying and an oscillatory signal would be represented by two points, both to the left of the ω axis (to give the decay) and symmetrically above and below the σ axis (to give the oscillation). A filter response can be described in a similar way. The points described above are called poles and are represented by crosses in the S-plane. There are also points called zeroes which often lie on the ω axis, and these are represented by small circles in the S-plane. These describe a zero response, that is, no output, at certain frequencies. Given a pole-zero diagram it is possible to predict the frequency response of a circuit.

A powerful image of the S-plane is given by the analogy of tents used in camping. The poles in the S-plane are like those used to hold up a canvas. The zeroes are like pegs that hold the canvas down, except that it has to be imagined that the edge of the canvas is held down far away from the tent's center (infinity, actually). The pegs (zeroes) hold down the canvas at discrete points along a straight line, so there are dips in the canvas around the pegs. Perhaps the canvas is more like a rubber sheet, so that it stretches near the pegs. (See Figure 3.5.)

The Pole-Zero "Tent"

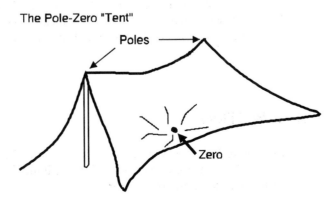

Figure 3.5

The S-Plane "Tent"

Now, taking this image further, consider a tent with a single pole. Move the tent to the left of the frequency axis so that the pole is along the negative real axis. This represents a first-order filter. Measuring the height of the canvas, by moving up and down the frequency axis, describes the frequency response of the filter. See also Figure 3.6.

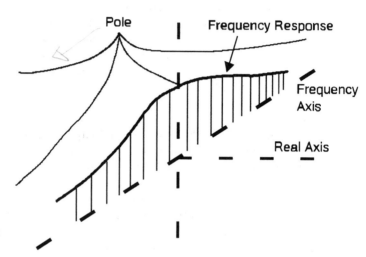

Figure 3.6

Frequency Response in the S-Plane

The Laplace Transform—Converting between Time and Frequency Domains

Reactive components have an impedance that can be expressed in terms of s. An inductor's impedance is sL, and a capacitor's impedance is 1/sC. Since reactive impedance is purely imaginary, the real part of S is equal to zero, or $\sigma = 0$, and the imaginary part has a magnitude equal to the frequency, or $j\beta = j\omega$. Having imaginary impedance means that if an AC signal is applied across the component, the current through it is 90° out of phase with the voltage.

First-Order Filters

First-order RC filters were described in Chapter 1 and are shown in Figure 3.7. The applications for a first-order filter are limited, but they are useful in developing analysis methods.

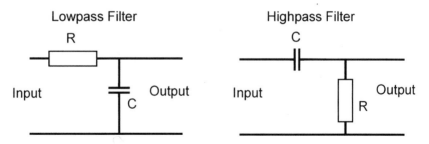

Figure 3.7

First-Order Filters

The frequency response of these circuits can be easily described in terms of the S-plane because they are simple potential dividers. Taking the lowpass filter first, the output voltage is given by:

$$V_0 = V_i \frac{1/sC}{R + 1/sC}$$

multiply top and bottom by sC

$$V_0 = V_i \frac{1}{1 + sCR}$$

The equation $1/(1 + sCR)$ is the transfer function for this filter. Since it is a first-order filter it only has one pole located on the negative real axis. In fact, a pole exists when the denominator is equal to zero, and this occurs when $sCR = -1$. In other words, $S = -1/CR$. Now, intuitively, if R or C is decreased in value the cutoff frequency is raised. This agrees with the pole location negative and moving along the real axis in the S-plane, away from the origin. The transfer function also has a zero, located at infinity. This can be explained because when $S = \infty$; the denominator is equal to infinity and therefore the equation is equal to zero.

The Laplace Transforms can be used to determine the time domain response of the RC filter. From published tables of time and frequency domain equivalents (many sources) is given:

Time domain ae^{-bt} = Frequency domain $a/(s + b)$

The transfer function needs manipulating to make it suitable for transformation. Letting $a = 1/CR$ and $b = 1/CR$ results in:

$$\frac{V_0}{V_i} = \frac{1/CR}{s + 1/CR} = \frac{1}{sCR + 1}$$

In the time domain Laplace Transforms give the response for an impulse. The impulse has a unit area and infinitely narrow width. A far more practical response is that obtained following the application of a step voltage. A step of amplitude "a" units—a.u(t)—in the time domain has a function a/s in the frequency domain, so multiplying the transfer function by $1/s$ (assuming a unit step) gives the desired result. Using a step input voltage, the frequency domain response becomes:

$$a/s(s+b).$$

$$\frac{a}{s(s+b)} = \frac{A}{s} + \frac{B}{s+b}$$

By the cover-up rule, $A = a/b[s = 0]$

$$B = a/-b[s = -b]$$

So this gives: $\dfrac{a}{bs}$ and $\dfrac{-a/b}{s+b}$

This equates to a step of $a/b - a/be^{-bt}$. Since a and b both equal $1/CR$, the equation simplifies and the time domain output voltage is, $V(t) = 1 - e^{\frac{-t}{CR}}$.

The time domain response for a lowpass filter is given in Figure 3.8. Decreasing RC reduces the decay period in the time domain, as well as raising the cutoff point in the frequency domain.

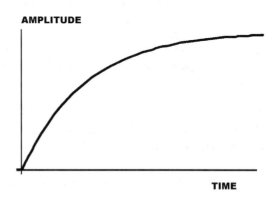

AMPLITUDE

TIME

Figure 3.8

Time Domain Response of
Lowpass Filter with Step Input

The highpass network can be analyzed in a similar way. Taking the highpass RC filter gives:

$$V_0 = V_i \frac{R}{R + 1/sC}$$

Divide by R, top and bottom

$$V_0 = V_i \frac{1}{1 + 1/sCR}$$

Alternatively, $\dfrac{V_0}{V_i} = \dfrac{sCR}{1 + sCR}$

By dividing top and bottom by CR, this becomes, $\dfrac{V_0}{V_i} = \dfrac{s}{1/CR + s}$.

Here the pole occurs at $S = -1/CR$, the same as for the lowpass filter, but now there is an S in the numerator. The equation is equal to zero when $S = 0$, so engineers refer to this as having a zero at the origin. The zero placed at infinity in the lowpass filter example has moved to the origin in the highpass case. This zero at $S = 0$ is intuitively correct since there is no output at zero frequency or DC.

The transfer function is similar to the lowpass filter case, except for the S that has appeared in the numerator. Multiplying a frequency domain equation by S means that the time domain equation must be differentiated. This is expressed mathematically as:

$$s.F(s) - f(0) = \frac{df(t)}{dt}, \text{ where } f(0) \text{ is the initial time domain condition.}$$

However, analysis of a step response requires division by S, so the equation for a highpass transfer function with a step input is simplified to:

$$\frac{V_0}{V_i} = \frac{1}{1/CR + s}$$

In the Laplace Transform $a/(s + b)$, $b = 1/CR$, and a = 1, so in the time domain:

$$h(t) = ae^{-bt} = e^{\frac{-t}{CR}}$$

The filter responds immediately to the fast rising edge of the step input, giving a step output followed by an exponential decay. This is shown in Figure 3.9.

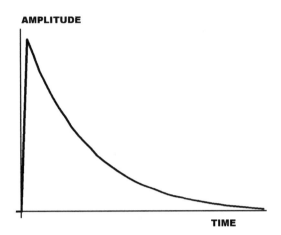

Figure 3.9

Time Domain Response of
Highpass Filters with Step Input

The simple lowpass and highpass filter examples given are simply to illustrate a point: that time domain and frequency domain responses are related. Generally, filter designers do not need to consider the time domain's step response.

Table 3.1, which follows, contains Laplace Transforms that may be useful.

Signal or Response	f(t)	F(s)
Exponential decay	ae^{-bt}	$\dfrac{a}{b+s}$
Critical damping	ate^{-bt}	$\dfrac{a}{(b+s)^2}$
Sine wave	$a.\sin \omega t$	$\dfrac{a\omega}{s^2+\omega^2}$
Cosine wave	$a.\cos \omega t$	$\dfrac{as}{s^2+\omega^2}$
Damped sine wave	$ae^{-bt}.\sin \omega t$	$\dfrac{a\omega}{(s+b)^2+\omega^2}$
Damped cosine wave	$ae^{-bt}.\cos \omega t$	$\dfrac{a(s+b)}{(s+b)^2+\omega^2}$

Table 3.1

Laplace Transforms

Pole and Zero Locations

Butterworth Poles

As briefly described above, the poles of the Butterworth response all lie on the unit circle; because of this they are the easiest to find out of all the filter designs.

The equations in the Appendix give the normalized pole positions for a Butterworth response with a 3 dB cutoff point at $\omega = 1$.

Using the formula given, pole positions have been obtained and are listed in Table 3.2.

Order, n	Real Part, $-\sigma$	Imaginary Part, $\pm j\omega$
1	1.0000	
2	0.7071	0.7071
3	0.5000	0.8660
	1.0000	
4	0.9239	0.3827
	0.3827	0.9239
5	0.8090	0.5878
	0.3090	0.9511
	1.0000	
6	0.9659	0.2588
	0.7071	0.7071
	0.2588	0.9659
7	0.9010	0.4339
	0.6235	0.7818
	0.2225	0.9749
	1.0000	
8	0.9808	0.1951
	0.8315	0.5556
	0.5556	0.8315
	0.1951	0.9808
9	0.9397	0.3420
	0.7660	0.6428
	0.5000	0.8660
	0.1737	0.9848
	1.0000	
10	0.9877	0.1564
	0.8910	0.4540
	0.7071	0.7071
	0.4540	0.8910
	0.1564	0.9877

Table 3.2

Butterworth Pole Positions

Bessel Poles

Bessel response poles also lie on a circle. However, when the poles are scaled to produce a response with a 3 dB cutoff frequency, the circle does not have a radius of unity and its center is not at the origin of the S-plane. The natural pole positions for a Bessel response are found for a filter that has a transmission delay of one second. In other words, they are normalized for their delay characteristics rather than their frequency response. The poles are not placed at equal angular distances from one another; they are spaced at approximately equal distances in the imaginary axis only. This is illustrated in Figure 3.10.

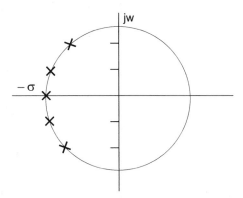

Figure 3.10

Bessel Pole Zero Diagram

Bessel response poles can be used to produce a filter with a 3 dB cutoff frequency if their positions are scaled. A table of pole positions for the Bessel response with a 3 dB cutoff frequency is provided here in Table 3.4. These values were found by re-normalizing the pole positions given by Thomson, which were normalized for a one-second delay. The frequency normalization process required the division of Thomson's values by a factor that was approximately equal to: $\sqrt{(2n-1).\ln 2}$. The actual factors used to normalize Thomson's values are given in Table 3.3.

Order, n	Normalizing Factor
1	1
2	1.36
3	1.75
4	2.13
5	2.42
6	2.7
7	2.95
8	3.17
9	3.39
10	3.58

Table 3.3

Bessel Normalizing Factors

Pole locations to produce the Bessel response are given in Table 3.4.

Order, n	Real Part, $-\sigma$	Imaginary Part, $\pm j\omega$
1	1.0000	
2	1.1030	0.6368
3	1.0509	1.0025
	1.3270	
4	1.3596	0.4071
	0.9877	1.2476
5	1.3851	0.7201
	0.9606	1.4756
	1.5069	
6	1.5735	0.3213
	1.3836	0.9727
	0.9318	1.6640
7	1.6130	0.5896
	1.3797	1.1923
	0.9104	1.8375
	1.6853	
8	1.7627	0.2737
	0.8955	2.0044
	1.3780	1.3926
	1.6419	0.8253
9	1.8081	0.5126
	1.6532	1.0319
	1.3683	1.5685
	0.8788	2.1509
	1.8575	
10	1.9335	0.2424
	0.8684	2.2996
	1.8478	0.7295
	1.6669	1.2248
	1.3649	1.7388

Table 3.4

Normalized Bessel Pole Positions

Chebyshev Pole Locations

The Chebyshev response has ripple in its passband. This is because the transfer function has poles that lie on an ellipse, rather than on a circle like the Butterworth response. The positions of the poles are related to Butterworth pole locations by hyperbolic trigonometric functions: $\sinh(x)$ and $\cosh(x)$. In general terms, poles move away from the real axis by a constant multiplying factor. They also move towards the imaginary axis by a different constant multiplying factor. This is shown in the S-plane diagram, in Figure 3.11.

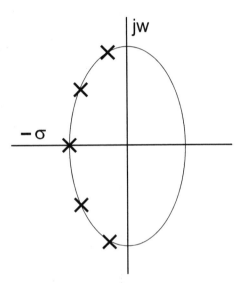

Figure 3.11

Chebyshev Pole Locations

Pole locations for the normalized Chebyshev response with a 3 dB cutoff point are given in Tables 3.5 to 3.9. The passband ripple values used to produce these tables are 0.01 dB, 0.1 dB, 0.25 dB, 0.5 dB, and 1 dB; these are the most popular values. You may notice that in all these tables, a first-order response pole is always real and positioned at −1.0. This should not be a great surprise since this is the same for all responses.

To keep the purists happy, pole locations for the normalized Chebyshev response with a "natural" cutoff frequency are given in Tables 3.10 to 3.14. If the natural cutoff frequency is at $\omega = 1$, the 3 dB attenuation frequency is at

$\omega = \cosh\left(\dfrac{1}{n}.\cosh^{-1}\dfrac{1}{\varepsilon}\right)$ where $\varepsilon = \sqrt{10^{0.1R} - 1}$ and R is the passband ripple in dB

and n is the filter order. The 3 dB attenuation frequency is always greater than one, provided that the passband ripple is less than 3 dB.

Filter Order	Real Part	Imaginary Part
1	1.00000	
2	0.67434	0.70750
3	0.42334	0.86631
	0.84668	
4	0.28009	0.92407
	0.67620	0.38276
5	0.19556	0.95120
	0.51199	0.58787
	0.63285	
6	0.14296	0.96603
	0.39057	0.70718
	0.53353	0.25885
7	0.10850	0.97501
	0.30401	0.78189
	0.43931	0.43392
	0.48760	
8	0.08490	0.98085
	0.24178	0.83152
	0.36185	0.55561
	0.42683	0.19510
9	0.06812	0.98486
	0.19613	0.86607
	0.30049	0.64282
	0.36860	0.34204
	0.39226	
10	0.05579	0.98773
	0.16191	0.89104
	0.25217	0.70714
	0.31776	0.45401
	0.35224	0.15644

Table 3.5

Chebyshev Poles with 3 dB
Bandwidth (0.01 dB Ripple)

Filter Order	Real Part	Imaginary Part
1	1.00000	
2	0.61042	0.71065
3	0.34896 0.69792	0.86837
4	0.21775 0.52570	0.92541 0.38332
5	0.14676 0.38423 0.47493	0.95211 0.58843
6	0.10494 0.28670 0.39165	0.96668 0.70766 0.25902
7	0.07850 0.21996 0.31785 0.35279	0.97550 0.78229 0.43414
8	0.06082 0.17321 0.25922 0.30577	0.98123 0.83185 0.55582 0.19518
9	0.04845 0.13952 0.21375 0.26221 0.27903	0.98516 0.86634 0.64302 0.34214
10	0.03948 0.11458 0.17846 0.22487 0.24927	0.98798 0.89127 0.70731 0.45412 0.15648

Table 3.6

Chebyshev Poles with 3 dB
Bandwidth (0.1 dB Ripple)

Filter Order	Real Part	Imaginary Part
1	1.00000	
2	0.56212	0.71536
3	0.30618	0.87122
	0.61236	
4	0.18646	0.92719
	0.45015	0.38405
5	0.12402	0.95330
	0.32469	0.58917
	0.40134	
6	0.08799	0.96754
	0.24040	0.70829
	0.32840	0.25925
7	0.06550	0.97613
	0.18354	0.78280
	0.26522	0.43442
	0.29437	
8	0.05058	0.98172
	0.14405	0.83226
	0.21559	0.55610
	0.25430	0.19528
9	0.04021	0.98555
	0.11577	0.86668
	0.17736	0.64327
	0.21757	0.34228
	0.23153	
10	0.03271	0.98830
	0.09491	0.89155
	0.14783	0.70754
	0.18628	0.45427
	0.20649	0.15653

Table 3.7

Chebyshev Poles with 3 dB
Bandwidth (0.25 dB Ripple)

Filter Order	Real Part	Imaginary Part
1	1.00000	
2	0.51291	0.72247
3	0.26829	0.87532
	0.53659	
4	0.16042	0.92970
	0.38728	0.38509
5	0.10570	0.95497
	0.27672	0.59020
	0.34205	
6	0.07459	0.96871
	0.20378	0.70915
	0.27837	0.25957
7	0.05534	0.97701
	0.15505	0.78350
	0.22406	0.43481
	0.24869	
8	0.04264	0.98240
	0.12143	0.83284
	0.18173	0.55648
	0.21436	0.19541
9	0.03384	0.98609
	0.09743	0.86715
	0.14928	0.64362
	0.18311	0.34247
	0.19487	
10	0.02750	0.98873
	0.07979	0.89195
	0.12428	0.70785
	0.15660	0.45447
	0.17360	0.15660

Table 3.8

Chebyshev Poles with 3 dB
Bandwidth (0.5 dB Ripple)

Filter Order	Real Part	Imaginary Part
1	1.00000	
2	0.45077	0.73514
3	0.22568	0.88230
	0.45135	
4	0.13251	0.93388
	0.31991	0.38683
5	0.08653	0.95772
	0.22654	0.59190
	0.28002	
6	0.06076	0.97066
	0.16599	0.71057
	0.22675	0.26009
7	0.04494	0.97845
	0.12591	0.78466
	0.18194	0.43545
	0.20194	
8	0.03455	0.98350
	0.09840	0.83377
	0.14727	0.55711
	0.17371	0.19563
9	0.02738	0.98697
	0.07885	0.86793
	0.12080	0.64420
	0.14818	0.34277
	0.15769	
10	0.02223	0.98945
	0.06451	0.89259
	0.10047	0.70837
	0.12660	0.45480
	0.14034	0.15671

Table 3.9

Chebyshev Poles with 3 dB
Bandwidth (1.0 dB Ripple)

Filter Order	Real Part	Imaginary Part
1	20.82774	
2	2.22776	2.33729
3	0.79469 1.58937	1.62621
4	0.41087 0.99192	1.35553 0.56148
5	0.25251 0.66109 0.81715	1.22820 0.75907
6	0.17147 0.46845 0.63992	1.15867 0.84820 0.31046
7	0.12426 0.34818 0.50313 0.55844	1.11664 0.89548 0.49695
8	0.09429 0.26852 0.40187 0.47404	1.08934 0.92350 0.61706 0.21668
9	0.07405 0.21321 0.32665 0.40070 0.42641	1.07060 0.94147 0.69879 0.37182
10	0.05971 0.17329 0.26991 0.34011 0.37701	1.05720 0.95371 0.75687 0.48594 0.16744

Table 3.10

Chebyshev Poles with Ripple
Bandwidth (0.01dB Ripple)

Filter Order	Real Part	Imaginary Part
1	6.55220	
2	1.18618	1.38095
3	0.48470	1.20616
	0.96941	
4	0.26416	1.12261
	0.63773	0.46500
5	0.16653	1.08037
	0.43599	0.66771
	0.53891	
6	0.11469	1.05652
	0.31335	0.77343
	0.42804	0.28309
7	0.08384	1.04183
	0.23492	0.83549
	0.33947	0.46366
	0.37678	
8	0.06398	1.03218
	0.18220	0.87504
	0.27268	0.58468
	0.32165	0.20531
9	0.05044	1.02551
	0.14523	0.90182
	0.22251	0.66935
	0.27294	0.35616
	0.29046	
10	0.04079	1.02071
	0.11837	0.92080
	0.18437	0.73075
	0.23232	0.46917
	0.25753	0.16166

Table 3.11

Chebyshev Poles with Ripple
Bandwidth (0.1dB Ripple)

Filter Order	Real Part	Imaginary Part
1	4.10811	
2	0.89834	1.14325
3	0.38361	1.09155
	0.76722	
4	0.21252	1.05678
	0.51306	0.43773
5	0.13503	1.03788
	0.35350	0.64145
	0.43695	
6	0.09339	1.02689
	0.25515	0.75173
	0.34854	0.27515
7	0.06845	1.02001
	0.19178	0.81798
	0.27714	0.45395
	0.30760	
8	0.05232	1.01545
	0.14900	0.86085
	0.22299	0.57520
	0.26304	0.20198
9	0.04130	1.01227
	0.11890	0.89018
	0.18217	0.66071
	0.22347	0.35156
	0.23781	
10	0.03342	1.00998
	0.09700	0.91112
	0.15108	0.72307
	0.19037	0.46424
	0.21102	0.15997

Table 3.12

Chebyshev Poles with Ripple
Bandwidth (0.25 dB Ripple)

Filter Order	Real Part	Imaginary Part
1	2.86278	
2	0.71281	1.00404
3	0.31323	1.02193
	0.62646	
4	0.17535	1.01625
	0.42334	0.42095
5	0.11196	1.01156
	0.29312	0.62518
	0.36232	
6	0.07765	1.00846
	0.21214	0.73824
	0.28979	0.27022
7	0.05700	1.00641
	0.15972	0.80708
	0.23080	0.44789
	0.25617	
8	0.04362	1.00500
	0.12422	0.85200
	0.18591	0.56929
	0.21929	0.19991
9	0.03445	1.00400
	0.09920	0.88291
	0.15199	0.65532
	0.18644	0.34869
	0.19841	
10	0.02790	1.00327
	0.08097	0.90507
	0.12611	0.71826
	0.15891	0.46115
	0.17615	0.15890

Table 3.13

Chebyshev Poles with Ripple
Bandwidth (0.5dB Ripple)

Filter Order	Real Part	Imaginary Part
1	1.96523	
2	0.54887	0.89513
3	0.24709	0.96600
	0.49417	
4	0.13954	0.98338
	0.33687	0.40733
5	0.08946	0.99011
	0.23421	0.61192
	0.28949	
6	0.06218	0.99341
	0.16988	0.72723
	0.23206	0.26618
7	0.04571	0.99528
	0.12807	0.79816
	0.18507	0.44294
	0.20541	
8	0.03501	0.99645
	0.09970	0.84475
	0.14920	0.56444
	0.17600	0.19821
9	0.02767	0.99723
	0.07967	0.87695
	0.12205	0.65090
	0.14972	0.34633
	0.15933	
10	0.02241	0.99778
	0.06505	0.90011
	0.10132	0.71433
	0.12767	0.45863
	0.14152	0.15803

Table 3.14

Chebyshev Poles with Ripple
Bandwidth (1.0dB Ripple)

Inverse Chebyshev Pole and Zero Locations

As suggested by their name, Inverse Chebyshev filters are derived from Chebyshev filters. The pole positions are the inverse of those given for Chebyshev filters. The frequency response of Chebyshev filters was described in Chapter 2. There are ripples in the passband with a smoothly decaying response in the stopband. Inverting the pole positions produces a filter with a smooth passband. The zeroes produce ripple in the stopband. Equations for finding Inverse Chebyshev poles are given in the Appendix.

Inverse Chebyshev Zero Locations

The zero frequency locations for any order of Inverse Chebyshev filter are provided in equations in the Appendix. Inverse Chebyshev zero locations found using these equations should be used with pole locations for the natural (normalized to stopband) response. The Inverse Chebyshev response can be normalized to have 3 dB passband attenuation. The zero locations for this response can be found by modifying these values. I have shown that the poles move away from the origin by a frequency-scaling factor (see Appendix for more details).

This same frequency factor has to be applied to zeroes, too. The zero locations move away from the origin, so the whole pole-zero diagram is scaled equally. Tables 3.16, 3.18, and 3.20 give the scaling factor and the new zero locations for Inverse Chebyshev filters with a 3 dB passband and with 20 dB, 30 dB, and 40 dB stopband attenuation, respectively. Tables 3.15, 3.17, and 3.19 give the corresponding pole locations.

Using these tables, a seventh-order pole-zero plot is given in Figure 3.12. The poles in high-order Inverse Chebyshev filters tend to be placed so that they lie

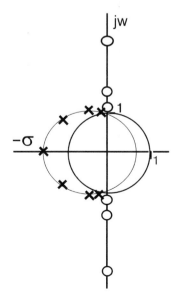

Figure 3.12

Seventh-Order Inverse Chebyshev
Pole Zero Plot

Filter Order	Real Part	Imaginary Part
2	0.70196	0.77604
3	0.42457 1.31299	0.96677
4	0.26575 1.19546	1.01172 0.78086
5	0.17766 0.81236 1.86437	1.01995 1.10097
6	0.12597 0.53765 1.67756	1.01931 1.16569 0.97457
7	0.09360 0.37271 1.13417 2.47872	1.01680 1.15880 1.35424
8	0.07215 0.27180 0.74868 2.17693	1.01423 1.13747 1.39884 1.21084
9	0.05725 0.20676 0.51888 1.45009 3.11628	1.01201 1.11619 1.35705 1.64504
10	0.04651 0.16264 0.37901 0.94794 2.68527	1.01018 1.09807 1.30383 1.66155 1.46308

Table 3.15

Inverse Chebyshev Poles with 3 dB Bandwidth and 20 dB Stopband Attenuation

Order	3dB Frequency	Zero 1	Zero 2	Zero 3	Zero 4	Zero 5
2	0.42738	3.30906				
3	0.65000	1.77646				
4	0.77384	1.39873	3.37683			
5	0.84462	1.24489	2.01428			
6	0.88763	1.16634	1.59325	4.35285		
7	0.91533	1.12060	1.39737	2.51797		
8	0.93408	1.09155	1.28757	1.92698	5.48758	
9	0.94730	1.07191	1.21893	1.64226	3.08645	
10	0.95696	1.05800	1.17281	1.47782	2.30176	6.67998

Table 3.16

Inverse Chebyshev Zero Locations with 3dB Bandwidth and 20dB Stopband Attenuation

Filter Order	Real Part	Imaginary Part
2	0.70658	0.72929
3	0.46668	0.91703
	1.13432	
4	0.31549	0.98080
	1.08499	0.57871
5	0.22153	1.00244
	0.84874	0.90666
	1.47021	
6	0.16191	1.00925
	0.62325	1.04100
	1.44056	0.64472
7	0.12265	1.01080
	0.45962	1.08414
	1.14673	1.03879
	1.88098	
8	0.09576	1.01049
	0.34762	1.09199
	0.85081	1.19326
	1.82236	0.76086
9	0.07667	1.00958
	0.27040	1.08743
	0.63200	1.23129
	1.44140	1.21808
	2.32176	
10	0.06269	1.00855
	0.21578	1.07916
	0.48091	1.22546
	1.06482	1.38252
	2.21852	0.89538

Table 3.17

Inverse Chebyshev Poles with 3dB Bandwidth and 30dB Stopband Attenuation

Order	3dB Frequency	Zero 1	Zero 2	Zero 3	Zero 4	Zero 5
2	0.24766	5.71025				
3	0.47234	2.44466				
4	0.63008	1.71786	4.14729			
5	0.73314	1.43419	2.32056			
6	0.80101	1.29247	1.76554	4.82355		
7	0.84702	1.21098	1.51006	2.72104		
8	0.87924	1.15962	1.36787	2.04716	5.82983	
9	0.90252	1.12510	1.27942	1.72376	3.23961	
10	0.91980	1.10075	1.22019	1.53753	2.39476	6.94985

Table 3.18

Inverse Chebyshev Zero Locations with 3 dB Bandwidth and 30 dB Stopband Attenuation

Filter Order	Real Part	Imaginary Part
2	0.70705	0.71416
3	0.48497	0.89059
	1.06023	
4	0.34458	0.95850
	1.01575	0.48477
5	0.25195	0.98715
	0.84806	0.78437
	1.27301	
6	0.18956	0.99931
	0.66693	0.94206
	1.27912	0.48413
7	0.14654	1.00433
	0.51754	1.01521
	1.10765	0.83445
	1.56433	
8	0.11606	1.00620
	0.40517	1.04570
	0.89101	1.02668
	1.57316	0.53963
9	0.09389	1.00667
	0.32256	1.05619
	0.70181	1.11330
	1.36537	0.93950
	1.89050	
10	0.07736	1.00652
	0.26151	1.05767
	0.55545	1.14468
	1.10003	1.15506
	1.88559	0.61545

Table 3.19

Inverse Chebyshev Poles with 3 dB Bandwidth and 40 dB Stopband Attenuation

Order	3dB Frequency	Zero 1	Zero 2	Zero 3	Zero 4	Zero 5
2	0.14072	10.04963				
3	0.33229	3.47501				
4	0.49672	2.17910	5.26081			
5	0.61882	1.69913	2.74925			
6	0.70627	1.46583	2.00236	5.47055		
7	0.76901	1.33382	1.66325	2.99707		
8	0.81470	1.25149	1.47623	2.20934	6.29166	
9	0.84865	1.19652	1.36063	1.83318	3.44524	
10	0.87438	1.15793	1.28357	1.61740	2.51915	7.31086

Table 3.20

Inverse Chebyshev Zero Locations with 3 dB Bandwidth and 40 dB Stopband Attenuation

in a circular pattern, with the axes centered on the real axes of the S-plane. As the pole positions move left along the real axis, their imaginary coordinate component increases rapidly to start with, but then slows as it reaches a maximum value. Moving further left, the pole's imaginary coordinate decreases again and approaches the negative real axis. One pole of an odd-order filter is on the negative real axis.

For those of you not wishing to use values normalized for a 3 dB cutoff point, Tables 3.21 to 3.23 give pole locations for the natural (normalized to stopband) Inverse Chebyshev responses. The tables give values for filters with 20 dB, 30 dB, and 40 dB stopband attenuation, respectively.

Zero locations have been found using the equations given in the Appendix. These are listed in Table 3.24 for the natural (normalized to stopband) Inverse Chebyshev response.

Filter Order	Real Part	Imaginary Part
2	0.17499	0.18062
3	0.22043	0.43315
	0.53578	
4	0.19879	0.61798
	0.68363	0.36464
5	0.16241	0.73493
	0.62225	0.66471
	1.07787	
6	0.12969	0.80842
	0.49923	0.83385
	1.15390	0.51642
7	0.10389	0.85616
	0.38931	0.91829
	0.97130	0.87987
	1.59322	
8	0.08420	0.88846
	0.30564	0.96013
	0.74807	1.04916
	1.60230	0.66898
9	0.06920	0.91116
	0.24405	0.98143
	0.57039	1.11126
	1.30089	1.09934
	2.09543	
10	0.05766	0.92766
	0.19848	0.99260
	0.44234	1.12717
	0.97942	1.27164
	2.04059	0.82357

Table 3.21

Inverse Chebyshev Poles with
20dB Stopband Bandwidth

Filter Order	Real Part	Imaginary Part
2	0.09950	0.10050
3	0.16115	0.29593
	0.35230	
4	0.17116	0.47610
	0.50454	0.24079
5	0.15592	0.61087
	0.52480	0.48539
	0.78777	
6	0.13388	0.70579
	0.47103	0.66535
	0.90341	0.34193
7	0.11269	0.77234
	0.39799	0.78070
	0.85179	0.64169
	1.20298	
8	0.09456	0.81975
	0.33009	0.85193
	0.72591	0.83644
	1.28166	0.43964
9	0.07968	0.85431
	0.27374	0.89634
	0.59559	0.94480
	1.15872	0.79730
	1.60437	
10	0.06764	0.88008
	0.22865	0.92480
	0.48568	1.00088
	0.96184	1.00996
	1.64872	0.53814

Table 3.22

Inverse Chebyshev Poles with
30dB Stopband Bandwidth

Order	Zero 1	Zero 2	Zero 3	Zero 4	Zero 5
2	1.41421				
3	1.15470				
4	1.08239	2.61313			
5	1.05146	1.70130			
6	1.03528	1.41421	3.86370		
7	1.02572	1.27905	2.30477		
8	1.01959	1.20269	1.79995	5.12583	
9	1.01543	1.15470	1.55572	2.92380	
10	1.01247	1.12233	1.41421	2.20269	6.39245

Table 3.23

Inverse Chebyshev Poles with 40 dB Stopband Bandwidth

Filter Order	Real Part	Imaginary Part
2	0.30000	0.33166
3	0.27597	0.62840
	0.85345	
4	0.20565	0.78291
	0.92509	0.60426
5	0.15005	0.86147
	0.68614	0.92991
	1.57469	
6	0.11182	0.90476
	0.47723	1.03469
	1.48905	0.86506
7	0.08568	0.93071
	0.34115	1.06068
	1.03813	1.23957
	2.26884	
8	0.06739	0.94737
	0.25388	1.06249
	0.69933	1.30663
	2.03342	1.13102
9	0.05423	0.95868
	0.19586	1.05738
	0.49154	1.28554
	1.37368	1.55835
	2.95207	
10	0.04451	0.96669
	0.15564	1.05081
	0.36270	1.24771
	0.90714	1.59003
	2.56969	1.40010

Table 3.24

Zero Locations for Inverse Chebyshev Filters with Natural Bandwidth

Cauer Pole and Zero Locations

Tables of pole and zero locations for some Cauer or elliptic function filters have been produced by Zverev,[2] but more extensive tables are given by Huelsman,[3] and by Stephenson.[4] These tables require the passband ripple, stopband attenuation, and the passband to stopband frequency ratio. For Cauer filters, the passband edge has the same attenuation as the ripple value; the response is not normalized to the 3 dB attenuation point. The reason is simply that the 3 dB point is difficult to calculate.

Pole and zero locations have been produced using these equations. Tables 3.25 to 3.30 give details of responses with passband ripple of 0.1 dB and 1 dB, and minimum stopband attenuation values of 30 dB, 40 dB, and 50 dB. For each passband ripple value, pole and zero locations have been tabulated to give stopband frequencies of 1.1, 1.2, 1.3, 1.4, 1.5, and 2.0. The tables show the stopband attenuation (loss) achieved, the filter order, and the pole and zero locations.

Stopband Frequency	Filter Order	Attenuation (dB)	Zero	Pole Real Part	Pole Imaginary Part
1.1	7	39	1.874772	0.3726101	0.7068724
			1.234481	0.1291187	0.9574289
			1.110913	0.0282793	1.0182750
				0.5996378	
1.2	6	39	3.598982	0.5476350	0.4296885
			1.495323	0.2354315	0.9076983
			1.222716	0.0545950	1.0342950
1.3	5	34	1.936892	0.3997024	0.8289210
			1.342284	0.0925559	1.0585590
				0.7155148	
1.4	5	39	2.138080	0.4105364	0.7979046
			1.450162	0.1048577	1.0630270
				0.6759573	
1.5	5	43	2.331876	0.4170394	0.7757674
			1.557406	0.1141299	1.0661520
				0.6497566	
2.0	4	41	4.922113	0.6704486	0.5356409
			2.143189	0.2162558	1.1168230

Table 3.25

Cauer Pole and Zero Locations (0.1 dB Ripple and 30 dB Stopband Attenuation)

Stopband Frequency	Filter Order	Attenuation (dB)	Zero	Pole Real Part	Pole Imaginary Part
1.1	8	49	3.886673	0.4667635	0.3448176
			1.542858	0.2464161	0.7959576
			1.194614	0.0923402	0.9640234
			1.108280	0.0222051	1.0136280
1.2	7	50	2.228609	0.3711950	0.6271911
			1.393320	0.1614481	0.9285521
			1.216500	0.0410804	1.0244980
				0.5197208	
1.3	6	46	4.130255	0.5189843	0.3893137
			1.664290	0.2561887	0.8800071
			1.328862	0.0660706	1.0392380
1.4	6	52	4.618428	0.5006298	0.3655174
			1.825298	0.2688599	0.8608322
			1.434274	0.0742502	1.0425360
1.5	5	43	2.331876	0.4170394	0.7757674
			1.557406	0.1141299	1.0661520
				0.6497566	
2.0	4	41	4.922113	0.6704486	0.5356409
			2.143189	0.2162558	1.1168230

Table 3.26

Cauer Pole and Zero Locations (0.1dB Ripple and 40dB Stopband Attenuation)

Stopband Frequency	Filter Order	Attenuation (dB)	Zero	Pole Real Part	Pole Imaginary Part
1.1	9	58	2.302010	0.3417308	0.5448127
			1.392020	0.1731486	0.8472668
			1.170604	0.0695129	0.9697934
			1.106502	0.0178438	1.0105670
				0.4482750	
1.2	7	50	2.228609	0.3711950	0.6271911
			1.393320	0.1614481	0.9285521
			1.216500	0.0410804	1.0244980
				0.5197208	
1.3	7	59	2.533631	0.3667336	0.5862146
			1.540439	0.1790991	0.9100119
			1.320986	0.0493328	1.0282150
				0.4818349	
1.4	6	52	4.618428	0.5006298	0.3655174
			1.825298	0.2688599	0.8608322
			1.434274	0.0742502	1.0425360
1.5	6	57	2.331876	0.4170394	0.7757674
			1.557406	0.1141299	1.0661520
				0.6497566	
2.0	5	58	3.250805	0.4290917	0.7213293
			2.089247	0.1389126	1.0735670
				0.5909335	

Table 3.27

Cauer Pole and Zero Locations (0.1dB Ripple and 50dB Stopband Attenuation)

Stopband Frequency	Filter Order	Attenuation (dB)	Zero	Pole Real Part	Pole Imaginary Part
1.1	5	30	1.480909	0.2021778	0.8048071
			1.122194	0.0346257	1.0002280
				0.4466498	
1.2	5	38	1.722895	0.2175734	0.7481695
			1.233340	0.0480848	0.9984797
				0.3915898	
1.3	4	32	2.845330	0.3773649	0.5212809
			1.368223	0.0887571	0.9976692
1.4	4	36	3.169408	0.3702850	0.4971396
			1.480785	0.0977832	0.9955948
1.5	4	39	3.478406	0.3649968	0.4806942
			1.592342	0.1044116	0.9939388
2.0	3	34	2.270068	0.2170489	0.9815897
				0.5400008	

Table 3.28

Cauer Pole and Zero Locations (1dB Ripple and 30dB Stopband Attenuation)

Stopband Frequency	Filter Order	Attenuation (dB)	Zero	Pole Real Part	Pole Imaginary Part
1.1	6	40	2.970935	0.3150958	0.4092459
			1.309230	0.1187325	0.8745158
			1.115061	0.0239272	0.9994161
1.2	6	50	3.598982	0.2894673	0.3598284
			1.495323	0.1365805	0.8342558
			1.222716	0.0332612	0.9983043
1.3	5	44	1.936892	0.2237550	0.7166589
			1.342284	0.0564516	0.9971266
				0.3649640	
1.4	5	49	2.138080	0.2269681	0.6961513
			1.450162	0.0622602	0.9960793
				0.3487910	
1.5	5	53	2.331876	0.2288748	0.6816781
			1.557406	0.0665407	0.9952537
				0.3378465	
2.0	4	51	4.922113	0.3512734	0.4424978
			2.143189	0.1214786	0.9891762

Table 3.29

Cauer Pole and Zero Locations (1dB Ripple and 40dB Stopband Attenuation)

Stopband Frequency	Filter Order	Attenuation (dB)	Zero	Pole Real Part	Pole Imaginary Part
1.1	8	59	3.886673	0.2486642	0.3003523
			1.542858	0.1396883	0.7377446
			1.194614	0.0547358	0.9343149
			1.108280	0.0133884	0.9992589
1.2	6	50	3.598982	0.2894673	0.3598284
			1.495323	0.1365805	0.8342558
			1.222716	0.0332612	0.9983043
1.3	6	57	4.130255	0.2757091	0.3356090
			1.664290	0.1454597	0.8107855
			1.328862	0.0390806	0.9974829
1.4	6	63	4.618428	0.2669266	0.3208161
			1.825298	0.1508535	0.7950983
			1.434274	0.0431288	0.9968596
1.5	5	53	2.331876	0.2288748	0.6816781
			1.557406	0.0665407	0.9952537
				0.3378465	
2.0	4	51	4.922113	0.3512734	0.4424978
			2.143189	0.1214786	0.9891762

Table 3.30

Cauer Pole and Zero Locations (1 dB Ripple and 50 dB Stopband Attenuation)

In some cases the values given in one table are the same as another. This occurs when the stopband attenuation achieved in producing one table exceeds that required for the next.

Cauer Pole Zero Plot

Cauer filters have a pole pattern similar to that of Chebyshev filters. The poles are placed in an elliptical pattern, but Cauer filters also have zeroes on the imaginary axis. An example of this is given in Figure 3.13, which shows the pole zero diagram for a fifth-order Cauer filter.

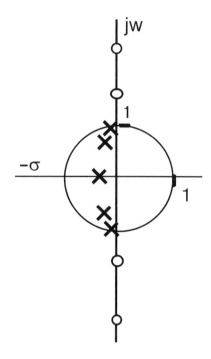

Figure 3.13

Fifth-Order Cauer Pole Zero Plot

References

1. Maddock, Robin. *Poles and Zeros in Electrical and Control Engineering*. London: Holt, Rinehart, Winston, 1982.

2. Zverev, Antonov. *Handbook of Filter Synthesis*. New York: John Wiley & Sons, 1967.

3. Huelsman, Laurance. *Active and Passive Analog Filter Design*. New York: McGraw-Hill, 1993.

4. Stephenson, Frederick. *RC Active Filter Design Handbook*. New York: John Wiley & Sons, 1985.

Exercises

3.1 What is the impulse response of a circuit that has a pole on the negative real axis?

3.2 Poles that have an imaginary component appear as a pair in the S-plane. If one such a pole has coordinates –0.3 + j0.67, what are the coordinates of the other pole?

3.3 On which axis do zeroes appear?

3.4   What is the effect of a pair of zeroes just outside the unit circle?

3.5   Where are Butterworth poles located, relative to the origin of the S-plane?

3.6   Where are Chebyshev poles located, relative to the origin of the S-plane?

CHAPTER 4

ANALOG LOWPASS FILTERS

This chapter describes how to design active or passive lowpass filters to almost any desired specification. Formulae and examples of how to use them are given for the denormalization of component values previously given in Chapters 2 and 3.

Passive Filters

Passive filters are the simplest to design from the normalized model. The model itself is a lowpass design, although normalized for a passband that extends from DC to 1 rad/s and is terminated with a 1 Ω load resistance. Denormalization for a higher load impedance requires component values to be scaled to have a higher impedance. The impedance of an inductor is proportional to its inductance, but the impedance of a capacitor is inversely proportional to its capacitance. Thus, if the load resistance is a more practical 50 Ω, inductance values are increased fifty-fold and capacitance values are reduced fifty-fold (to increase their impedance).

As an example, let's see how the component values change with a fifth-order Butterworth filter. In Figure 4.1 is the normalized lowpass model.

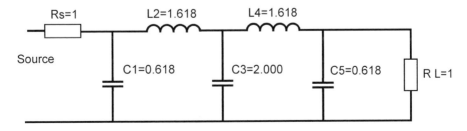

Figure 4.1

Fifth-Order Butterworth Normalized Model

Now scale these values, so that the source and load are terminated in 50 Ω. To do this you must multiply the inductance values by 50 and divide the capacitance values by 50. The result obtained by these calculations is shown in Figure 4.2.

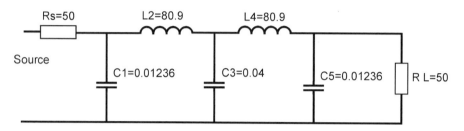

Figure 4.2

Fifth-Order Butterworth—Impedance Scaled

The component values are in units of Henries and Farads. Clearly these are not very practical values, but the filter design has a 1 rad/s cutoff frequency. So the next step is to frequency scale the design.

How do the values change when the cutoff frequency is scaled? Well, inductance values can be reduced because their impedance is proportional to frequency. As the signal frequency is raised, the inductor's reactance increases, so a lower value inductance can provide the same impedance as the inductor in the normalized filter. Capacitor values can also be reduced because as the signal frequency is raised, the capacitor's impedance decreases. To maintain the same performance at the new frequency the impedance must be increased. Since a capacitor's impedance is inversely proportional to the signal frequency, reducing the capacitance value raises the impedance and gives us the required result. Therefore, both capacitors and inductors are scaled by dividing their normalized values by the frequency scaling factor.

Since the normalized model has a 1 rad/s cutoff frequency, the scaling factor is $2 \pi F_C$ to convert the frequency into Hertz. Suppose a lowpass filter with a 4 MHz cutoff frequency and 50 Ω termination is wanted. The frequency scaling factor is $2\pi . 4 \times 10^6 = 25.133 \times 10^6$. In other words, the cutoff frequency required is 25.133×10^6 rad/s. All the inductor and capacitor values in the fifth-order lowpass filter (50 Ω version) must be divided by the frequency scaling factor. The result is shown in Figure 4.3 below.

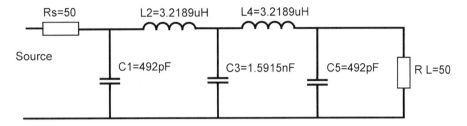

Figure 4.3

Fifth-Order Lowpass Filter Frequency Scaled to 4 MHz

Formulae for Passive Lowpass Filter Denormalization

I have described the process of passive filter denormalization. Now its time to write these as simple mathematical expressions:

$$L = \frac{RL*}{2\pi F_c}$$

$$C = \frac{C*}{2\pi F_c\, R}$$

$L*$ and $C*$ are the normalized lowpass component values. L and C are the final values after scaling. In practice, the design would be scaled for impedance and frequency in one step, by substitution of values into the given formulae.

A simple example will now be given. Suppose a fourth-order lowpass filter is required that has $600\,\Omega$ load impedance and a cutoff frequency of 3.4 kHz for telephone band speech. The filter is to be driven from a $0\,\Omega$ source (i.e., an ideal op-amp) and a 0.1 dB ripple Chebyshev response has been chosen.

The normalized values (refer to Chapter 2) are: $L1' = 1.51567$; $C2' = 1.77396$; $L3' = 1.45978$; $C4' = 0.67474$. The L' and C' here refer to the normalized component values given in the table. The apostrophe indicates that the ladder network begins with a series inductor. If the source had been of infinite impedance the ladder network would have begun with a capacitor and the values would have been for $C1$, $L2$, $C3$, and $L4$ respectively.

In the scaling formulae:

$$L = \frac{RL*}{2\pi F_c} \quad \text{and} \quad C = \frac{C*}{2\pi F_c\, R}$$

$R = 600$, $2\pi F_C = 21{,}363\,\text{rad/s}$.

The scaled component values for this filter are: $L1 = 42.57\,\text{mH}$; $C2 = 138.4\,\text{nF}$; $L3 = 41.0\,\text{mH}$; $C4 = 52.64\,\text{nF}$. Using these component values, the circuit given in Figure 4.4 is obtained.

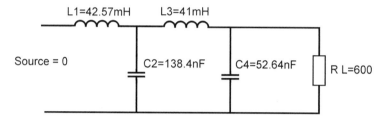

Figure 4.4

Passive Fourth-Order 0.1 dB Ripple Chebyshev Lowpass Filter (Scaled for 3.4 kHz and 600 Ω)

Denormalizing Passive Filters with Resonant Elements

Cauer (or elliptic function) and Inverse Chebyshev filters have series or parallel resonant circuits. These parallel resonant circuits provide a "zero" in the filter's stopband, which gives these filters a steep skirt response. But can the same denormalizing equations can be used for the resonant circuits?

The tuned frequency of an LC network is: $\omega_o = \dfrac{1}{\sqrt{LC}}$. If L and C are frequency scaled, by dividing them by a factor K $(=2\pi F_C)$, the equation becomes:

$$\omega_o = \frac{1}{\sqrt{\dfrac{L}{K} \cdot \dfrac{C}{K}}} = \frac{K}{\sqrt{LC}}$$

The tuned frequency has been multiplied by K, the scaling factor, which is exactly what was wanted. Therefore, the same denormalizing equations can be used with passive Cauer filters. Figures 4.5 and 4.6 give the circuit diagram for a Cauer filter having 0.1 dB ripple in the passband and 59 dB attenuation at twice the cutoff frequency. Note: Cauer filters have a cutoff point where the passband ripple is exceeded, which is at 0.1 dB in this case and not the 3 dB that I have been using up to now. The reason for not using the 3 dB point is the difficulty in scaling the component values. The normalized filter values were taken from Stephenson,[1] and a diagram of the circuit is given in Figure 4.5.

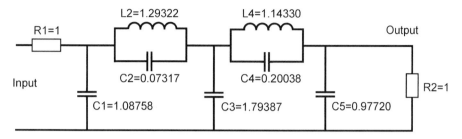

Figure 4.5

Normalized Cauer Lowpass Filter, 1 Rad/s Cutoff

If this design is now denormalized to have a $600\,\Omega$ load and a cutoff frequency of $10\,\text{kHz}$, then the design scaling factors are found using the formulae given.

$$L = \frac{RL*}{2\pi F_c} \text{ and } C = \frac{C*}{2\pi F_c\, R}$$

$L = L*(9.5493 \times 10^{-3})$ and $C = C*(2.65258 \times 10^{-8})$. Therefore, $L2 = 1.29322 \times (9.5493 \times 10^{-3}) = 12.349\,\text{mH}$, and $C1 = 1.08758 \times (2.65258 \times 10^{-8}) = 28.85\,\text{nF}$, and so on. The final result is shown in Figure 4.6.

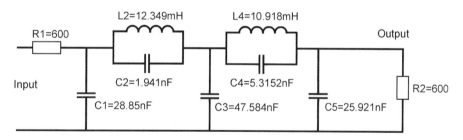

Figure 4.6

Denormalized Cauer Filter: $R = 600$; $F_c = 10\,\text{kHz}$

Mains Filter Design

Mains filters carry potentially high currents at dangerously high voltages, so care is essential in their design. The working voltage and current rating of components can be decided once the specification is known. The basic specification should include mechanical details such as the enclosure size, method of fixing, and any limit on its weight. The electrical specification should include the voltage and current rating. In addition the EMC performance and the allowable leakage current should be specified. The electrical specification must comply with national safety standards.

Filters work on the principal of providing a large discontinuity in the characteristic impedance seen by an unwanted signal. The intention is to reflect most of this unwanted energy back to its source. In the case of mains supplies, the source and load impedance varies wildly with frequency. The source impedance is variable over time and can be anywhere from $2\,\Omega$ to $2000\,\Omega$. The actual impedance is dependent on the loads that are connected to it and the frequency of interest. The characteristic impedance of the mains lead to the load is around $150\,\Omega$, and the load itself may have a variable impedance.

Mains filters are tested with a $50\,\Omega$ source and load impedance because most RF test equipment has a characteristic impedance of $50\,\Omega$. This allows consistent test results and allows direct comparison between one design and another. However, because the source and load impedance is not generally $50\,\Omega$ in practical situations, the attenuation predicted for a design based on this specification is generally optimistic compared with its performance in working equipment.

Inductors resonate at some frequencies due to unwanted interwinding capacitance. Similarly, capacitors resonate at some frequencies due to unwanted lead inductance. In a filter the performance of the inductors and capacitors used can depend critically on the resonant frequencies and on the source and load impedance.

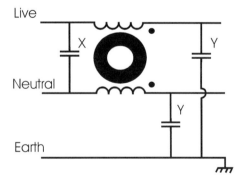

Figure 4.7

Typical Single-Stage Mains Filters

Mains filters with a single stage, such as those in Figure 4.7, are very sensitive to source and load impedance. This type of filter can easily increase the level of unwanted signals, rather than reduce them, when operated with source and load impedance other than their specification. This often occurs in the 150 kHz to 10 MHz frequency band, and the apparent gain can be as high as 20 dB.

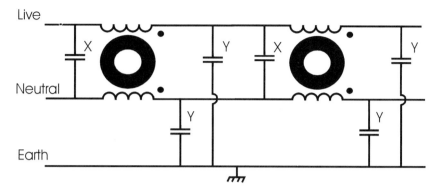

Figure 4.8

A Typical Two-Stage Mains Filter

Filters with two or more stages, such as the one found in Figure 4.8, are able to maintain an internal node at an impedance that is largely independent of the source and load impedance. This enables them to provide attenuation closer to the level specified for a $50\,\Omega$ source and load. These filters are larger and more expensive than the single-stage type.

There are two modes of interfering signals. Common-mode signals have a current that travels along both mains wires in the same direction and returns through earth or ground. Differential signals have a current that travels along one mains wire and returns along the other; thus the sum of the current carried by the two wires is zero, as is the earth current. The mains power supply is a differential signal with a low frequency ($50\,Hz$ in Europe, $60\,Hz$ in the United States). Since the differential mains supply signal carries high current, the filter inductors must be designed so they do not saturate their magnetic cores.

Most mains filters use common-mode chokes that are wound so that no magnetic flux is produced in the core by a purely differential signal. This is achieved by using an inductor with two windings and arranging for the go and return current to flow through them in opposing directions. Since no magnetic flux is produced, there is no inductive reactance. A common-mode current that flows in the same direction through both supply wires will generate a magnetic flux in the core and will thus have an inductive reactance. The common-mode choke thus appears as having a high series impedance to common-mode signals, but low series impedance to differential signals.

Differential-mode signals are presented with low impedance between the go and return wires by so-called "X capacitors." These X capacitors provide some degree of attenuation to the unwanted signals, but if high levels of attenuation are required, differential-mode chokes may have to be used. Because they must handle mains current these inductors tend to have low values of differential inductance and are physically quite large.

Most mains filters use so-called "Y capacitors" connected between earth and the go and return wires. These Y capacitors typically have values of around a few nano-farad (larger values would exceed earth leakage limits imposed by the relevant safety authorities).

The earth leakage limits imposed on medical equipment, especially if patient-connected, usually makes it impossible to use any reasonable size of Y capacitor. Instead, such filters have to use better inductors and more filter stages. To avoid this large and costly filter, the patient-connected end of the equipment is often made battery-powered and communicates with the mains-powered equipment through an electrically isolated path, such as an opto-coupler or fiber optic link.

Active Lowpass Filters

In Chapter 3, I stated that active filters are designed using pole and zero locations, which are determined from the frequency response's transfer function. This is not possible in passive filter designs because all the components interact with each other. However, in active filters the operational amplifier (op-amp), the "active" part of the circuit, buffers one stage from the next so there is no interaction. Each stage can therefore be designed to provide the frequency response of one pair of complex poles, or a single real pole, or sometimes both. When all the stages are connected in series, the desired overall response is produced.

Now that I have set the scene, I will describe some active filter designs and see how the pole and zero locations are used to find component values.

First-Order Filter Section

The first-order section is a simple structure comprising a lowpass RC network, followed by a buffer, as shown in Figure 4.9. The buffer serves to provide a high input impedance, so that the voltage at the connection node of the RC network is transferred to the buffer's output without being loaded by following stages. A simple RC network on its own would be loaded by following stages and therefore not have the expected frequency response.

The first-order section is an all-pole network, because it cannot produce zeroes in its frequency response. In fact, the first-order section has one real pole at $-\sigma$.

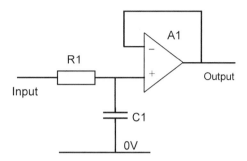

Figure 4.9

First-Order Active Filter

Letting $R1$ equal $1\,\Omega$ in the normalized lowpass model, calculation of $C1$ is simple:

$$C1 = \frac{1}{\sigma}, \text{ where } \sigma \text{ is the pole position on the negative real axis of}$$
the S-plane.

Sallen and Key Lowpass Filter

The Sallen and Key filter provides a second-order all-pole response and is a simple active lowpass design. It can be used for Bessel, Butterworth, or Chebyshev responses. High-order filters can be produced by cascading second-order sections. Odd-order filters can be produced by using a series of second-order sections and then adding a first-order section at the end.

The Sallen and Key filter uses an amplifier (which may be connected as a unity gain buffer) with a network of resistors and capacitors at the input. Capacitive feedback from the output is also used, and this can give rise to peaking in the frequency response. Peaking is required in second-order circuits where the Q is greater than unity and occurs due to phase shifts around the feedback loop. If the Q is large, say $Q = 10$, for example, the amplifier is providing a gain of 10 that restricts its bandwidth to 0.1 of the gain-bandwidth product. The diagram in Figure 4.10 shows the circuit.

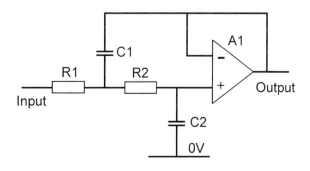

Figure 4.10

Sallen and Key Lowpass Filter
(Second-Order)

By letting $R1$ and $R2$ equal $1\,\Omega$ in the normalized design, the values of $C1$ and $C2$ can easily be calculated.

$$C1 = \frac{2Q}{\omega_n} = \frac{1}{\sigma} \quad \text{and} \quad C2 = \frac{1}{2\omega_n Q} = \frac{\sigma}{\sigma^2 + \omega^2}$$

In the case of Butterworth filters, $\omega_n = 1$ and $C2 = \sigma$, that is, the reciprocal of $C1$.

For example, the first pair of poles of a Butterworth fourth-order filter are $0.9239 \pm j0.3827$. A Sallen and Key filter section that has the same pole locations has $C1 = 1.0824$ and $C2 = 0.9239$.

The second filter section capacitors will number in sequence, being C3 and C4 and calculated from the same formula by substituting for C1 and C2, respectively. With poles at $0.3827 \pm j0.9239$, this filter section has capacitor values of $C3 = 2.613$ and $C4 = 0.3827$. The diagram in Figure 4.11 illustrates the whole circuit.

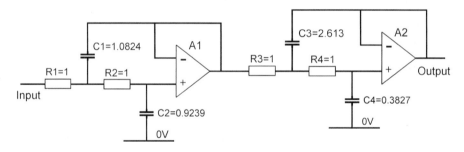

Figure 4.11

Fourth-Order Filter

The Sallen and Key lowpass filter is good if the requirements are not too demanding, with section Q factors below 50. In particular the gain-bandwidth product of the op-amps can limit the filter's cutoff frequency. I previously described this phenomenon in a magazine article,[2] in which I showed that the cutoff frequency limit was given by the empirical expressions:

$$\text{Butterworth passband frequency limit} = \frac{\text{Gain} - \text{Bandwidth Product}}{(\text{filter order})^2}$$

$$\text{Chebyshev (1dB) passband frequency limit} = \frac{\text{Gain} - \text{Bandwidth Product}}{(\text{filter order})^{3.2}}$$

As an example of how these formulae are used, consider a fifth-order filter using amplifiers with a 1 MHz gain-bandwidth product. If the filter is to have a

Butterworth response, its maximum passband frequency is 1 MHz/25 = 40 kHz. If, instead, a 1 dB Chebyshev response is wanted, the maximum passband frequency is limited to 1 MHz/172.5 = 5.8 kHz.

These frequency limits are for a maximum error in the passband of 2 dB. If no error is acceptable, the frequency limit will be much lower. Although the frequency limit can be raised by using an amplifier having a greater gain-bandwidth product, it can lead to instability. Usually, amplifiers with a high gain-bandwidth product have a minimum gain for stability. For example, the OP37 amplifier has a gain-bandwidth product of 63 MHz, but at a minimum gain of five.

Denormalizing Sallen and Key Filter Designs

In active filter designs the resistor values used should all be in the range 1 kΩ to 100 kΩ where possible. If resistor values are lower than 1 kΩ there may be a problem with loading of op-amp stage outputs. Loading can cause distortion and increases the supply current. If resistor values are much higher than 100 kΩ there may be problems with noise pickup. High impedance circuits can capacitively couple with external electric fields. These unwanted signals can then interfere with the wanted signal. Also, thermal noise voltage generated by the circuit's resistors increases in proportion to their resistance.

Active filters are based on a lowpass normalized filter model, using 1 Ω source and load resistors and a cutoff frequency of 1 rad/s. Denormalization is quite simple: (1) scale the impedance; the input impedance will tend towards 1 Ω as the frequency approaches the passband cutoff point; and (2) scale for frequency by denormalizing the capacitance value.

Impedance scaling is simply multiplying the resistor values by a value that gives a suitable input impedance. If you are driving the filter from a 600 Ω source it is probably better to make the input impedance high, say 56 kΩ (about 100 times 600 Ω), and then provide a separate 600 Ω resistive termination to match the source. This makes the input impedance correct for all frequencies. If 600 Ω resistors were used in the filter, the impedance would only be correct close to the cutoff frequency. The input impedance of an active filter changes with frequency because of the shunt and feedback capacitance.

Scaling the capacitor values can now be carried out using the following equation: $C = \dfrac{C'}{2\pi F_c R}$. Where C' is the normalized value calculated earlier, and R is the denormalized value chosen to give a suitable input impedance.

An example of denormalizing a Sallen and Key lowpass filter will now be given. Let us just consider a single second-order stage from a fifth-order Chebyshev filter having 0.5 dB passband ripple. The pole locations for this stage are 0.1057 ± j0.95497. The frequency scaling factor for this filter to have a 10 kHz cutoff frequency (F_C) is $2\pi F_C = 62,832$ rad/s.

The equations to find the capacitor values for a normalized 1 rad/s cutoff frequency are: $C'1 = \dfrac{1}{\sigma}$ and $C'2 = \dfrac{\sigma}{\sigma^2 + \omega^2}$. In these equations, substituting $\sigma = 0.1057$ and $\omega = 0.95497$ gives the result $C'1 = 9.46074$ and $C'2 = 0.1145$.

To find the frequency scaled component values use $C = \dfrac{C'}{2\pi F_c\, R}$. If $R = 10\,\text{k}\Omega$ (suitable for a source of about $100\,\Omega$), and the frequency scaling factor $2\pi F_C = 62,832$ rad/s, are substituted into the above equation, this gives:

$$C1 = \frac{9.46074}{62,832.10^4} = 15.057\,\text{nF} \approx 15\,\text{nF}$$

$$C2 = \frac{0.1145}{62,832.10^4} = 182.23\,\text{pF} \approx 180\,\text{pF}$$

Finally, $R1 = R2 = R = 10\,\text{k}\Omega$.

State Variable Lowpass Filters

This circuit design has a lower sensitivity to the op-amp's gain-bandwidth product limitation, and section Q factors of up to 200 are possible. It does, however, need three op-amps, as shown in Figure 4.12.

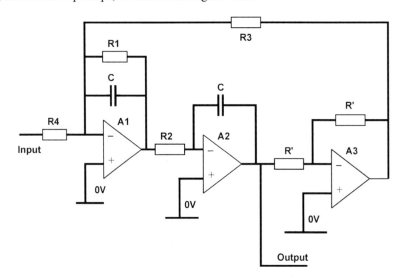

Figure 4.12

State Variable
Lowpass
(All-Pole)

Note that the output is in-phase with the input (subject to phase shifts due to the filter's response). The output could have been taken from A3, but would have been inverted.

The equations for this filter allow the arbitrary choice of capacitor, C.

$$R1 = \frac{1}{2\sigma C}$$

$$R2 = R3 = R4 = \frac{1}{C\sqrt{\sigma^2 + \omega^2}} = \frac{1}{\omega_n C}$$

A circuit gain of greater than unity can be achieved if the value of $R4$ is reduced. Dividing the value of $R4$ given in the last equation by a factor K gives the circuit a gain. The gain is equal to K.

Cauer and Inverse Chebyshev Active Filters

To design a Cauer or Inverse Chebyshev filter a different circuit topology is required. The Cauer response has zeroes outside the passband, so a notch circuit is required. This can be achieved using a circuit that is an extension of the state variable filter and is known as a biquad. This circuit is illustrated in Figure 4.13.

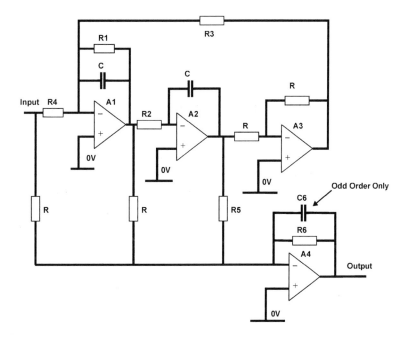

Figure 4.13

The Biquad Filter

The following equations give component values for the active biquad filter. As in the case of the state variable, the value of C can be chosen as any suitable value, then resistor values calculated from the equations. First compute the section's frequency from the pole location:

$$\omega_n = \sqrt{\sigma^2 + \omega^2}$$

$$R1 = R4 = \frac{1}{2\sigma C}$$

$$R2 = R3 = \frac{1}{\omega_n C}$$

$$R5 = \frac{2\sigma\omega_n R}{\omega_Z^2 - \omega_n^2}$$

ω_Z = the normalized zero frequency.

$$R6 = \left(\frac{\omega_n}{\omega_z}\right)^2 . AR$$

The gain at DC and low frequencies is represented by "A" in the equation. The resistors labeled R can be any arbitrary value; a typical value may be in the range $1\,k\Omega$ to $100\,k\Omega$, say $10\,k\Omega$. Odd-order filter sections can be implemented by adding a capacitor across $R6$. The value of this capacitor is given by the equation below:

$$C6 = \frac{1}{\sigma R6} \text{ where } \sigma \text{ is the value of the pole on the S-plane negative}$$

real axis.

Denormalizing State Variable or Biquad Designs

I have shown that the normalized component values used in passive filters, and in Sallen and Key active filters, can be scaled for different frequencies. However, the simplest approach with state variable and biquad filters is to start by frequency scaling the poles (and zeroes in the biquad case). Scaling pole and zero locations is easy: simply multiply them by the frequency scaling factor, $2\pi F_C$. The frequency scaled pole and zero locations can then be used in the design equations for state variable and biquad filters. These were given in the previous two sections.

Frequency scaling pole and zero locations can be visualized by considering the S-plane diagram. Frequency scaling moves the poles outward on a line that extends from the S-plane origin. To picture this, think of a pole at, say, $s = -0.75 + j1.2$ in a normalized response. If this is scaled for a frequency of $10\,Hz$, the scaling factor is $2\pi F_C = 62.83\,rad/s$, and the pole moves to $-47.12 + j75.396$. This is shown in the diagram of Figure 4.14 (not to scale).

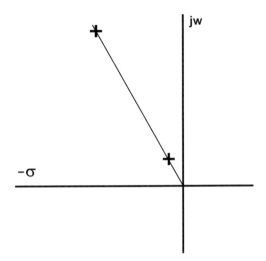

Figure 4.14

Frequency Scaling of Pole
Location in S-Plane

Each pole has a certain natural frequency (ω_n) and a certain magnifying factor (Q). The Q depends on the angle of the line from the S-plane origin to the pole location. As the pole-zero diagram is scaled for a higher cutoff frequency, the pole moves along the line from the S-plane origin to the pole location. This means that the value of Q remains unchanged as the pole location is scaled for frequency. The natural frequency ω_n is dependent upon the "σ" coordinate (real part), and this changes in proportion to the scaling of the diagram. More detail of frequency scaling of poles is given in the Appendix.

Zeroes are located on the imaginary axis, so scaling is simple. They are moved along this axis in proportion to the scaling frequency.

Choose a capacitor value and then use the equations given here to find the resistor values. If the resistor values are very small or very large, select a new capacitor value and try again. Again, aim to keep the resistor values between 1 kΩ and 100 kΩ. Here is an example for a biquad filter.

For example, design a second-order biquad filter, based on an Inverse Chebyshev design. The filter should have a passband of 1 kHz and a 30 dB stopband attenuation. Using the pole and zero location in Tables 3.17 and 3.18 given in Chapter 3, for a 3 dB passband attenuation at 1 rad/s, the zero is at 5.71025 and the poles are at 0.70658 ± j0.72929.

To scale these for a 1 kHz passband, multiply the pole and zero locations by the frequency scaling factor $2\pi F_C = 6283$ rad/s. Hence $F_Z = 35,877.5$ rad/s. The scaled poles are located at 4439.44 ± j4582.13 ($\sigma = 4439.44$ and $\omega = 4582.13$). The natural frequency of this pair of poles is given by

$$\omega_n = \sqrt{\sigma^2 + \omega^2} = 6380 \text{ rad/s}.$$

Component values can now be found by choosing an arbitrary value capacitor, C. Let $C = 100\,pF$.

$$R1 = R4 = \frac{1}{2\sigma C} = \frac{1}{2.4439.44.10^{-7}} = 1.126\,k\Omega$$

$$R2 = R3 = \frac{1}{\omega_n C} = \frac{1}{6380.10^{-7}} = 1.567\,k\Omega$$

$$R5 = \frac{2\sigma\omega_n R}{\omega_Z^2 - \omega_n^2} = \frac{56647254.R}{1.2465.10^9} = 0.0454454\,R$$

Letting $R = 10\,k\Omega$ gives $R5 = 454\,\Omega$. This is too low, so let $R = 33\,k\Omega$. Now $R5 = 1500\,\Omega$.

ω_z = the denormalized zero frequency of 35,877.5 rad/s. Let gain $A = 1$.

$$R6 = \left(\frac{\omega_n}{\omega_z}\right)^2 .AR = \left(\frac{6380}{35,877.5}\right)^2 .33k\Omega.\text{Hence } R6 = 1k\Omega.$$

Frequency Dependent Negative Resistance (FDNR) Filters

Frequency dependent negative resistance (FDNR) circuits can be used to make an active filter based on a passive ladder filter design. In applications where an elliptical lowpass filter is required and an active filter is possible, FDNR filters can be used as an alternative to a biquad filter. For example, a third-order elliptic lowpass filter requires a biquad design with four op-amps, ten resistors, and three capacitors. The same design using an FDNR requires two op-amps, eight resistors, and four capacitors. An obvious advantage is the reduction of op-amps from four down to two. Halving the number of op-amps required for the filter halves the supply current, assuming that the same type of op-amp would be required in both circuits.

However, there is a catch. In order for the circuit to work as specified, the source impedance should be zero. This can be compensated for by simply reducing the value of a series resistor in the design (more on this later). The greater problem is the output load. The load must be high impedance for the circuit to work properly. Of course, in multistage filters such as a seventh-order elliptic filter, a biquad design would require three biquad stages connected in series (twelve op-amps). A similar FDNR filter would require six op-amps, seven including a buffer at the output.

The most significant advantage of doubly terminated lossless ladder circuits is the low sensitivity to component tolerances. However, inductors are bulky and are difficult to obtain. Low value inductors for radio applications are reasonably easy to find, but audio frequency applications require much larger values. High-value inductors often have to be specially wound in order to obtain the required inductance.

Replacing the inductors and capacitors by resistors and FDNRs gives the same low sensitivity to component tolerances. If there are two signal paths in a system that must be closely matched in terms of amplitude and phase, an FDNR filter is the better choice. For all these reasons there is some advantage in using the FDNR for "all-pole" designs, such as Butterworth or Chebyshev. So now I have convinced you, I hope, that in some application, FDNR filters are a "Good Thing." But what are FDNRs?

The schematic symbol for an FDNR looks like a capacitor with four plates instead of the usual two and is assigned a letter D. The FDNR is also known as a D-element. A frequency dependent negative resistance (FDNR) is an active circuit that behaves like an unusual capacitor. In a lowpass RC circuit, the voltage drop across the shunt capacitor falls with increasing frequency. Beyond the passband, doubling the frequency halves the voltage across the capacitor. In a lowpass RD circuit, in which the FDNR has replaced the capacitor, the voltage drop across the FDNR falls at double the rate. Thus, above the passband, doubling the frequency quarters the output signal amplitude.

In decibel terms, a signal applied to an RC network has a rate of fall of 6 dB/octave (a first-order filter). The same signal applied to an RD network has a rate of fall of 12 dB/octave of a capacitor. This double rate of fall is the reason for the four plates in the D-element symbol, rather than the two in a capacitor symbol. The circuit of an FDNR is given in Figure 4.15.

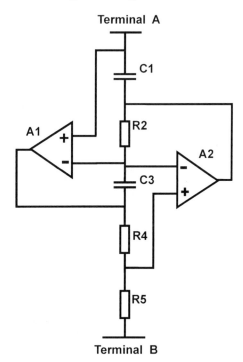

Figure 4.15

Circuit Diagram of an FDNR

In a simple approach where all resistors are equal to $1\,\Omega$ and all capacitors are equal to $1\,F$, the circuit behaves like a negative resistance of $-1\,\Omega$. The equation for the negative resistance is:

$$D = \frac{R2.R4.C1.C3}{R5}$$

If $C1 = C3 = 1\,F$ and $R4 = R5$, the negative resistance equals $R2$.

Now I have shown what an FDNR looks like. How do you use it? Transformation of the passive components is needed. FDNR elements are used to replace the capacitors in passive lowpass filters. Resistors are used to replace the inductors. This allows the filter size to be reduced, and a miniature hybrid circuit is possible. The design begins with a conventional double terminated lowpass LC filter design, in the T configuration. This has resistors (for the source and load), series and shunt inductors, and shunt capacitors. Figure 4.16 shows a normalized elliptic lowpass LC filter.

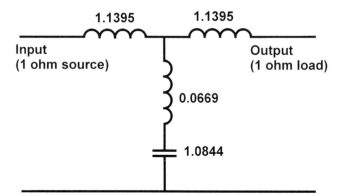

Figure 4.16

Circuit of Normalized
Lowpass LC Filter

To convert the passive design into an FDNR design, the resistors are replaced by capacitors, the inductors are replaced by resistors, and the capacitors are replaced by FDNRs. If the source and load resistor are $1\,\Omega$, these are replaced by capacitors of $1\,F$. Generally, the capacitor value is $1/R$, so if the load was $0.2\,\Omega$ the capacitor would be $5\,F$.

Inductors are replaced by resistors. A $1\,H$ inductor becomes a $1\,\Omega$ resistor. Generally, $R = L$, so a $1.1395\,H$ inductor would be replaced by a $1.1395\,\Omega$ resistor.

Capacitors are replaced by FDNRs. In an FDNR, the resistors are normalized to $1\,\Omega$ and the capacitors are normalized to $1\,F$, to replace a $1\,F$ capacitor. If the normalized capacitor is not $1\,F$, the value of $R2$ (in Figure 4.15) is scaled in proportion. Generally, $R1 = C$. Thus a $1.0844\,F$ capacitor is replaced by an FDNR that has $R2 = 1.0844\,\Omega$.

The conversion process is displayed in Figure 4.17.

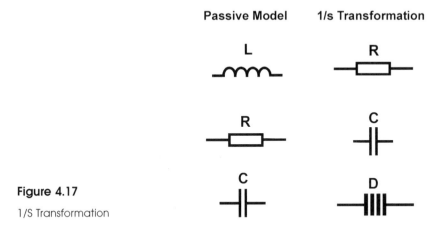

Figure 4.17

1/S Transformation

Applying these simple rules to the normalized lowpass design given in Figure 4.16 gives the FDNR equivalent design, illustrated in Figures 4.18 and 4.19.

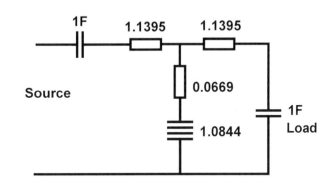

Figure 4.18

Lowpass Filter with D-Element

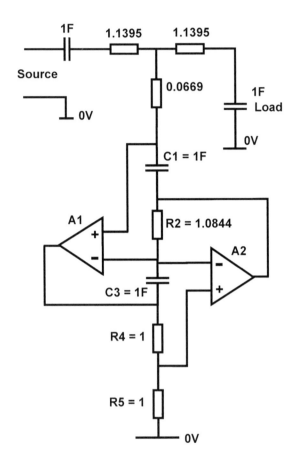

Figure 4.19

Normalized Lowpass FDNR Filter

Denormalization of FDNR Filters

Now apply frequency scaling to obtain practical component values. I will now design a third-order filter that has a passband of 15 kHz. The normalized design has a passband of 1 rad/s, so the frequency scaling factor is $2.\pi.F$. The frequency scaling factor is 94,247.78 in this case. All capacitor values must now be divided by 94,247.78, which makes each one equal to $10.6103\,\mu F$. This value is a little too large and must be reduced to a more convenient value. Let us divide the capacitor value by 1061.03, so that all capacitors in the circuit are now 10 nF. Each resistor must now be multiplied by this scaling factor. Resistor values of $1.061\,k\Omega$ are now required for R4 and R5.

Before redrawing the filter, the value of $R2$ in the FDNR circuit must be defined. If the normalized capacitor is not 1 F, the value of $R2$ is given by $1.061\,k\Omega$ multiplied by the normalized capacitor value. If, for example, the capacitor in the passive filter has a value of $1.0844\,F$, the value of $R2$ in the FDNR will be $1.061\,k \times 1.084 = 1.15\,k\Omega$.

Finally, a DC path from the source to the load must be allowed. This will give 6 dB insertion loss, the same as a terminated lossless ladder filter. The output load should be a high value, compared with the other series components; a value of 100 kΩ is often used. The input capacitor must be bypassed by a resistor that has a value less than 100 kΩ. The bypass resistor value should be 100 kΩ minus the sum of other series resistors. Suppose the other series resistors (replacing series inductors in the passive filter) sum to 2.416 kΩ, the bypass resistor should have a value of (100 − 2.416) kΩ or 97.584 kΩ.

Figure 4.20 gives the circuit diagram of the final FDNR lowpass filter.

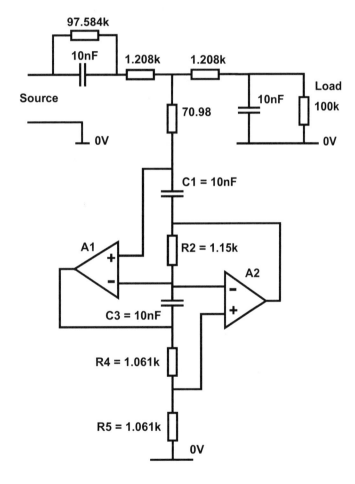

Figure 4.20

FDNR Lowpass Filter

An important point is that the common rail of the filter should be connected to the 0 V rail of the supply. The op-amp should then be powered from positive and negative supply rails.

References

1.   Stephenson, F. W. *RC Active Filter Design Handbook*. New York: John Wiley & Sons, 1985.

2.   Winder, S. "The Real Choice for Active Filters." *Electronics World and Wireless World*, September 1993.

Exercises

4.1   A normalized inductor value is 0.8212. Denormalize this for a passive lowpass filter having a load resistance of 50 ohms and a cutoff frequency of 20 kHz. What is its denormalized value?

4.2   A normalized capacitor value is 0.5532. Denormalize this for a passive lowpass filter having a load resistance of 600 ohms and a cutoff frequency of 100 kHz. What is its denormalized value?

4.3   What happens to pole locations in the S-plane as the frequency is scaled?

4.4   A second-order Butterworth filter has poles at $-0.7071 \pm j0.7071$. What are the two capacitor values for a normalized Sallen and Key active filter (let $R1 = R2 = 1\,\Omega$)?

4.5   For the active filter in Exercise 4.4, let the resistor values equal $1\,k\Omega$. What are the capacitors' values ($C1$ and $C2$) if the cutoff frequency is 10 kHz?

CHAPTER 5

HIGHPASS FILTERS

This chapter describes how to design an analog active or passive highpass filter having almost any desired specification. This chapter, like the previous one, uses information from Chapters 1, 2, and 3. Examples for most types of highpass filter are given. Formulae will be presented for the denormalization of component values given in previously presented tables.

Passive Filters

Passive highpass filters are designed using the normalized lowpass model. The model is normalized for a passband that extends from DC to 1 rad/s and is terminated with a 1 Ω load resistance. The first part of the process is to carry out the conversion to a highpass model; this can then be scaled for the desired load impedance and cutoff frequency. The highpass model has a passband that extends from 1 rad/s to infinity (in theory, at least). In practice, parasitic components exist to reduce the upper frequency response. These parasitic components are, for example, capacitance between wires in an inductor's windings or inductance in the leads of a capacitor. More details on these are given in Chapter 10.

Converting the lowpass model into a highpass equivalent is not too demanding in all-pole filters, like Butterworth or Chebyshev types. The process requires replacing each inductor in the lowpass model by a capacitor. Similarly, each capacitor in the lowpass model has to be replaced by an inductor.

In Cauer or Inverse Chebyshev filters there are series or parallel resonant LC networks. For these components, replacing inductors in the lowpass model by capacitors and replacing capacitors in the lowpass model by inductors would appear to give no change. The net result is a series or parallel resonant circuit as before. However, when each component is replaced by one with an opposite reactance, the replacement will have a value that is the reciprocal of its value in the lowpass model. Thus, the inductance value will be the reciprocal of the capacitance value that it replaced. Also, the capacitance value will be the reciprocal of the inductance value that it replaced. The LC network will then resonate at the reciprocal of its lowpass frequency.

Figures 5.1 and 5.2a show the component-replacing process for a simple all-pole filter. More complex filters, such as Cauer, will be described further later on in the chapter.

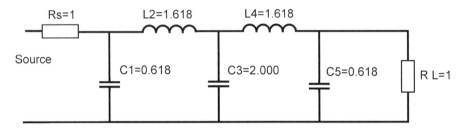

Figure 5.1

Normalized Fifth-Order Butterworth Lowpass Model

Converting this into a highpass model, gives the result in Figure 5.2a.

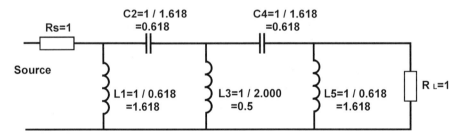

Figure 5.2a

Normalized Fifth-Order Butterworth Highpass Model

This is not a minimum inductor design any longer. However, a circuit with an entirely equal response is given in Figure 5.2b in which shunt inductors have been replaced by series capacitors of the same value. Also, shunt inductors replace series capacitors of the same value.

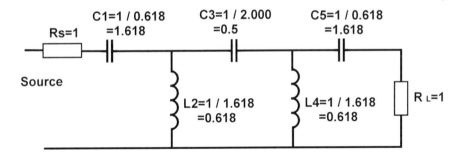

Figure 5.2b

Minimum Inductor Fifth-Order Butterworth Highpass Model

Denormalization of the highpass model for higher load impedance requires component values to be scaled to have higher impedance. This is an identical process to that of denormalizing a lowpass filter. The impedance of an inductor is proportional to its inductance, but the impedance of a capacitor is inversely proportional to its capacitance. Thus, if the load resistance is a more practical 600 Ω, inductance values are increased 600-fold and capacitance values are reduced 600-fold.

As an example, let's see how the component values change in the fifth-order Butterworth highpass model given in Figure 5.2a. Now, let's scale these values so that the source and load are terminated in 600 Ω. By multiplying the inductance values by 600 and dividing the capacitance values by 600, the result shown in Figure 5.3 is obtained.

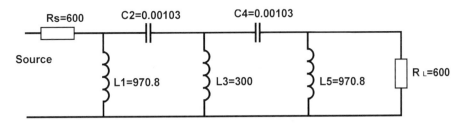

Figure 5.3

Fifth-Order Butterworth Impedance Scaled

The component values are in Henries and Farads. As found with the lowpass denormalization, these are not very practical values. The cutoff frequency is still 1 rad/s, so the next step is to frequency scale the design.

How do the values change when the cutoff frequency is scaled? In exactly the same way that lowpass values change, by reducing both capacitance and inductance values by 2π times the cutoff frequency (in Hertz). Inductance values can be reduced because their impedance is proportional to frequency. To maintain the same impedance at a higher frequency requires less inductance. Capacitor values can also be reduced because a capacitor's impedance is inversely proportional to the frequency, to have the same impedance at a higher frequency requires less capacitance.

Since the normalized model has a 1 rad/s cutoff frequency, the scaling factor is $2\pi F_c$ to convert the frequency into Hertz. Let's design a highpass filter with a 100 kHz cutoff frequency and 600 Ω termination. The frequency scaling factor is $2\pi . 100 . 10^3 = 628.32 . 10^3$. In other words, the cutoff frequency required is $628.32 . 10^3$ rad/s. All the inductor and capacitor values in the fifth-order highpass filter shown in Figure 5.3 (which has already been scaled for a 600 Ω source and load) must be divided by the frequency-scaling factor. The result is shown in Figure 5.4.

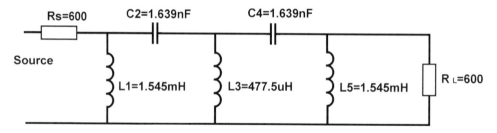

Figure 5.4

Fifth-Order Highpass Filter Frequency Scaled to 100 khz

Formulae for Passive Highpass Filter Denormalization

The process of filter denormalization for highpass filters has an addition step compared with the process used in denormalizing lowpass filters. The mathematical expressions are similar to those used in the lowpass case, except that now the inductance value is proportional to the inverse of the normalized lowpass capacitance value. Similarly, the capacitance value is proportional to the inverse of the normalized lowpass inductance value.

$$L = \frac{R}{2\pi F_c C*}$$

$$C = \frac{1}{2\pi F_c RL*}$$

$L*$ and $C*$ are the normalized lowpass component values. L and C are the final values after scaling. Usually the design would be scaled for impedance and frequency in one step, using the given formulae.

Here is a simple example using the formulae. Design a third-order highpass filter having a 0.25 dB Chebyshev response, a 20 kHz cutoff frequency, infinite (open circuit) load, and source impedance of 150 Ω. Note that the source impedance is quoted, but the load is infinite. Referring to Tables 2.16 to 2.20 in Chapter 2 for Chebyshev passive filters, you will notice that they are for filters having a zero or infinite source impedance. The normalized values of a 0.25 dB Chebyshev lowpass model with infinite load impedance can be found by reversing the order of the elements.

From Table 2.18 in Chapter 2, the element values, as given, are C1 = 1.53459, L2 = 1.52828, and C3 = 0.81651. Reversing these to give an infinite load impedance gives C1 = 0.81651, L2 = 1.52828, and C3 = 1.53459, as shown in Figure 5.5. Note that a shunt capacitor is needed at the load end to terminate the filter; a series inductor connected to the output would have no effect.

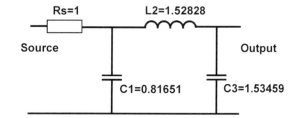

Figure 5.5

Normalized Lowpass Filter with
Infinite Load Impedance

The filter design in Figure 5.5 is a normalized lowpass and, using the formulae, can be converted to a highpass denormalized design in one step. Applying the formulae to scale and convert to highpass gives:

$$L = \frac{R}{2\pi F_c C*} \qquad C = \frac{1}{2\pi F_c R L*}$$

$$2\pi F_c = 2\pi . 20\,\text{kHz} = 125{,}664\,\text{rad/s}$$

$$L1 = 150/(125{,}664 . C1*) = 1.4619\,\text{mH}$$

$$C2 = 1/(125{,}664 . L2* . 150) = 34.713\,\text{nF}$$

$$L3 = 150/(125{,}664 . C3*) = 0.7778\,\text{mH}$$

The filter now has a shunt inductor across the output, replacing the shunt capacitor in the lowpass model. A minimum inductor design is not possible in this case. This circuit is shown in Figure 5.6.

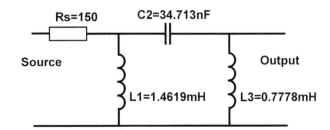

Figure 5.6

Highpass Filter with Infinite Load
Impedance

Highpass Filters with Transmission Zeroes

Cauer (or elliptic function) and Inverse Chebyshev filters are more complicated
than the ladder filters already described. They have series or parallel resonant
circuits. The lowpass model can have either parallel tuned LC circuits in the
series arms (replacing the inductors in the ladder circuit) or series tuned LC
circuits in the shunt arms (replacing the capacitors in the ladder circuit). In
Chapter 4, I showed that the same denormalizing equations can be used for the
resonant circuits; now I am also converting to highpass, so the element values
must be swapped as well. An example of this follows.

The normalized component values for a 0.1 dB passband ripple Cauer filter were
taken from a table given in Stephenson.[1] This circuit has 30 dB stopband attenuation, starting at 2.5 times the cutoff frequency; a diagram of the circuit is
given in Figure 5.7.

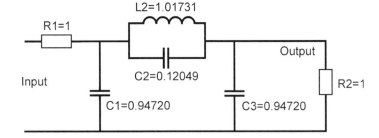

Figure 5.7

Normalized Cauer
Lowpass Filter, 1 rad/s
Cutoff

An alternative design uses two series inductors between source and load, with
a series resonant circuit between their connecting node and the common rail. In
that case the value of $L2$ in Figure 5.7 becomes the value of $C2$ in the alternative design. Similarly, the value of $C2$ in Figure 5.7 becomes that of $L2$ in the
alternative design.

Converting the minimum inductor design into a highpass circuit is straightforward. The shunt capacitor of the lowpass prototype becomes a shunt inductor

in the highpass design. This also applies to the component values in the parallel tuned circuit. The parallel tuned arm between the shunt capacitors of the lowpass prototype is also present in the highpass design. However, in the highpass circuit, the value of the inductor is derived from the capacitor value in the prototype, and the value of the capacitor is derived from the inductor value in the prototype. The value of the highpass component is the inverse of the lowpass design. This uses the same equations as before.

$$L = \frac{R}{2\pi F_c C^*} \qquad C = \frac{1}{2\pi F_c RL^*}$$

This highpass circuit is illustrated in Figure 5.8

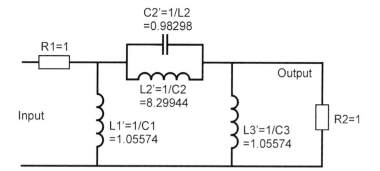

Figure 5.8

Third-Order Highpass
Filter

Generally, circuits with a minimum of inductors are preferred for ease of manufacture. Conversion from this design into minimum inductor highpass filters is straightforward, though. A shunt inductor becomes a series capacitor with the same element value. A parallel tuned circuit in the series arm becomes a series tuned circuit in the shunt arm. The value of the series arm capacitor is used for the value of the shunt arm inductor, and the value of the series arm inductor is used for the shunt arm capacitor.

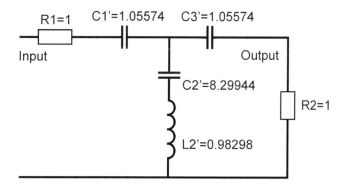

Figure 5.9

Minimum Inductor Highpass
Conversion

Equations to convert from the minimum inductor lowpass model to the minimum inductor highpass filter are given by:

$$L = \frac{R}{2\pi F_c L^*} \qquad C = \frac{1}{2\pi F_c R C^*}$$

Note that the inductor and capacitor values in this circuit are given by the reciprocal of the inductor and capacitor values, respectively, in the normalized lowpass. Previously the capacitor values were determined by the reciprocal of the lowpass inductor values. The reason for the change is that now the position of capacitors in the lowpass model coincides with the position of capacitors in the highpass model. The same is true for inductors.

Active Highpass Filters

Active filters use pole and zero locations from the frequency response's transfer function. Tables of pole and zero values were given in Chapter 3. The operational amplifier (op-amp), the "active" part of the circuit, buffers one stage from the next so there is no interaction. Each stage can therefore be designed to provide the frequency response of one pair of complex poles, or a single real pole, or sometimes both. When all the stages are connected in series the overall response is that which is desired.

A lowpass to highpass translation is required to find the highpass normalized pole and zero locations. Normalized lowpass response pole and zero locations are used as a starting point in the following formulae:

$$\sigma_{HP} = \frac{\sigma}{\sigma^2 + \omega^2}$$

$$\omega_{HP} = \frac{\omega}{\sigma^2 + \omega^2}$$

For a real pole at σ, the imaginary component is zero ($\omega = 0$ in the above equation). Simplifying the equation gives $\sigma_{HP} = 1/\sigma$, which means that the highpass pole is located at the reciprocal of the pole location in the lowpass prototype. Similarly, for a zero on the (imaginary) frequency axis, the real component is zero, so $\sigma = 0$ in the above equation. Simplifying the equation gives $\omega_{ZHP} = 1/\omega_Z$, which means that the highpass zero is located at the reciprocal of the zero location in the lowpass prototype.

So, what does the S-plane diagram look like now? In Chapter 4 an example of a fourth-order lowpass filter was given. This had a Butterworth response, with poles on a unit circle at $-0.9239 \pm j0.3827$ and $-0.3827 \pm j0.9239$. Since the poles are on a unit circle, the denominators in the equations are equal to one. There-fore the poles are in the same place, as shown in Figure 5.10.

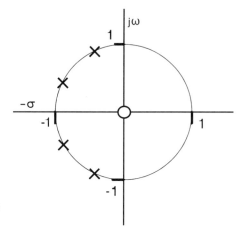

Figure 5.10

Fourth-Order Butterworth Highpass
Pole Locations

The difference is that the zeroes that were at infinity in the lowpass design have now moved to the S-plane origin. In other words, the filter does not pass DC.

Scaling normalized highpass pole and zero locations is easy: simply multiply them by $2\pi F_c$. The zeroes stay at the origin, but the poles move outwards away from the origin just as they did in the case of lowpass filters. This is shown in Figure 5.11.

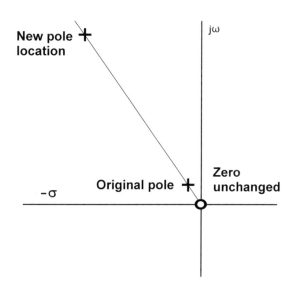

Figure 5.11

Frequency Scaling of Pole
Location in S-Plane

Important factors, which are related to the pole locations, are ω_n and Q. The values of these for the highpass filter are found in the same way as for the lowpass filter. The method is repeated here. The natural frequency ω_n is dependent upon σ, and this changes in proportion to the scaling of the diagram. The origin to pole distance is equal to ω_n. The value of Q is given by the distance from the pole to the origin divided by twice the real coordinate. Thus Q depends on the ratio of ω/σ. As the pole-zero diagram is scaled for a higher cutoff frequency, the value of Q remains unchanged.

Now that I have set the scene, let's take a look at some basic highpass active filter designs and see how the pole and zero locations are used to find component values. I shall return to the S-plane later when discussing active Cauer and Inverse Chebyshev filters: these types both have zeroes in the stopband.

First-Order Filter Section

The first-order section is a simple structure comprising a highpass RC network, followed by a buffer, as shown in Figure 5.12. The buffer serves to provide a high input impedance, so that the voltage at the connection node of the RC network is transferred to the buffer's output and prevents the RC network from being loaded by following stages. A simple RC network on its own would not have the expected frequency response if additional resistance is added in parallel with the shunt resistor.

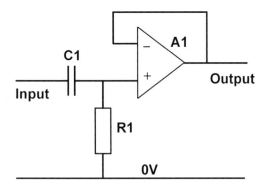

Figure 5.12

First-Order Highpass Active Filter

The first-order section is called an all-pole network, because zeroes cannot be placed on the frequency axis in its frequency response. In fact, the first-order highpass section has one real pole at $-1/\sigma$.

Letting $C1$ equal 1 Farad in the normalized highpass model enables simple calculation of $R1$.

$$R1 = \frac{1}{\sigma_{HP}} = \sigma_{LP},$$

where σ is the pole position on the negative real axis of the S-plane. As the cut-off frequency increases, the highpass pole σ_{HP} moves further from the origin. The denormalization process requires the value of σ_{HP} to be multiplied by $2\pi F_C$, hence the normalized value of $R'1$ must be divided by the frequency scaling factor. Thus, for a given capacitor value, the resistor value must decrease to raise the cutoff frequency.

Does this make sense? Well, intuitively, you may be able to see that by reducing the value of R the potential at the node between C and R will be lower at a given frequency. Increasing this frequency lowers the capacitor's reactance and restores the potential to what it was at the original frequency. In other words, to maintain a certain potential (for example, the 3 dB point of 0.7071 volts) at a higher frequency requires a reduction in the value of R.

Sallen and Key Highpass Filter

The Sallen and Key filter produces a second-order all-pole response and is a simple active highpass design. It can be used for Bessel, Butterworth, or Chebyshev responses. Cascading second-order sections can produce high-order filters. Odd-order filters can be produced by using a series of second-order sections and then adding a first-order section at the end.

The Sallen and Key filter uses an amplifier (which may be connected as a unity gain buffer) with a network of resistors and capacitors at the input. Resistive feedback from the output is also used, and this can give rise to peaking in the frequency response. Peaking is required in second-order circuits where the Q is greater than unity, and occurs due to phase shifts around the feedback loop. If the Q is large, say $Q = 15$, the amplifier is providing a gain of 15, which restricts its bandwidth to 0.0666 of the gain-bandwidth product. The diagram in Figure 5.13 shows the circuit.

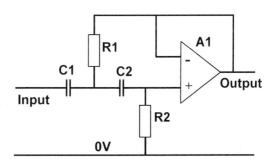

Figure 5.13

Sallen and Key Highpass Filter
(Second-Order)

Using Lowpass Pole to Find Component Values

By letting $C1$ and $C2$ equal 1F in the normalized design, the values of $R1$ and $R2$ can easily be calculated from the lowpass pole locations.

$$R1 = \frac{\omega_{nLP}}{2Q_{LP}} = \sigma_{LP} \quad \text{and} \quad R2 = 2\omega_{nLP}Q_{LP} = \frac{{\sigma_{LP}}^2 + {\omega_{LP}}^2}{\sigma_{LP}}$$

Lowpass pole positions have been used because they are readily available in tables. Thus it is not necessary to convert to highpass pole positions first. Note that in the case of Butterworth filters, $\omega_n = 1$ (for highpass and lowpass).

For example, given that the locations of the first pair of lowpass poles of a Butterworth fourth-order filter is $0.9239 \pm j0.3827$. A Sallen and Key filter section, having the same pole locations, has resistor values $R1 = 0.9239$ and $R2 = 1.0824$. As previously stated, to use the simplified equations, the normalized highpass has capacitor values of 1 Farad.

The numbering of resistors in the next filter section follows the number sequence and are labeled $R3$ and $R4$. The value of $R3$ and $R4$ can be calculated from the same equations that were used to find $R1$ and $R2$. Substitute $R3$ for $R1$ and $R4$ for $R2$. With poles at $0.3827 \pm j0.9239$ this filter section has resistor values of $R3 = 0.3827$ and $R4 = 2.613$.

The diagram in Figure 5.14 illustrates the whole circuit.

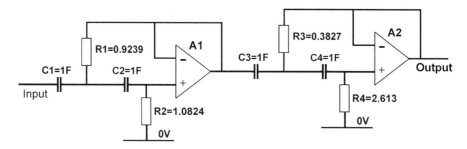

Figure 5.14

Fourth-Order Filter

Using Highpass Poles to Find Component Values

If you want to design a Sallen and Key highpass filter from its highpass pole positions, the following equations should be used:

$$R1 = \sigma_{LP} = \frac{\sigma_{HP}}{\sigma_{HP}^2 + \omega_{HP}^2} = \frac{1}{2Q_{HP}.\omega_{nHP}}$$

$$R2 = \frac{\sigma_{LP}^2 + \omega_{LP}^2}{\sigma_{LP}} = \frac{1}{\sigma_{HP}} = \frac{2Q_{HP}}{\omega_{nHP}}$$

The relationship between the equations using highpass pole locations, and those previously presented using lowpass pole locations, can be seen. Note that for both resistors, the equations have a frequency-dependent factor in the denominator. Frequency scaling can therefore be achieved by dividing the normalized highpass resistor values by $2\pi F_c$.

Operational Amplifier Requirements

Sallen and Key highpass filters are good if the requirements are not too demanding, with section Q factors below 50. As with lowpass designs, the gain-bandwidth product of the op-amps can limit the filter's cutoff frequency. The lowpass cutoff frequency limit was given by the empirical expressions:

$$\text{Butterworth passband frequency limit} = \frac{\text{Gain} - \text{Bandwidth Product}}{(\text{filter order})^2}$$

$$\text{Chebyshev (1dB) passband frequency limit} = \frac{\text{Gain} - \text{Bandwidth Product}}{(\text{filter order})^{3.2}}$$

These equations can also be used for highpass filters, by letting the passband frequency limit equal the highest frequency to be passed (i.e., do not use the $-3\,\text{dB}$ cutoff frequency). Remember that if several amplifiers are cascaded, the gain-bandwidth product of each one has to be higher than what is required overall. This is because each one contributes to high frequency roll-off as the gain-bandwidth frequency is approached.

The passband frequency limit for a given amplifier gain-bandwidth product is for a maximum of 2 dB amplitude error in the passband. A lower passband frequency limit must be set if no amplitude error is acceptable. Although using an amplifier having a greater gain-bandwidth product can raise the passband frequency limit, it can lead to instability. Amplifiers that have a high gain-bandwidth product are often unstable in a unity gain configuration.

Denormalizing Sallen and Key or First-Order Designs

In active filter designs the resistor values used should all be in the range $1\,\text{k}\Omega$ to $100\,\text{k}\Omega$ where possible. If resistor values are lower than $1\,\text{k}\Omega$ there may be a problem with loading of op-amp stage outputs. Remember, as stated in the previous chapter, loading can cause distortion and increase the power supply current. If resistor values are much higher than $100\,k\Omega$ there may be problems with noise pickup. High impedance circuits capacitively couple with the electric field from other circuits. This coupling could cause the pickup of noise and other unwanted signals, which may interfere with the wanted signal. Also, thermal noise voltage increases in proportion to the resistance.

The normalized highpass active filter model uses 1F capacitors between the filter input and the op-amp input. The normalized design is based on a cutoff frequency of 1 rad/s. Denormalization is quite simple: (1) scale the impedance; (2) scale for frequency by denormalizing the capacitance value.

Impedance scaling is simply dividing the input capacitor(s) value to give suitable input impedance. The input impedance of an active filter will tend towards $1\,\Omega$ as the frequency approaches the normalized cutoff frequency of 1 rad/s, since the series capacitor $C = 1F$ and its reactance is $X_C = 1/\omega C$. The input impedance will therefore change with frequency. To reduce this effect, capacitors with a reactance of about 100 times the desired filter input impedance could be used. A separate terminating resistor could then be used to provide the correct load impedance at all frequencies.

Scaling the resistor values can now be carried out using the following equation:

$$R = \frac{R'}{2\pi F_c C}$$

Where R' is the normalized value calculated earlier and C is the denormalized value chosen to give a suitable input impedance.

For example, suppose you want a second-order Butterworth filter using a high-pass Sallen and Key design with an input impedance of $600\,\Omega$ and a cutoff frequency $F_c = 4\,kHz$. The normalized lowpass poles are located at $0.7071 \pm j0.7071$.

Scaling the capacitor for a $60\,k\Omega$ reactance at $4\,kHz$, gives $X_c = 60,000 = \dfrac{1}{2\pi F_c C}$.

Thus $C = \dfrac{1}{120,000\pi F_c} = 663\,pF$.

This is a nonstandard value, so let $C1$ (and $C2$) = $680\,pF$. A smaller value (higher reactance) could have been used to increase the filter's input impedance.

The normalized resistor values for highpass Sallen and Key designs are related to lowpass pole locations by, $R'1 = \sigma_{LP}$ and $R'2 = \dfrac{\sigma_{LP}^2 + \omega_{LP}^2}{\sigma_{LP}}$.

Hence the normalized $R'1 = 0.7071$ and the normalized $R'2 = 1/0.7071 = 1.4142$.

Frequency scaling should now be carried out using $R = \dfrac{R'}{2\pi F_c C}$, where R' is the normalized value. Thus $R1 = 41{,}374\,\Omega$ and $R2 = 82{,}749\,\Omega$.

State Variable Highpass Filters

The state variable circuit is actually a simple type of biquad. It provides a second-order stage suitable for use in all-pole filter designs. The state variable circuit has a lower sensitivity to the op-amp's gain-bandwidth product limitation, and section Q factors of up to 200 are possible. The penalty for having this good performance is that it needs three op-amps and associated passive components, as shown in Figure 5.15.

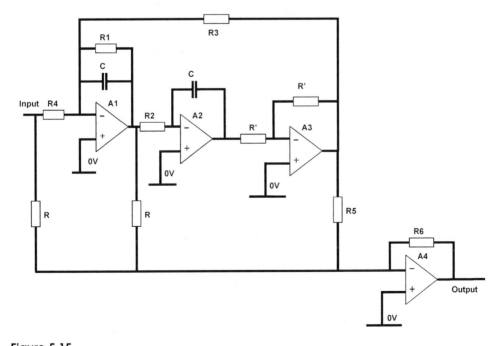

Figure 5.15

State Variable Highpass (All-Pole)

The equations for this filter require the use of normalized highpass pole locations. They allow the arbitrary choice of capacitor, C.

$$R1 = R4 = \frac{1}{2\sigma_{HP}C}$$

$$R2 = R3 = \frac{1}{C\sqrt{\sigma^2 + \omega^2}} = \frac{1}{\omega_{nHP}C}$$

$$R5 = \frac{2\sigma_{HP}R}{\omega_{nHP}}$$

The value of $R6$ determines the gain: $R6 = KR$, where K = gain. The value of R' is arbitrary, but a typical value could be $10\,k\Omega$.

An odd-order filter is made from second-order sections connected in series, followed by a first-order section. The second-order sections are all as just described, requiring four op-amps for each pole pair. The first-order section is usually added at the end of the second-order sections, and comprises a CR network followed by an op-amp.

Cauer and Inverse Chebyshev Active Filters

Let's return to the S-plane and take a look at how the Inverse Chebyshev pole-zero diagram changes when going from lowpass to highpass response. Normalized pole and zero positions of 20 dB Inverse Chebyshev with 3 dB cutoff frequency are given in Table 5.1.

Filter Order	Real Part	Imaginary Part
7	0.09360	1.01680
	0.37271	1.15880
	1.13417	1.35424
	2.47872	

Order	Zero 1	Zero 2	Zero 3
7	1.12060	1.39737	2.51797

Table 5.1

Pole and Zero Locations of Seventh-Order Inverse Chebyshev Lowpass Response

These poles and zeroes were plotted in Chapter 3 and are repeated in Figure 5.16.

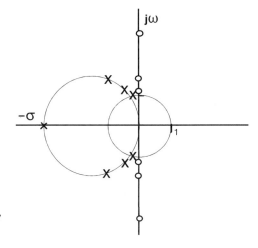

Figure 5.16

Seventh-Order Inverse Chebyshev
Lowpass Pole Zero Plot

Using the lowpass to highpass conversion equations given earlier:

$$\sigma_{HP} = \frac{\sigma}{\sigma^2 + \omega^2}$$

$$\omega_{HP} = \frac{\omega}{\sigma^2 + \omega^2}$$

Where σ and ω are the real and imaginary parts of the lowpass response. Applying these to the first pair of poles gives:

$$\sigma_{1HP} = \frac{0.0936}{0.0936^2 + 1.0168^2} = 0.089772$$

$$\omega_{1HP} = \frac{1.0168}{0.0936^2 + 1.0168^2} = 0.975214$$

This process can be repeated to find the other highpass pole locations, as shown in Table 5.2.

Table 5.2

Highpass Pole Locations for
20 dB Inverse Chebyshev
Response

Pole Number	Real Part	Imaginary Part
1	0.089772	0.975214
2	0.251537	0.782059
3	0.363480	0.434010
4	0.403434	0

The highpass zero locations are the reciprocal of the lowpass locations, as shown in Table 5.3.

Table 5.3

Highpass Zero Locations for 20 dB Inverse Chebyshev Response

Pole Number	Pole Location
1	0.89238
2	0.71563
3	0.39715

When these are all put together, they form the highpass pole-zero diagram as shown in Figure 5.17.

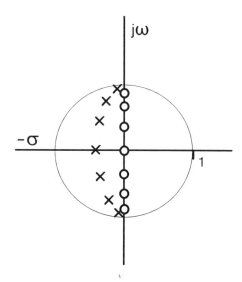

Figure 5.17

Seventh-Order Inverse Chebyshev Highpass Pole-Zero Plot

In converting from a lowpass to a highpass Inverse Chebyshev response the poles and zeroes in the S-plane have moved inside the unit circle (unlike those in the Butterworth case). This is because the pole positions are now in a similar position to a lowpass Cauer response. The zero positions are now inside the unit circle.

To design a Cauer or Inverse Chebyshev filter, a different circuit topology is required. The Cauer response has zeroes outside the passband, so a notch-generating circuit is needed. This can be achieved using a circuit that is an extension of the state variable filter and is known as a biquad. This circuit, illustrated in Figure 5.18, is exactly the same as the state variable circuit previously given for all-pole highpass filters, except that different component values are required.

Also note that in the highpass biquad $R5$ is connected to a different node from that used in the lowpass biquad.

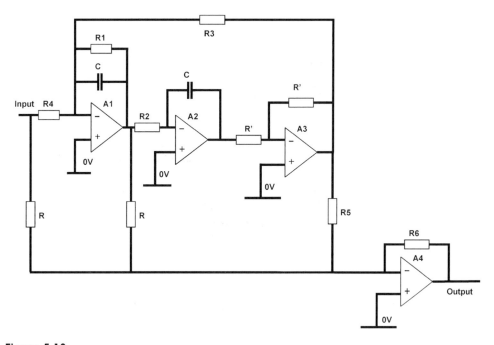

Figure 5.18

The Biquad Filter

The following equations give component values. As in the case of the state variable, the value of C can be chosen as any suitable value, then resistor values calculated from the equations. First compute the section's frequency from the pole location: $\omega_{nHP} = \sqrt{\sigma_{HP}^2 + \omega_{HP}^2}$

$$R1 = R4 = \frac{1}{2\sigma_{HP}C}$$

$$R2 = R3 = \frac{1}{\omega_{nHP}C}$$

$$R5 = \frac{2\sigma_{HP}\omega_{nHP}R}{\omega_{nHP}^2 - \omega_{ZHP}^2}$$

ω_z = the normalized zero frequency.

$R6 = AR$

The gain at DC is denoted by the symbol "A." The resistors labeled R and R' can be any arbitrary value; a typical value may be in the range $1\,k\Omega$ to $100\,k\Omega$, say $10\,k\Omega$. The resistors labeled R have an effect on the input impedance of the filter section.

Denormalizing State Variable or Biquad Designs

The simplest approach with state variable and biquad filters is to scale the poles (and zeroes, in the biquad case) before using the design equations. Choose a capacitor value, and then use the equations to find the resistor values. If the resistor values are very small or very large, select a new capacitor value and try again. Again, aim to keep the resistor values between $1\,k\Omega$ and $100\,k\Omega$.

Here's an example of denormalizing a biquad highpass filter design. Design a second-order Inverse Chebyshev filter that gives $40\,dB$ stopband attenuation. This uses normalized lowpass pole locations $0.70705 \pm j0.71416$ and a zero location 10.04963. The passband cutoff frequency is $1\,kHz$. Let $R = 10\,k\Omega$ and let the reactance of $C \approx 10\,k\Omega$ at $1\,kHz$ ($C \approx 1/[2\pi F_C . 10^3]$). Thus, using the nearest $E6$ preferred value, $C = 150\,nF$.

First compute the pole locations from the normalized lowpass values:

$$\sigma_{HP} = \frac{\sigma}{\sigma^2 + \omega^2}$$

$$\omega_{HP} = \frac{\omega}{\sigma^2 + \omega^2}$$

Thus, $\sigma_{HP} = 0.700102$ and $\omega_{HP} = 0.707142$.

The frequency scaling factor at $1\,kHz$ is $2\pi F_C = 6283\,rad/s$. Multiplying the pole locations by this factor gives:

$$\sigma_{HP} = 0.700102 . 6283 = 4398.87$$

and $\omega_{HP} = 0.707142 . 6283 = 4443.10$.

Now compute the section's frequency from the normalized highpass pole location

$$\omega_{nHP} = \sqrt{\sigma_{HP}^2 + \omega_{HP}^2} = 0.995084 . 6283 = 6252.29.$$

Now compute the highpass zero location, using $\omega_{ZHP} = 1/\omega_Z$ and the normalized lowpass zero location of 10.04963, which gives the highpass zero at

0.099506. Multiplying this by the frequency scaling factor of 6283 gives $\omega_{ZHP} = 625.197\,\text{rad/s}$.

Component values can now be found after substituting capacitor value, $C = 150\,\text{nF}$ and $R = 10\,\text{k}\Omega$.

$$R1 = R4 = \frac{1}{2\sigma_{HP}C} = 757\Omega \quad (750\Omega \text{ is the nearest E24 series preferred value})$$

$$R2 = R3 = \frac{1}{\omega_{nHP}C} = 1500\Omega \quad (\text{a standard value})$$

$$R5 = \frac{2\sigma_{HP}\omega_{nHP}R}{\omega_{nHP}^2 - \omega_{ZHP}^2} = 14{,}213\Omega \quad (14.3\text{k}\Omega \text{ is the nearest 1\% tolerance value})$$

ω_Z = the normalized zero frequency.

$$R6 = A\,R$$

Let the gain at DC be unity, $A = 1$. Let the resistors $R = R' = 10\,\text{k}\Omega$. High precision resistors are sometimes necessary to achieve values close to those calculated.

Gyrator Filters

Gyrators are related to the FDNR circuits described in Chapter 4 and are used to replace inductors. The gyrator uses two op-amps, four resistors, and a capacitor. The gyrator can be smaller than the inductor it replaces, especially if surface-mount components are used. Other advantages of using a gyrator instead of an inductor are that using suitable components can reduce temperature effects and that the component value can be adjusted easily.

The gyrator has the same structure as the FDNR: two op-amps connected to a chain of passive elements. The gyrator only has one capacitor, instead of the two used in the FDNR. All remaining passive components are resistors. The gyrator has a capacitor in place of the fourth element instead of in place of the first and third element.

A circuit diagram for the gyrator is given in Figure 5.19.

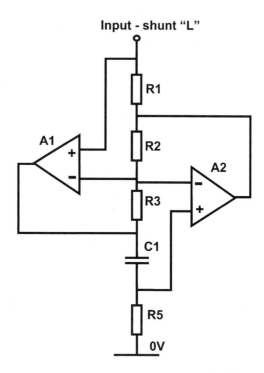

Figure 5.19

Gyrator Circuit

The gyrator behaves like a shunt inductor whose value is given by:

$$L = \frac{C.R1.R3.R5}{R2}$$

If $C1$, $R1$, $R3$, and $R2$ are all normalized to unity, then $L = R5$. If all resistors in the gyrator circuit are equal to R, $L = R^2C$.

Suppose you wish to design a highpass filter. First you should obtain the normalized lowpass passive filter component values. You should then convert the design into a normalized highpass circuit by replacing inductors (that have a value L) by capacitors that have a value of $1/L$.

Also you need to replace capacitors (that have a value of C) by inductors that have a value of $1/C$. The gyrator circuit now replaces the inductor, so $R5$ in the gyrator circuit has a value of $1/C$.

Finally, all component values are normalized. This means that all capacitor values in the final circuit are divided by $Z . 2\pi Fc$ and all resistor values are multiplied by Z.

For example, suppose you wish to design a third-order highpass filter using a gyrator. The filter should have a passband cutoff frequency of 10 kHz with input and output impedance of 600 Ω.

A passive filter must be designed first, and then the gyrator used to replace the inductor. The normalized lowpass model has two inductors in series with a central shunt capacitor. The component values are: $L1 = 1.4328$; $C2 = 1.5937$; and $L3 = 1.4328$. This is shown in Figure 5.20.

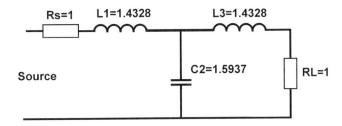

Figure 5.20

Lowpass Model

The normalized lowpass model is converted into a highpass equivalent by replacing the series inductors by series capacitors; thus $L1$ becomes $C1$, and so on. The capacitor values in the highpass model are the inverse of the inductor values in the lowpass model. In this case, $C1 = 1/1.4328 = 0.697934$. Due to symmetry, $C3 = 0.697934$. The shunt capacitor in the normalized lowpass model becomes a shunt inductor in the highpass model. The value of the shunt inductor is the inverse of the shunt capacitor in the lowpass model, so $C2$ becomes $L2$. The value of $L2 = 1/1.5937 = 0.627471$. This is illustrated in Figure 5.21.

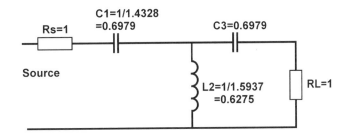

Figure 5.21

Highpass Model

In order to replace $L2$ with a gyrator, as shown in Figure 5.19, the value of $R5$ becomes 0.627471, with $R1 = R2 = R3 = 1 \, \Omega$, and $C2$ of the gyrator circuit equals 1F.

To denormalize the filter, all resistor values must be multiplied by the load impedance of 600 Ω. Resistors, $R1$, $R2$, and $R3$ all become 600 Ω. $R5$ becomes

$376\,\Omega$. The capacitor values must all be divided by the load impedance and by the cutoff frequency in radians $(2\pi Fc)$. Thus, capacitors $C1$ and $C3$ become 18.5133 nF and $C2$ becomes 26.5258 nF. The circuit is given in Figure 5.22.

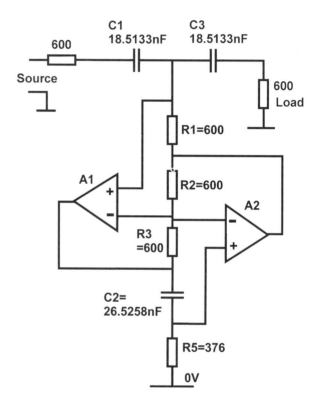

Figure 5.22

Gyrator Highpass Filter

The gyrator resistors all have a low value, which could be a problem for op-amp drive capability. Although most op-amps do have a reasonable output drive performance, low power devices do not. To overcome this, the resistance values of $R1$, $R2$, $R3$, and $R5$ can be increased, provided that the combined multiplying factor of $R1$, $R3$, and $R5$ is equal to the multiplying factor of $R2$.

Suppose, for example, that $R1$, $R3$, and $R5$ were all multiplied by 2. The value of $R2$ would have to be multiplied by 8 to restore the balance of the equation. The modified component values are then: $R1 = R3 = 1.2\,\text{k}\Omega$, $R5 = 752\,\Omega$, and $R2 = 4.8\,\text{k}\Omega$. The value of $C2$ was unchanged for this modification, but it could be reduced so that the value of $R2$ would not have to increase by such a large factor. The highpass filter circuit with revised component values is given in Figure 5.23.

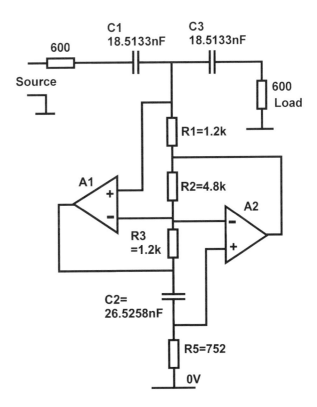

Figure 5.23

Revised Highpass Filter

The secret is to design the filter as initially described, and then modify component values in order to make them practical. Remember to keep the equation for the gyrator inductance (equivalent to the value of *L2*) balanced. In practical circuits, the value of *C2* would probably have to be produced by two or more capacitors wired in parallel. Standard capacitor values are usually in the *E6* range, which is coarsely spaced. It is unlikely that the gyrator capacitor would just happen to fall on one of these *E6* values. Fortunately, it is easier to find resistor values that are close tolerance and finely spaced, so a single component can usually be used.

Reference

1.   Stephenson, F. W. *RC Active Filter Design Handbook*, Chapter 13. New York: John Wiley & Sons, 1985.

Exercises

5.1 A normalized inductor value is 0.6834. Denormalize this for a passive highpass filter having a load resistance of 100 ohms and a cutoff frequency of 12 kHz. What is its denormalized value?

5.2 A normalized capacitor value is 0.7490. Denormalize this for a passive highpass filter having a load resistance of 75 ohms and a cutoff frequency of 10 kHz. What is its denormalized value?

5.3 A second-order filter has lowpass poles at $-0.6205 \pm j0.8075$. What are the two resistor values for a normalized highpass Sallen and Key active filter (let $C1 = C2 = 1F$)?

5.4 For the active filter in Exercise 5.3, let the capacitor values equal 1 nF. What are the resistors values ($R1$ and $R2$) if the cutoff frequency is 15 kHz?

CHAPTER 6

BANDPASS FILTERS

There are two categories of bandpass filters: wideband and narrowband. Filters are classified as **wideband** if their upper and lower passband cutoff frequencies are more than an octave apart. This is when the upper frequency is over twice that of the lower frequency. Wideband filters are ideally constructed from lowpass and highpass filters connected in series. The denormalization and scaling process for these has already been described in Chapters 4 and 5. This chapter describes how to design **narrowband** analog active or passive bandpass filters. Narrowband filters have upper and lower frequencies that are an octave or less apart.

Passive bandpass filter designs will be based on the tables of normalized lowpass component values in Chapter 2. Formulae will be given for the denormalization and scaling of these component values to produce a bandpass design. The equations are more complex than for lowpass or highpass transformations, but examples of their application will be given.

Active bandpass filter designs will be based on the normalized lowpass pole and zero locations, given in Chapter 3. Formulae will be given for denormalizing this pole and zero information, which will allow component values to be obtained. The equations are complex, but they are broken down into easier steps in order to simplify the process and reduce the chance of errors. Examples of how to use the equations will be given.

Lowpass to Bandpass Transformation

There is a close relationship between the bandwidth of a bandpass filter and the normalized lowpass filter from which it is derived. The bandwidth of a lowpass filter is from DC to the cutoff frequency, and the bandwidth of a bandpass filter is between the lower and upper cutoff frequencies. To obtain a particular bandwidth in a bandpass filter, first scale the normalized lowpass design to have this bandwidth, and then transform this into a bandpass filter design. The resultant bandpass filter bandwidth will be the same as the lowpass filter from which it was derived. Figure 6.1 illustrates this.

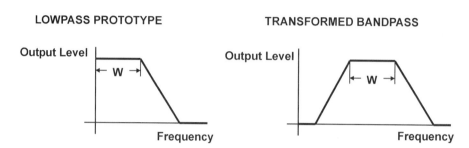

Figure 6.1

Lowpass to Bandpass Response Transformation

The relationship between the bandpass filter and its lowpass prototype does not only apply to the −3 dB bandwidth. The width of the skirt in the bandpass filter response, at any given amount of attenuation, will be equal to the width of the skirt in the lowpass filter response frequency at which the same attenuation is achieved.

For example, suppose a bandpass filter with a center frequency of 10 kHz is desired. This filter must have a −3 dB bandwidth of 6.8 kHz and 40 dB attenuation at $Fc \pm 10$ kHz, that is, the width of the skirt response at 40 dB attenuation is 20 kHz. The bandpass filter must be based on a lowpass filter design that produces the same response. That is, it must have 40 dB attenuation at a frequency of 20 kHz. The normalized stopband-to-passband frequency ratio of the lowpass filter is the same as that of the bandpass filter: 20 kHz divided by 6.8 kHz, which gives a ratio of 2.94. Thus, in a normalized lowpass prototype with a 1 rad/s passband frequency, 40 dB attenuation is required at a frequency of 2.94 rad/s. The normalized lowpass attenuation curves given in Chapter 2 can be examined to find the filter order required to achieve this response.

Passive Filters

Passive bandpass filters are derived from the normalized lowpass model. The model is normalized for a passband that extends from DC to 1 rad/s and is terminated with a 1 Ω load resistance. The first process that you must carry out is to scale the lowpass model for the desired cutoff frequency, transform it into a bandpass filter, and, finally, scale for the correct load impedance.

The design process starts with identifying the lowpass prototype. This may be Butterworth, Chebyshev, or another design. The filter order must also be determined. Starting with the specification given in the introduction, you need a filter with a 6.8 kHz 3 dB bandwidth and with 40 dB attenuation at ±10 kHz. In addition, let the filter have a center frequency, Fo, of 198 kHz. Design a Butterworth bandpass filter that achieves this specification.

The stopband-to-passband ratio is 20/6.8 = 2.94, as explained in the previous example. Referring to the attenuation versus frequency curves for Butterworth filters, you can see that a fifth-order filter will provide the required performance. Start with a lowpass prototype, as shown in Figure 6.2.

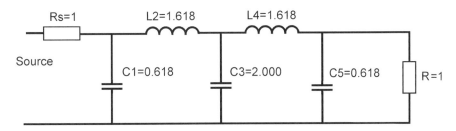

Figure 6.2

Normalized Fifth-Order Butterworth Lowpass Model

The lowpass model must be frequency scaled to have a cutoff frequency of 6.8 kHz. This is done in the same way that lowpass filters are scaled, that is, the inductors and capacitors are divided by $2\pi Fc$, where Fc is the cutoff frequency. The divisor factor is therefore 42,725.66; results in the component values are shown in Figure 6.3.

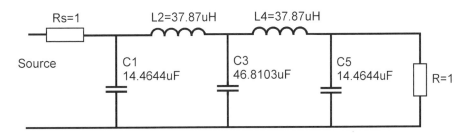

Figure 6.3

Scaled Fifth-Order Butterworth Lowpass Filter

To frequency translate the scaled lowpass prototype into a bandpass model you must resonate each branch of the ladder at the center frequency, *Fo*. Series inductors become series *LC* circuits, and shunt capacitors become parallel tuned *LC* circuits. The capacitor and inductor values in the lowpass model are unchanged.

Remember that for a tuned circuit at resonance, $F_O = \dfrac{1}{2\pi\sqrt{LC}}$. The inductor and capacitor values can be found by manipulating this equation. Hence the inductor required to tune the lowpass capacitor becomes $L_{BP} = \dfrac{1}{4\pi^2 Fo^2 C_{LP}}$, and the capacitor required to tune the lowpass inductor becomes $C_{BP} = \dfrac{1}{4\pi^2 Fo^2 L_{LP}}$.

For the bandpass filter tuned to 198 kHz, the frequency translating factor is $4\pi^2 Fo^2 = 1.547712 \times 10^{12}$. Using this information the bandpass circuit component values are given in Table 6.1.

Lowpass Component	Lowpass Value	Bandpass Component	Bandpass Value
C1	14.4644×10^{-6}	L1	44.669×10^{-9}
L2	37.87×10^{-6}	C2	17.0614×10^{-9}
C3	46.8103×10^{-6}	L3	13.8028×10^{-9}
L4	37.87×10^{-6}	C4	17.0614×10^{-9}
C5	14.4644×10^{-6}	L5	44.669×10^{-9}

Table 6.1

Bandpass Component Values

Putting these components into the circuit, you now have the bandpass filter shown in Figure 6.4.

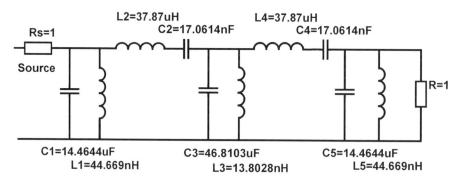

Figure 6.4

Bandpass Filter, with 1 Ω Load Resistance

The capacitor and inductor values given are for a normalized 1 Ω load. Denormalization of the bandpass model for higher load impedance requires component values to be scaled to have higher impedance. This is done in exactly the same way that lowpass or highpass filters are scaled. Inductor values increase in proportion to the load impedance and capacitor values reduce in inverse proportion to the load. Capacitor values reduce because their impedance is inversely proportional to their capacitance values. As the load impedance increases, all the reactances in the circuit must increase in order to have the same response as the model.

The filter you have been designing is intended to provide a filter for a simple radio receiver, to pick up a carrier at 198 kHz. This requires a 50 Ω source and load impedance, to match the radio frequency components at its input and output (50 Ω is the standard impedance for *RF* circuits; 75 Ω is standard for television picture transmission). Impedance scaling is achieved by multiplying the inductor values by 50 and dividing the capacitor by 50. Finally, the filter circuit given in Figure 6.5 is obtained.

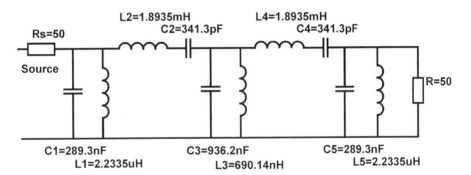

Figure 6.5

Bandpass Filter, Denormalized with 50 Ω Load Resistance

This circuit is one of two possible configurations. This configuration was developed from the minimum inductor prototype and had two series resonant arms. Three parallel resonant shunt arms were connected, to the common rail at one end and to either the source, the load, or the central node at their other end. This design gives low impedance outside the passband because the shunt arms have low impedance at DC and at frequencies above resonance.

If the design were, instead, developed from the minimum capacitor prototype, the end result would have used the same number of capacitors and inductors. The difference would have been that the filter would have had three series resonant arms between the source and load. Also, there would have been two parallel resonant shunt arms connected between the nodes of the series arms and the common rail. The alternative circuit is shown in Figure 6.6. This circuit was designed by *FILTECH*, which calculates the normalized element values and then scales them using double precision floating-point arithmetic. Although the transfer function of this filter is identical to the previous version, the input and output impedances of this version are high outside the filter's passband.

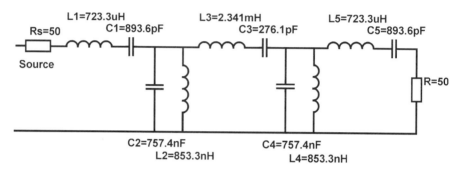

Figure 6.6

Bandpass Filter, Denormalized with 50Ω Load Resistance

Having gone through this long-winded process, readers will be pleased to know that there are formulae that allow the whole process to be completed in one step. Of course there are slight complications: because of the different circuit topologies there are a number of formulae and the difficulty is knowing which to use. I will give guidance on this subject, with examples in this chapter.

Formula for Passive Bandpass Filter Denormalization

$$C_{Series} = \frac{F_U - F_L}{2\pi F_U F_L RX}$$

$$L_{Series} = \frac{RX}{2\pi.(F_U - F_L)}$$

$$C_{Parallel} = \frac{X}{2\pi.(F_U - F_L).R}$$

$$L_{Parallel} = \frac{(F_U - F_L).R}{2\pi F_U F_L X}$$

The series and parallel subscripts indicate which circuit element is being considered. In the equations, the factor X is the normalized lowpass element value taken from the tables in Chapter 2. The same value of X must be used for both components in a single branch. This is because each branch in the lowpass filter has one component, while branches in the bandpass have two components that are either series or parallel resonant. Both components in a single branch are related to a single component value in the lowpass prototype.

It may be helpful to redesign the fifth-order Butterworth filter to illustrate the use of these formulae. Since it is a symmetrical design, only the first three branches need to be calculated. As before, $R = 50$, $Fu = (198 + 3.4)\,\text{kHz} = 201.4\,\text{kHz}$, $Fl = (198 - 3.4)\,\text{kHz} = 194.6\,\text{kHz}$.

The first branch has a value $X = 0.618$ and could be a series arm or a shunt arm. Taking the shunt arm case first (parallel resonant) gives:

$$C_{Parallel} = \frac{X}{2\pi.(F_U - F_L).R} = 0.618/2.136283 \times 10^6 = 289.3 pF$$

$$L_{Parallel} = \frac{(F_U - F_L).R}{2\pi F_U F_L X} = 340 \times 10^3/15.218466 \times 10^{10} = 2.23413\mu H$$

The second branch has a value $X = 1.618$. Since the first arm was chosen to be a shunt arm, this arm must be connected in series. Calculating the values gives:

$$C_{Series} = \frac{F_U - F_L}{2\pi F_U F_L RX} = 6.8 \times 10^3/1.992189 \times 10^{13} = 341.3 pF$$

$$L_{Series} = \frac{RX}{2\pi.(F_U - F_L)} = 80.9/42,725.66 = 1.8935 mH$$

The third branch is a parallel shunt arm, the same as the first branch. This time the value of X is 2.0. Let's cheat by using the results of the first branch and multiplying them by a ratio of X_3 to X_1.

$$C_3 = 289.3 \times 2.0/0.618 = 936.2\, pF$$

$$L_3 = 2.23413 \times 0.618/2.0 = 0.69035\mu H, \text{ or } 690.35 nH$$

The differences between these results and those obtained in Figure 6.5 are due to round-off errors in the tables of normalized values and during the calculations. The calculations were done by hand using a calculator. Floating-point arithmetic in a computer program such as *FILTECH* achieves more accurate results.

To obtain the circuit given in Figure 6.6, it is necessary to calculate the series arm first. This will use a value of $X = 0.618$.

$$C_{Series} = \frac{F_U - F_L}{2\pi F_U F_L RX} = 6.8 \times 10^3 / 7.609233 \times 10^{12} = 893.6 \, pF$$

$$L_{Series} = \frac{RX}{2\pi . (F_U - F_L)} = 30.9 / 42,725.66 = 723.2 \, \mu H$$

A shunt arm must be calculated next, using $X = 1.618$, followed by another series arm, using $X = 2.0$. Because of symmetry, the final two arms will have the same component values as previously calculated for the first two arms. The last arm will have the same component values as the first arm. The one-before-last arm will have the same component values as the second arm.

Passive Cauer and Inverse Chebyshev Bandpass Filters

So far, procedures for designing all-pole bandpass filters have been explained. However, Cauer and Inverse Chebyshev responses have zeroes in the stopband, so their circuit topology must be more complex. I have shown in earlier chapters that designing for lowpass or highpass Cauer filters is straightforward. This is because the zeroes are scaled outward from the S-plane origin in the lowpass case. Zeroes are inverted and then scaled to be less than the cutoff frequency in the highpass case. Zeroes in the resultant passive filter are produced by parallel resonant circuits in the series arm, or series resonant circuits in the shunt arm.

When it comes to designing Cauer bandpass filters, two zeroes are required for each zero in the lowpass prototype, one above and one below the passband frequency range. This means that two resonant circuits are required in the bandpass filter for each one in the lowpass prototype. The procedure for finding these component values will follow. Consider the third-order Cauer lowpass prototype given in Figure 6.7.

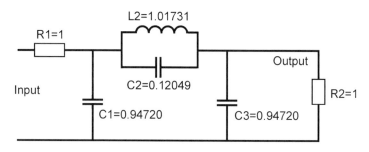

Figure 6.7

Normalized Cauer Lowpass Filter, 1 Rad/s Cutoff

The zero-producing series branch, L_2 and C_2, is a parallel resonant circuit. In the bandpass filter derived from this prototype two zeroes need to be produced. Therefore it is necessary to replace L_2 and C_2 with two parallel resonant circuits that are connected in series. In the transformed circuit, one resonant circuit comprises L_a and C_a, the other L_b and C_b. The resonant circuit comprising L_a and C_a gives a zero above the passband; the resonant circuit comprising L_b and C_b gives a zero below the passband.

The zero frequencies are given by $\omega_{\infty,a} = \sqrt{\beta}$ and $\omega_{\infty,b} = 1/\omega_{\infty,a}$, where β is given by:

$$\beta = 1 + \frac{1}{2L_2C_2} + \sqrt{\frac{1}{4L_2{}^2C_2{}^2} + \frac{1}{L_2C_2}} = 10.0588$$

From this the transformed circuit pairs can be found:

$$L_a = \frac{1}{C_2(\beta+1)} = 0.75048$$

but $L_b = \beta L_a = 7.54896$

$$C_a = \frac{1}{L_b} = 0.13247 \quad \text{and} \quad C_b = \frac{1}{L_b} = 1.33248.$$

These component values must then be normalized. Multiply inductor values by R and divide them by $2\pi F_o$; divide capacitor values by $2\pi F_o R$. Notice that only the value of C_2 is required in these equations, in addition to β. This also applies to equations that will be given later, which convert the lowpass prototype directly into a bandpass design.

Now, you may be wondering what to do with the shunt capacitors that are on either side of the parallel tuned circuit in my example. The answer is exactly the same as with all-pole filters: resonate each shunt capacitor with a parallel tuned inductor at the passband center frequency. The final filter topology is shown in Figure 6.8.

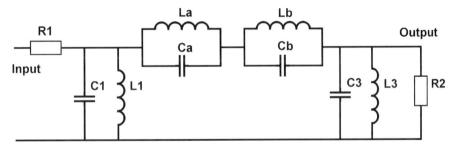

Figure 6.8

Normalized Cauer Lowpass Filter, 1 Rad/s Cutoff

There are equations that allow direct conversion from the parallel tuned circuit elements of the normalized Cauer lowpass prototype. The result is pairs of tuned circuits for the denormalized bandpass filter. These are given below:

$$L_a = \frac{(F_U - F_L).R}{(\beta + 1)2\pi F_U F_L X}$$

$$L_b = \frac{\beta.(F_U - F_L).R}{(\beta + 1)2\pi F_U F_L X}$$

$$C_a = \frac{(\beta + 1)X}{\beta.R.2\pi.(F_U - F_L)}$$

$$C_b = \frac{X.(\beta + 1)}{2\pi.(F_U - F_L).R}$$

Where X is the normalized lowpass series arm capacitor value (C_2 in this case). As I pointed out earlier, the inductor value is not needed. The inductor value is, however, used to derive β. The function β is the squared resonant frequency of the parallel tuned circuit in the normalized lowpass design. It can be derived from the series arm capacitor and inductor values.

Active Bandpass Filters

Active filters can be designed using pole and zero locations, which are derived from the frequency response's transfer function. Operational amplifiers (op-amps) are the "active" part of the circuit. These are used to buffer one stage from the next, which prevents interaction between stages. Each stage can therefore be designed to provide the frequency response of one pair of complex poles. Zeroes are also required, above and below the passband. Active networks used in bandpass filter circuits also produce zeroes. Because each filter stage is buffered from the next, the overall response is correct when all the stages are connected in series.

Bandpass Poles and Zeroes

Normalized **lowpass** filter response's pole and zero locations are used as a starting point. Frequency translation is then required to convert these into normalized **bandpass** pole and zero locations. Frequency translation in both transfer functions and the S-plane are made by replacing s with s'' as given by the following equation:

$$s'' = \frac{\omega_0}{BW}\left[\frac{s}{\omega_0} + \frac{\omega_0}{s}\right]$$

The passband center frequency is $\omega_0 = \sqrt{\omega_U . \omega_L}$ and BW is the bandwidth, given by the difference between the upper and lower passband frequencies, $\omega_U - \omega_L$. This is not particularly easy to evaluate. However, Williams[1] has published equations for finding the Q and resonant frequency, f_R, of each stage of a bandpass filter from a lowpass model. These are all that are needed to design active bandpass filters. I have manipulated Williams' equations slightly, to be consistent with those used to design bandstop filters. Bandstop filter equations will be given in the next chapter.

To start with you need to know the Q of bandpass filter, Q_{BP}, and the real and imaginary parts of the lowpass prototype pole location, σ and ω. The pole positions can be found by using the formulae or referring to tables given in Chapter 3. The bandpass Q is the center frequency, f_0, divided by the bandwidth.

$$m = \frac{\sigma}{Q_{BP}}$$

$$J = \frac{\omega}{Q_{BP}}$$

$$n = m^2 + J^2 + 4$$

The required $Q = \sqrt{\dfrac{n + \sqrt{n^2 - 16m^2}}{8m^2}}$

This gives the frequency scaling factor, $W = Qm + \sqrt{Q^2 m^2 - 1}$

And the frequencies are $f_{R1} = \dfrac{f_0}{W}$ and $f_{R2} = Wf_0$.

These are the pole transformation equations. Now the zero locations are needed, and, in an all-pole filter such as Chebyshev or Butterworth response, these are at the S-plane origin and at infinity. In Cauer and Inverse Chebyshev filters the zero locations have to be calculated, as follows:

$$k = \frac{\omega_\infty}{Q_{BP}}$$

$$h = \frac{k^2}{2} + 1$$

The zero scaling factor can now be found, $z = \sqrt{h + \sqrt{h^2 - 1}}$

The bandpass zero frequencies are then $f_{\infty,1} = \dfrac{f_0}{z}$ and $f_{\infty,2} = zf_0$.

What does the S-plane diagram look like now? An example of a fourth-order lowpass filter was given in Chapter 4, Figure 4.11. This had a Butterworth response, with poles on a unit circle at $-0.9239 \pm j0.3827$ and $-0.3827 \pm j0.9239$. Suppose the filter is required to have a passband between 9 rad/s and 11 rad/s ($BW = 2$, this is for illustration only and not intended to be a practical value). This gives $BW = 2$, $\omega_0 = 9.95$ rad/s, and $Q_{BP} = 4.975$. Notice that the geometric center frequency (9.95 rad/s) is not the same as the arithmetic center frequency (10 rad/s). Taking one pole from the first pair: $s = -0.9239 + j0.3827$, $\sigma = 0.9239$, and $\omega = 0.3827$.

The two bandpass poles produced from this are found from the following equations:

$$m = \frac{\sigma}{Q_{BP}} = 0.18571$$

$$J = \frac{\omega}{Q_{BP}} = 0.076925$$

$$n = m^2 + J^2 + 4 = 4.0404056$$

$$Q = \sqrt{\frac{n + \sqrt{n^2 - 16m^2}}{8m^2}} = 5.388756$$

$$W = Qm + \sqrt{Q^2 m^2 - 1} = 1.039375$$

The frequencies are $f_{R1} = \dfrac{f_0}{W} = 9.57306$ and $f_{R2} = Wf_0 = 10.34178$.

The second pair of poles can be found in a similar way. Due to symmetry $\sigma = 0.3827$ and $\omega = 0.9239$:

$$m = \frac{\sigma}{Q_{BP}} = 0.076925$$

$$J = \frac{\omega}{Q_{BP}} = 0.18571$$

$$n = m^2 + J^2 + 4 = 4.0404056$$

$$Q = \sqrt{\frac{n + \sqrt{n^2 - 16m^2}}{8m^2}} = 13.0556778$$

$$W = Qm + \sqrt{Q^2 m^2 - 1} = 1.09723$$

The frequencies are $f_{R1} = \dfrac{f_0}{W} = 9.068286$ and $f_{R2} = Wf_0 = 10.917444$.

In order to help you visualize what has happened to the poles, I provide a pole-zero diagram in Figure 6.9. This diagram only shows the positive frequency poles; there are symmetrical negative frequency poles, but these have been omitted for clarity. Also, note that for a given Q the poles lie on a line that passes through the origin. The two poles just calculated both had a Q of about 5.4.

The other poles had a Q of about 13, but are further from the bandpass filter's center frequency, Fc. Remember that the Q of a pole is given by the equation:

$$Q = \frac{\sqrt{\sigma^2 + \omega^2}}{2\sigma}$$

The Q of a bandpass pole is approximately $\dfrac{2\omega_0 Q_{LP}}{BW}$ where Q_{LP} is the normalized lowpass pole Q. Figure 6.9 only shows the zeroes at the origin; there are also zeroes at infinity that cannot be shown (!).

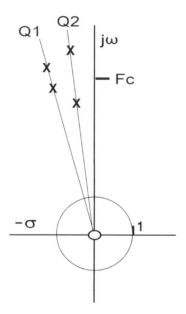

Figure 6.9

Fourth-Order Butterworth
Bandpass Pole Locations

The scene has been set. I will now take a look at some basic bandpass active filter designs and show how the pole and zero locations are used to find component values. I shall return to the S-plane later when discussing active Cauer and Inverse Chebyshev filters: these types both have zeroes in the stopband.

Bandpass Filter Midband Gain

One of the main features of a bandpass filter is its center frequency, f_0. However, each stage of a bandpass filter has a resonant frequency, f_R, which could be above or below f_0. The gain of each stage is measured at these two important frequencies, f_0 and f_R, which gives gain G_0 and G_R respectively. The gain of all stages are added together to give the overall filter gain at any particular frequency. Since the frequency response is symmetrical about the center frequency, there will be an equal number of stages resonant above and below the center frequency. In the example frequency response, illustrated by the graph in Figure 6.10, f_R is below f_0.

The gain of the filter at its center frequency can be found from the following equation, which also requires the stage's Q to be known. The terms f_R and Q can be found from the bandpass pole positions and using the relationship $G_R = 2Q^2$. The bandpass filter center frequency, f_0, is found from the filter's specification.

$$G_0 = \frac{G_R}{\sqrt{1+Q^2\left(\dfrac{f_0}{f_R} - \dfrac{f_R}{f_0}\right)}}, \text{ which simplifies to } G_0 = \frac{2Q^2}{\sqrt{1+Q^2\left(\dfrac{f_0}{f_R} - \dfrac{f_R}{f_0}\right)^2}}$$

This equation gives the midband gain of the stage being designed. Suppose the bandpass filter design is required to have unity gain in the passband. The simplest way to do this is to have unity gain at the passband center frequency (f_0) in each stage, then $G_{RR} = \dfrac{G_R}{G_0}$. Suppose that $G_0 = 10$ and $G_R = 15$. Since I want a center frequency gain of 1, not 10, the revised gain at resonance, G_{RR}, has to be scaled to be a tenth of G_R. In this case, $G_{RR} = \dfrac{G_R}{G_0} = \dfrac{15}{10} = 1.5$. This means that the stage will need a potential divider, usually at its input, to reduce the "natural" gain of the stage from 15 to 1.5.

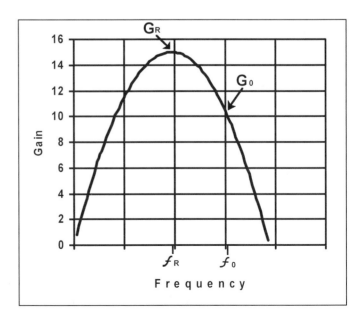

Figure 6.10

Gain versus Frequency for a Single Stage

If the desired midband gain is greater than unity, given by factor k, then G_{RR} must also be scaled by factor k: $G_{RR} = \dfrac{kG_R}{G_0}$. To achieve this scaling, the potential divider is modified to allow a greater proportion of the input signal into the filter stage.

If a number of stages are used, the overall midband gain will be the product of all the separate stage gains: $G_P = G1 * G2 * G3 *$, and so on. If each stage has a gain that is not unity at the filter center frequency, an inverting amplifier following the filter stages with a gain of $1/G_P$ could be used to restore the overall filter gain to unity.

Multiple Feedback Bandpass Filter

One of the simplest bandpass filters is the Multiple Feedback Bandpass (MFBP) circuit. It is suitable for producing an all-pole response. This filter stage looks like lowpass and highpass Sallen and Key stages combined into one and is illustrated in Figure 6.11.

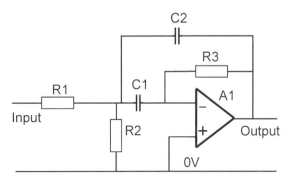

Figure 6.11

Multiple Feedback Bandpass (MFBP) Filter

The MFBP circuit is typically limited to applications where the pole's Q value is less than 20. This limitation restricts its use considerably, but for simple applications it is easy to use. The performance of the MFBP circuit depends mainly on the op-amp employed. The gain-bandwidth product of the device should be well in excess of the resonant frequency multiplied by the resonant gain. In mathematical terms: $GBW \gg G_R f_R$. The gain at the circuit resonant frequency is given by: $G_R = 2Q^2$. This can be used later in the equation to find the passband center frequency gain. Therefore the op-amp's $GBW \gg 2Q^2 f_R$.

Input resistors $R1$ and $R2$ form a potential divider network to allow gain adjustment. However, the impedance seen from the remaining circuitry is a parallel combination of $R1$ and $R2$, so adjusting only one will affect more than just the gain. This can be worked out from the design equations:

$$R3 = \frac{Q}{\pi f_R C}$$

The parallel combination of $R1$ and $R2$ ($R1 // R2$) is given by:

$$\frac{R1.R2}{R1+R2} = \frac{R3}{4Q^2}$$

If $R2$ is omitted, $R1 = \dfrac{R3}{4Q^2}$.

If the revised gain at the resonant frequency, G_{RR}, is known (it can be calculated from the formulae previously given in this chapter), in the section "Bandpass Filter Midband Gain" there are equations for calculating the values of $R1$ and $R2$:

$$R1 = \frac{R3}{2.G_{RR}} \quad \text{and} \quad R2 = \frac{R3}{4Q^2 - 2G_{RR}}.$$

If the filter stage's center frequency gain G_0 is less than unity it will not be possible to scale each stage's gain using a potential divider network. Instead, the gain produced by all the filter stages must be added together to find the overall gain, and then one or two stages should have the gain scaling circuit added. If all stages have a less than unity gain, a separate gain stage will have to be added to amplify the signals and give the overall filter a gain of 1.

Denormalizing MFBP Active Filter Designs

In active filter designs, the resistor values used should all be in the range $1\,k\Omega$ to $100\,k\Omega$ where possible. If resistor values are lower than $1\,k\Omega$ they may load the op-amp and cause distortion. If resistor values are much higher than $100\,k\Omega$ there may be problems with noise pickup. High impedance circuits capacitively couple with electric fields from other circuits and these can interfere with the wanted signal. Thermal noise may also cause problems in high impedance circuits because noise voltage increases in proportion to the resistance.

A MFBP filter stage can be designed using the poles found earlier for a bandpass filter with a passband from $9\,rad/s$ to $11\,rad/s$. The first pair of poles were found to be $\sigma = 0.9239$ and $\omega = 0.3827$.

From before, $Q = \sqrt{\dfrac{n + \sqrt{n^2 - 16m^2}}{8m^2}} = 5.388756$

$W = Qm + \sqrt{Q^2m^2 - 1} = 1.039375$

The frequencies are $f_{R1} = \dfrac{f_0}{W} = 9.57306$ and $f_{R2} = Wf_0 = 10.34178$.

Consider the pole with $f_{R1} = 9.57306$ and $G_R = 2Q^2 = 58.07738$.

Let $C = 1\mu F$

$$R3 = \dfrac{Q}{\pi f_R C} = \dfrac{5.388756}{9.57306 \ 10^{-6} \ \pi} = 179.2 \text{ k}\Omega$$

Note this is somewhat higher than the nominal $100\,\text{k}\Omega$ maximum, so the value of C must be increased to allow the value of $R3$ to be reduced. However, you also have to take into account the suggested minimum resistance of $1\,\text{k}\Omega$ and the fact that $R1 = \dfrac{R3}{4Q^2}$ and $4Q^2 = 116.15476$. Thus, reducing the value of $R3$ to below $100\,\text{k}\Omega$ will also reduce the value of $R1$ below $1\,\text{k}\Omega$. If C is increased in value to $1.5\,\mu F$ (multiplying by a factor of 1.5), $R3$ is reduced in value to $119.453\,\text{k}\Omega$ and $R1$ then becomes $1.028\,\text{k}\Omega$. This is a reasonable compromise.

Now the gain, G_0, at the bandpass center frequency, f_0, needs to be found. The gain is given by:

$$G_0 = \dfrac{2Q^2}{\sqrt{1 + Q^2\left(\dfrac{f_0}{f_R} - \dfrac{f_R}{f_0}\right)^2}}$$

$G_R = 2Q^2 = 58.07738$, at $f_{R1} = 9.57306\,rad/s$.

The center frequency is given by $f_0 = \sqrt{f_U \cdot f_L}$ and is $9.95\,\text{rad/s}$. Substituting these values into the equation for the center frequency gain gives, $G_0 = 53.61636$.

The gain at resonance, G_R was previously found to be $G_R = 58.07738$. The gain needed at resonance in order to give a gain of 1 at the passband center frequency is given by:

$$G_{RR} = \dfrac{G_R}{G_0} = 1.0832.$$

Attenuation is required, so a potential divider formed from two resistors $R1$ and $R2$ is needed. The value of $R1$ has to be changed from the initial value calculated earlier, and resistor $R2$ has to be introduced.

Using the equations, $R1 = \dfrac{R3}{2.G_{RR}} = 55.139\,\text{k}\Omega$ and $R2 = \dfrac{R3}{4Q^2 - 2G_{RR}} = 1.048\,\text{k}\Omega$.

Dual Amplifier Bandpass (DABP) Filter

The dual amplifier bandpass filter is more complicated than the MFBP structure, but it has the advantage that much higher Q factors can be achieved. Q factors of up to 150 are possible. The DABP is a bandpass filter stage with an all-pole response. The circuit diagram for a DABP filter is given in Figure 6.12.

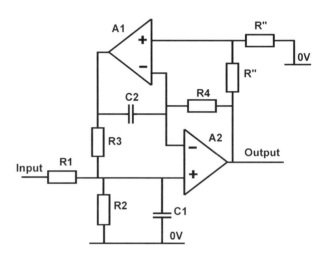

Figure 6.12

Dual Amplifier Bandpass Filter

The capacitance of $C1$ and $C2$ should be equal but may have any arbitrary value. In practice, the capacitors' values are chosen so that the resistors' values are all in a reasonable range, typically from $1\,\text{k}\Omega$ to $100\,\text{k}\Omega$. The limits are up to the designer, but remember that $R3$ and $R4$ load the output of op-amps $A1$ and $A2$. A high value of $R1$ will introduce noise, and may degrade the signals, because $R1$ is in series with the signal path.

Consider the equation: $R = \dfrac{1}{2\pi.f_R C}$, where $R = R3 = R4$. $R1' = Q.R$, assuming for the moment that $R2$ is open circuit (i.e., not there!). $R1$ is designated $R1'$ to show that this is for the case where $R2$ is not present. If the stage's Q is high, say 75, then $R1'$ will be 75 times $R3$ and $R4$. The value of C will have to be chosen so that R is close to the lower resistance limit. $R1'$ will then be close to the upper resistance limit.

Now, you will probably have realized that resistor $R2$ is used to adjust the gain. With $R2$ missing, the gain at resonance is 2. If lower gain at resonance is required, $R2$ must be in circuit. Resistors $R1$ and $R2$ form a potential divider, and their parallel resistance replaces $R1'$ in the equations given. The following equations use $R1'$ to determine $R1$ and $R2$:

$$R1 = \frac{2.R1'}{G_{RR}}$$

$$R2 = \frac{2.R1'}{2 - G_{RR}} \quad \text{(condition: } G_{RR} < 2)$$

The revised gain at resonance G_{RR} can be found from the equation $G_{RR} = \dfrac{G_R}{G_0}$.

In the DABP case the resonant frequency gain is always equal to 2, by default due to internal feedback. Hence $G_R = 2$ and this can be used to find G_0, given the overall filter center frequency f_0 and the pole characteristics f_R and Q.

$$G_0 = \frac{2}{\sqrt{1 + Q^2 \left(\dfrac{f_0}{f_R} - \dfrac{f_R}{f_0} \right)^2}}$$

Because the gain of each DABP stage at resonance is equal to 2, the gain at the filter center frequency may be less than unity. In this case, a separate amplifier stage may be needed if a unity gain bandpass filter is required.

This circuit has independent adjustment of resonant frequency and Q. The parallel combination of $R1$ and $R2$ adjust the Q at resonance. Resistor $R3$ determines the resonant frequency.

Denormalizing DABP Active Filter Designs

As discussed earlier in this chapter, the resistor values used should all be in the range $1\,\text{k}\Omega$ to $100\,\text{k}\Omega$ where possible. This will prevent overloading of the op-amp's output and reduce noise pickup.

Consider a DABP filter stage design that uses the poles found earlier in this chapter for a bandpass filter with a passband from $9\,\text{rad/s}$ to $11\,\text{rad/s}$. The first pair of poles were found to be $\sigma = 0.9239$ and $\omega = 0.3827$.

From before, $Q = \sqrt{\dfrac{n + \sqrt{n^2 - 16m^2}}{8m^2}} = 5.388756$

$$W = Qm + \sqrt{Q^2 m^2 - 1} = 1.039375$$

The frequencies are $f_{R1} = \dfrac{f_0}{W} = 9.57306$ and $f_{R2} = Wf_0 = 10.34178$.

Consider the pole with $f_{R2} = 10.34178$.

Let $C = 1\mu F$

$$R = R3 = R4 = \frac{1}{2\pi f_R C} = \frac{1}{2\pi\, 10.34178\, 10^{-6}} = 15.389\,k\Omega$$

$$R1' = Q.R = 82.928\,k\Omega.$$

Note that these values are less than the nominal $100\,k\Omega$ maximum and greater than the $1\,k\Omega$ minimum, so the value of C is suitable.

$$G_0 = \frac{2}{\sqrt{1 + Q^2\left(\dfrac{f_0}{f_R} - \dfrac{f_R}{f_0}\right)^2}} = 1.8463811.$$

Hence, $G_{RR} = \dfrac{G_R}{G_0} = \dfrac{2}{1.8463811} = 1.0832$

An attenuator is needed to reduce the gain at resonance from 2 to 1.0832. The following equations use $R1'$ to determine $R1$ and $R2$:

$$R1 = \frac{2.R1'}{G_{RR}} = \frac{165,856}{1.0832} = 153.117\,k\Omega$$

$$R2 = \frac{2.R1'}{2 - G_{RR}} = \frac{165,856}{0.9168} = 180.907\,k\Omega$$

The parallel combination of $R1$ and $R2$ equal the value of $R1'$. Thus, although the individual resistance values of $R1$ and $R2$ exceed the recommended maximum of $100\,k\Omega$, the effective resistance into the op-amp will be $82.928\,k\Omega$.

State Variable Bandpass Filters

The state variable design can be used for all-pole responses. It has a lower sensitivity to the op-amp's gain-bandwidth product limitation, and stage Q factors of up to 200 are possible. It does, however, need three op-amps, as shown in Figure 6.13.

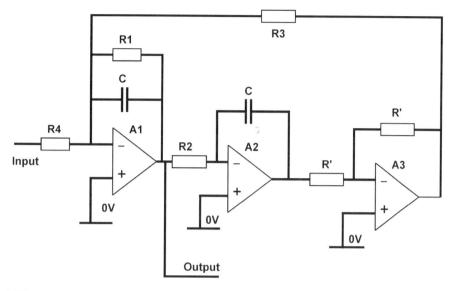

Figure 6.13

State Variable Bandpass (All-Pole)

The equations for this filter allow the arbitrary choice of capacitor, C.

$$R1 = \frac{Q}{2\pi f_R C}$$

$$R2 = R3 = \frac{R1}{Q}$$

$$R4 = \frac{R1}{G_{RR}}$$

The value of R' is arbitrary, but a typical value could be $10\,\text{k}\Omega$. G_{RR} is found from the equations given earlier in this chapter in the stage dealing with mid-band gain (Bandpass Filter Midband Gain).

Denormalization of State Variable Design

The second pair of poles of the fourth-order design considered earlier in this chapter were $\sigma = 0.3827$ and $\omega = 0.9239$. The overall bandpass filter had a center frequency $9.95\,\text{rad/s}$ and a Q_{BP} of 4.975.

$$m = \frac{\sigma}{Q_{BP}} = 0.076925$$

$$J = \frac{\omega}{Q_{BP}} = 0.18571$$

$$n = m^2 + J^2 + 4 = 4.0404056$$

$$Q = \sqrt{\frac{n + \sqrt{n^2 - 16m^2}}{8m^2}} = 13.0556778$$

$$W = Qm + \sqrt{Q^2 m^2 - 1} = 1.09723.$$

The frequencies are $f_{R1} = \dfrac{f_0}{W} = 9.068286$ and $f_{R2} = W f_0 = 10.917444$.

I will use $f_{R1} = 9.068286$ to find the filter stage gain, G_0, given that the gain at resonance will be $G_R = 2Q^2 = 340.9014456$.

$$G_0 = \frac{340.9014456}{\sqrt{1 + Q^2 \left(\dfrac{f_0}{f_R} - \dfrac{f_R}{f_0} \right)^2}} = 129.9005655.$$

$$G_{RR} = \frac{G_R}{G_0} = \frac{340.9014456}{129.9005655} = 2.624326109$$

Let capacitor, $C = 1 \mu F$.

$$R1 = \frac{Q}{2 \pi f_R C} = 229.136 \, k\Omega.$$

This value is too high, so let $C = 2.2 \, \mu F$.

$$R1 = 104.153 \, k\Omega.$$

$$R2 = R3 = \frac{R1}{Q} = 7.977 \, k\Omega.$$

$$R4 = \frac{R1}{G_{RR}} = 39.687 \, k\Omega.$$

Let the value of $R' = 10 \, k\Omega$.

Cauer and Inverse Chebyshev Active Filters

Designing bandpass filters with a Cauer or an Inverse Chebyshev response is slightly more difficult because each filter stage must provide both poles and zeroes close to the filter center frequency. Moreover, the pole and zero pairing must also be considered. A filter may have a number of poles and zeroes and, in principle, any zero could be associated with any pole. In practice the pole-zero pairing affects performance. Pole and zero pairing are illustrated in Figure 6.14.

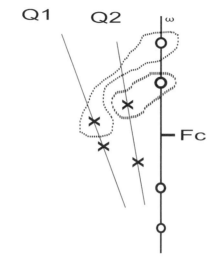

Figure 6.14

Cauer Pole and Zero Pairing

To design a Cauer, or Inverse Chebyshev, filter a different circuit topology is required. The Cauer response has zeroes outside the passband, so a notch-generating circuit is required. Notches can be produced using a circuit that is an extension of the state variable filter, and is known as a **biquad**. This circuit is illustrated in Figure 6.15.

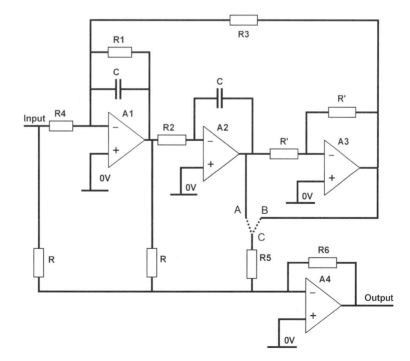

Figure 6.15

The Bandpass
Biquad Filter

Note that in the bandpass biquad shown in Figure 6.15, $R5$ is connected to different nodes, dependent on whether the zero is above or below the resonant frequency. If the zero frequency, f_Z, is above the resonant frequency, f_R, connect nodes A and C. If the zero frequency, f_Z, is below the resonant frequency, f_R, connect nodes B and C.

The following equations give component values.

$$R1 = R4 = \frac{Q}{2\pi f_R C}$$

$$R2 = R3 = \frac{1}{2\pi f_R C}$$

$$R5 = \frac{f_R^2 R}{Q\left| f_R^2 - f_Z^2 \right|}$$

If the filter stage is the last of an odd-order filter (i.e., no zero is required), $R5$ is not in circuit and $R6 = R$.

If a zero is required, $R5$ is in circuit and the value of $R6$ is given by the following equation.

$$R6 = \frac{R}{\sqrt{\left[\frac{f_0^2 - f_Z^2}{\sqrt{\left(f_R^2 - f_0^2 \right)^2 + \frac{f_R^2 f_0^2}{Q^2}}} \right]^2}}$$

The resistors labeled R and R' can be any arbitrary value. A typical value may be in the range $1\,\text{k}\Omega$ to $100\,\text{k}\Omega$, say $10\,\text{k}\Omega$. The resistors labeled R have an effect on the input impedance of the filter stage.

Denormalizing Biquad Designs

The simplest approach with biquad filters is to scale the poles and zeroes before using the design equations. Choose a convenient capacitor value, and then use the equations to find the resistor values required by the design. If the resistor values are very small or very large, select a new capacitor value and try again. Again, aim to keep the resistor values between $1\,\text{k}\Omega$ and $100\,\text{k}\Omega$.

Consider the filter stage design needed to produce a pole at $f_R = 10.255\,\text{rad/s}$, with $Q = 21$. The filter center frequency $f_0 = 9.1\,\text{rad/s}$ and a zero at $14.2\,\text{rad/s}$ is required.

Let $C = 10\,\mu F$.

$$R1 = R4 = \frac{Q}{2\pi f_R C} = \frac{21}{2\pi\,10.255\,10^{-5}} = 32.591\,k\Omega.$$

$$R2 = R3 = \frac{1}{2\pi f_R C} = 1.552\,k\Omega.$$

Let $R = 22\,k\Omega$.

$$R5 = \frac{f_R^{\,2}R}{Q\left|f_R^{\,2} - f_Z^{\,2}\right|} = 1.142\,k\Omega.$$

$$R6 = \frac{R}{\sqrt{\left[\dfrac{f_0^{\,2} - f_Z^{\,2}}{\sqrt{\left(f_R^{\,2} - f_0^{\,2}\right)^2 + \dfrac{f_R^{\,2}f_0^{\,2}}{Q^2}}}\right]^2}}$$

Substituting the values $f_0^{\,2} = 82.81$, $f_R^{\,2} = 105.165$, $f_Z^{\,2} = 201.64$ and $Q^2 = 441$, gives $R6 = 22\,k\Omega/5.2135792 = 4.21975\,k\Omega$.

Reference

1.   Williams, A., and F. J. Taylor. *Electronic Filter Design Handbook*, New York. McGraw-Hill, 1988.

Exercises

6.1   Starting with a passive normalized lowpass filter design, what step-by-step processes must be performed in order to produce a bandpass design?

6.2   A passive lowpass Cauer type filter has a parallel LC network in each series arm. A parallel LC circuit has high impedance at the resonant frequency and, when in a filter circuit, this produces a notch in the frequency response. This notch is mathematically equivalent to a "zero" in the S-plane. A bandpass filter has two zeroes in the S-plane for each lowpass zero. How is the parallel LC circuit of the lowpass filter translated into the bandpass design?

6.3   An active multiple feedback bandpass (MFBP) filter stage has a $Q = 10$ and resonant frequency, $f_R = 80\,kHz$. Given that $C = 220\,pF$ and the gain at resonance is 20, find the values of $R1$, $R2$, and $R3$.

6.4   An active filter stage is required to have $Q = 50$ at center frequency 35 kHz. A dual amplifier bandpass (DABP) design is necessary. The gain at 35 kHz is 1.5 and 4.7 nF capacitors are used. Find the values of $R1$, $R2$, $R3$, and $R4$.

CHAPTER 7

BANDSTOP FILTERS

There are two categories of bandstop filters: wideband and narrowband. Filters are classified as wideband if their upper and lower passband cutoff frequencies are several octaves apart. This is when the upper frequency is many times that of the lower frequency.

Wideband filters are ideally constructed from odd-order lowpass and highpass filters connected in parallel. Odd-order filters are necessary because, outside their passband, these have both high input impedance and high output impedance. High impedance in the stopband prevents loading of the parallel-connected filter. Otherwise impedance mismatches could occur that would lead to an incorrect overall frequency response. The denormalization and scaling process for lowpass and highpass filters has already been described (in Chapters 4 and 5).

This chapter describes how to design narrowband active and passive bandstop filters to almost any specification. Narrowband filters have upper and lower frequencies that are less than about three octaves apart. The design of these uses the normalized lowpass filter pole and zero or component values as a starting point. I use information from previous chapters and give examples where this helps in the understanding. I also provide formulae for passive designs in the denormalization and scaling of normalized component values previously given in Chapter 2, and describe the method of denormalizing pole and zero information, given in Chapter 3 for use with active filters.

Bandstop filter design starts with normalized component values, which are converted into normalized highpass values. These highpass values are then scaled to give a new cutoff frequency, W. The new cutoff frequency must be made equal to the difference between upper and lower cutoff frequencies for the desired bandstop filter. In mathematical terms, $W = f_U - f_L$. Figure 7.1 illustrates this.

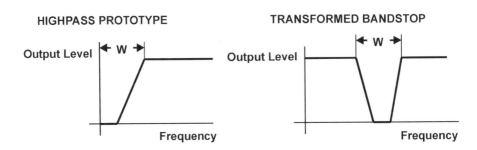

Figure 7.1

Lowpass to Bandstop Response Transformation

The highpass filter's stopband frequency, to give a certain level of attenuation, is made equal to the bandstop filter's stopband width, N. An example will help to explain this.

Let's say that the stopband width is N Hertz to give 40 dB attenuation. The highpass filter is required to have 40 dB attenuation at a frequency of N Hertz. To find the filter order needed to achieve this response, the frequencies must be normalized before using the graphs given in Chapter 2. The stopband where 40 dB attenuation occurs on the normalized frequency response curves is at W/NHz. Using graphs given in Chapter 2 for the normalized lowpass prototype, the filter order needed for the bandpass design can be found.

For example, suppose you want a bandstop filter where the difference between the upper and lower cutoff frequencies is 6.8 kHz and gives 40 dB attenuation at $F_0 \pm 1$ kHz, that is, the width of the skirt response at 40 dB attenuation is 2 kHz. Thus $W = 6.8$ kHz and $N = 2$ kHz. The normalized lowpass filter must give 40 dB attenuation at a normalized frequency ratio of 6.8 kHz divided by 2 kHz equals 3.4 rad/s. The normalized lowpass attenuation curves given in Chapter 2 can be examined to find the filter order.

Passive Filters

Passive bandstop filters are derived from the normalized lowpass model. The model is normalized for a passband that extends from DC to 1 rad/s and is terminated with 1 Ω load resistance. The first process that you must carry out is to convert the lowpass model into a highpass prototype, scaled for the desired cutoff frequency. Then transform the highpass prototype into a bandstop filter with the correct center frequency. Finally, scale for the correct load impedance.

As in the case of all filters, the design process starts with identifying the lowpass prototype. This may be Butterworth, Chebyshev, or another design. The filter order must also be determined. Suppose you need a filter with a 2.4 kHz bandwidth between the 3 dB points and with 40 dB attenuation at ±250 Hz (a 500 Hz stopband width). In addition the circuit is required to have a center frequency, F_0, of 320 kHz. Design a Butterworth bandstop filter that achieves this specification.

Passband to stopband ratio = 2.4/0.5 = 4.8. Referring to the normalized responses in Chapter 2, a third-order filter will just about achieve the required 40 dB attenuation at 4.8 rad/s. Start with a lowpass prototype, as shown in Figure 7.2.

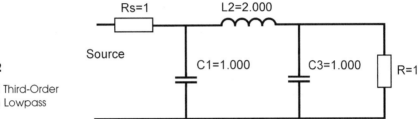

Figure 7.2

Normalized Third-Order Butterworth Lowpass Model

The lowpass model must be converted into a highpass model by replacing capacitors by inductors, and vice versa, using reciprocal values. In this case, the normalized highpass values of L_1 and L_3 remain equal to 1 Henry, but C_2 becomes 0.5 Farad. This normalized design is then frequency scaled to have a cutoff frequency of 2.4 kHz. This is done in the same way that lowpass filters are scaled. The inductors and capacitors are divided by $2\pi Fc$, where Fc is the cutoff frequency. The divisor factor is therefore 15,079.65 and results in the component values shown in Figure 7.3.

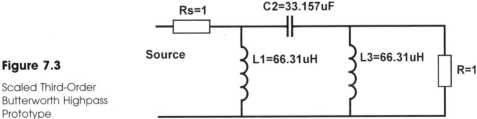

Figure 7.3

Scaled Third-Order Butterworth Highpass Prototype

To frequency translate into a bandstop model, resonate each branch of the ladder at the center frequency, Fo. Series capacitors become parallel tuned LC circuits. Shunt inductors become series tuned LC circuits. The capacitor and inductor values in the highpass prototype are unchanged.

Remember that, at resonance, $F_0 = \dfrac{1}{2\pi\sqrt{LC}}$, so the inductor required to tune

the highpass capacitor becomes $L_{BS} = \dfrac{1}{4\pi^2 {F_O}^2 C_{HP}}$, and the capacitor required

to tune the highpass inductor becomes $C_{BS} = \dfrac{1}{4\pi^2 {F_O}^2 L_{HP}}$.

For the bandstop filter tuned to 320 kHz, the frequency translating factor is $4\pi^2 F_O^2 = 4.04259 \times 10^{12}$. Using this information the bandstop circuit component values are given in Table 7.1.

Highpass Component	Highpass Value	Bandstop Component	Bandstop Value
L_1	66.31×10^{-6}	C_1	3.73045×10^{-9}
C_2	33.157×10^{-6}	L_2	7.46045×10^{-9}
L_3	66.31×10^{-6}	C_3	3.73045×10^{-9}

Table 7.1

Bandstop Component Values

Putting these components into the circuit gives the bandstop filter shown in Figure 7.4.

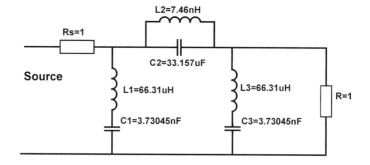

Figure 7.4

Bandstop Filter, with 1 Ω Load Resistance

Denormalization of the bandstop model for higher load impedance requires component values to be scaled to have higher impedance. This is done in exactly the same way that lowpass or highpass filters are scaled. Inductor values increase in proportion to the load impedance. Capacitor values reduce inversely proportional to the load. Capacitor values reduce because their impedance is inversely proportional to their capacitance value. As the load impedance increases, all the reactance values must increase their impedance in order to have the same response as the prototype model.

The filter design requires a 50 Ω source and load impedance to match the radio frequency components at its input and output. The normalized values of source and load impedance are increased 50-fold, therefore the impedance of the reactive components must also be increased 50-fold. Multiplying the inductor values by 50, and dividing the capacitor values by 50, results in the filter design shown in Figure 7.5.

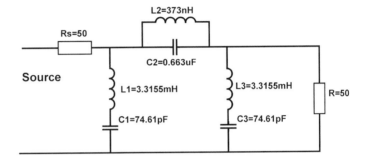

Figure 7.5

Bandstop Filter,
Denormalized with 50 Ω
Load Resistance

This gives one of two possible configurations. This design was developed from the minimum inductor prototype and has one series arm that is parallel resonant. It also has two shunt arms that are series resonant. The series resonant shunt arms are connected across the input and the output terminals, so the input impedance will be low in the stopband.

If the design were, instead, developed from the minimum capacitor prototype the end result would have used the same number of capacitors and inductors. The difference would have been that the filter would have had two parallel resonant arms wired in series between the source and load. Also, there would have been one shunt arm that was series resonant, connected between the common rail and the joining node of the two series arms.

The alternative circuit is shown in Figure 7.6 and was designed by *FILTECH* (a filter design program that I helped to develop). The *FILTECH* program calculates the normalized element values and then scales them using double precision floating-point arithmetic. The transfer function of this filter is identical to the previous version. However, the input and output impedance of this version are high in the filter's stopband.

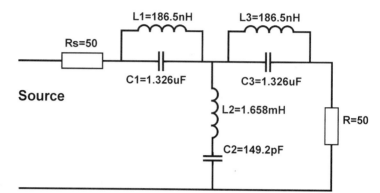

Figure 7.6

Bandstop Filter,
Denormalized with
50Ω Load Resistance

Having gone through this laborious process, readers will be pleased to know that there are formulae that allow the whole process to be completed in one step. These formulae are similar to those used in the bandpass filter design process. Care must be taken to use the correct formulae for each stage of the design.

Formula for Passive Bandstop Filter Denormalization

$$C_{Series} = \frac{1}{2\pi.[F_U - F_L]RX}$$

$$L_{Series} = \frac{[F_U - F_L].RX}{2\pi F_U.F_L}$$

$$C_{Shunt} = \frac{[F_U - F_L].X}{2\pi.F_U F_L.R}$$

$$L_{Shunt} = \frac{R}{2\pi.[F_U - F_L].X}$$

The series and shunt subscripts indicate which circuit element is being considered. A series subscript indicates the series arm (which is parallel resonant). A shunt subscript indicates the shunt arm (which is series resonant). In the equations, the factor X is the normalized lowpass element value taken from the tables in Chapter 2. The same value of X must be used for both components in a single branch. Remember that each branch in the all-pole lowpass filter has one component, while branches in the bandstop have two components that are either series or parallel resonant.

It may be helpful to redesign the third-order Butterworth filter to illustrate the use of these formulae. Since it is a symmetrical design, only the first three branches need to be calculated. As before $R = 50$, $F_u = (320 + 1.2)$ kHz = 321.2 kHz, $F_L = (320 - 1.2)$ kHz = 318.8 kHz.

The first branch has a value $X = 1.000$, and could be a series arm or a shunt arm. Taking the shunt arm case first (series resonant) gives:

$$C_{Shunt} = \frac{[F_U - F_L].X}{2\pi.F_U F_L.R} = 2.4 \times 10^3 / 3.2169 \times 10^{13} = 74.6 \, \text{pF}$$

$$L_{Shunt} = \frac{R}{2\pi.[F_U - F_L].X} = 50/15,079.65 = 3.3157 \, \text{mH}$$

The second branch has a value $X = 2.000$. Since the first arm was chosen to be a shunt arm, this arm must be series. Calculating the values gives:

$$C_{Series} = \frac{1}{2\pi.[F_U - F_L]RX} = 1/1,507,964.5 = 0.663 \, \mu\text{F}$$

$$L_{Series} = \frac{[F_U - F_L].RX}{2\pi F_U.F_L} = 24 \times 10^4 / 6.43389 \times 10^{11} = 373 \, \text{nH}$$

The third branch has the same prototype element values as the first branch. The filter is symmetrical, so the first and third branch component values will be the same. Symmetry is useful because if components have the same value, the cost of manufacturing is sometimes lower.

Differences between the results just obtained and those presented in Figure 7.5 are due to round-off errors, both in the tables of normalized values and during the calculations. The calculations were done by hand using a calculator. Floating-point arithmetic in a computer program achieves more accurate results.

To obtain the circuit given in Figure 7.6, it is necessary to calculate the series arm first. This will use a value of $X = 1.000$.

$$C_{Series} = \frac{1}{2\pi.[F_U - F_L]RX} = 1/753,982.2 = 1.32629 \, \mu\text{F}$$

$$L_{Series} = \frac{[F_U - F_L].RX}{2\pi F_U.F_L} = 12 \times 10^4 / 6.43389 \times 10^{11} = 186.51 \, \text{nH}$$

A shunt arm must be calculated next, using $X = 2.0$. Readers are invited to do the calculations themselves and compare their results with the values given in Figure 7.6. Because of symmetry, the final arm's component values are identical to those calculated above.

Passive Cauer and Inverse Chebyshev Bandstop Filters

The method for designing all-pole bandstop filters has been explained. However, unlike all-pole filters, Cauer and Inverse Chebyshev responses produce zeroes in the S-plane that are not at the center of the stopband. Odd-order filters have one zero at the center of the stopband. All the other zeroes are in the stopband, placed symmetrically either side of the center frequency. Even-order filters just have zeroes at stopband frequencies, symmetrically placed around the center. As you would expect, the circuit topologies of Cauer and Inverse Chebyshev filters are more complex. Their circuits are similar to those described for bandpass filters.

I have shown in earlier chapters that designing for lowpass or highpass Cauer filters is straightforward. Lowpass filter zeroes are scaled outward from the S-plane origin. Highpass filter zeroes are inverted, and then they are scaled to be in the stopband. Zeroes in the resultant passive filter are produced by parallel resonant circuits in the series arm or series resonant circuits in the shunt arm.

Cauer and Inverse Chebyshev lowpass responses both have zeroes in the stop-band and at infinity. During transformation into a bandstop filter, these zeroes change location. Zeroes at infinity in the lowpass filter's S-plane diagram move to the center of the stopband, just like those of all-pole filters. Zeroes in the lowpass filter's stopband become two zeroes in the bandstop filter's S-plane diagram. These are placed symmetrically about the stopband center frequency.

Physically each zero becomes a resonant circuit tuned to the zero's frequency. In the lowpass prototype a zero is produced by a parallel resonant circuit in the series arm. However, one zero in the lowpass prototype became two zeroes in the bandstop filter. Therefore, each resonant circuit in the series arm of the lowpass prototype becomes two resonant circuits in the bandstop filter. The two resonant circuits are connected in series and form a single arm of the filter. Each one resonates at a different frequency, one above and one below the stopband center frequency.

The first action in designing the bandstop filter is to take the required lowpass prototype and convert it into a highpass prototype. This must then be scaled by the bandpass Q factor before being converted into a normalized bandpass prototype. The parallel resonant series arms are then transformed into dual parallel resonant networks. These will create two stopband zeroes in the final frequency response. Frequency and impedance scaling are then used to find the final component values. Consider the third-order Cauer lowpass prototype given in Figure 7.7.

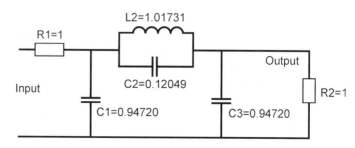

Figure 7.7

Normalized Cauer
Lowpass Filter, 1 Rad/S
Cutoff

This can be converted into a highpass prototype by replacing the capacitors with inductors of a reciprocal value. Inductor L2 must be replaced by a capacitor of a reciprocal value, as shown in Figure 7.8.

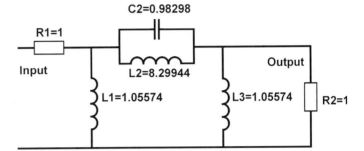

Figure 7.8

Normalized Highpass
Prototype Filter

The next step is to scale this filter by the Q factor of the desired filter. Using the same specification as in the previous example, $Q = F_o/[F_U - F_L] = 320/2.4 = 133.33$. This Q factor is very high, and it may be difficult to produce it. This is because the Q needed for the individual inductors will be much higher than the filter's Q value. All component values are multiplied by this value of Q. The resultant values are given in Figure 7.9.

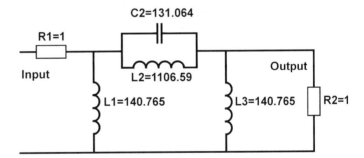

Figure 7.9

Scaled Highpass Filter

The zero producing series branch, L_2 and C_2, is a parallel resonant circuit. The bandstop filter derived from this prototype must replace this L_2 and C_2 with a pair of parallel resonant circuits, which are connected in series with each other. In the transformed circuit, one resonant circuit comprises L_a and C_a, the other L_b and C_b.

The zero frequencies are given by $\omega_{\infty,a} = \sqrt{\beta}$, and $\omega_{\infty,b} = 1/\omega_{\infty,a}$ where β is given by:

$$\beta = 1 + \frac{1}{2L_2C_2} + \sqrt{\frac{1}{4L_2{}^2C_2{}^2} + \frac{1}{L_2C_2}} = 1.000000009$$

From this the transformed circuit pairs can be found:

$$L_a = \frac{1}{C_2(\beta+1)} = 3.81493093 \times 10^{-3}$$

but $L_b = \beta L_a = 3.814930128 \times 10^{-3}$

$$C_a = \frac{1}{L_b} = 262.128012 \text{ and } C_b = \frac{1}{L_a} = 262.1279437$$

These component values must then be normalized. Multiply inductor values by R and divide them by $2\pi F_o$. Divide capacitor values by $2\pi F_o R$. The value of L_a then becomes 1.89738 nH. Notice that only the value of C_2 from the highpass prototype is required in these equations, in addition to β. In the equations that will be given later, to convert the lowpass prototype directly into a bandstop design, the value of L_2 in the lowpass prototype will be used.

Now you may be wondering what to do with the shunt inductors in the highpass prototype. These are connected onto either side of the parallel tuned circuit in my example. The answer is exactly the same as with all-pole filters; resonate each shunt inductor with a series tuned capacitor at the stopband center frequency. The final filter topology is shown in Figure 7.10.

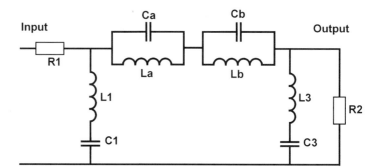

Figure 7.10

Cauer Bandstop Filter,
1 Rad/S Cutoff

Parallel tuned circuit elements in the normalized lowpass prototype can be converted directly into pairs of tuned circuits for the denormalized bandstop filter. Equations that allow this are given below:

$$L_a = \frac{(F_U - F_L).RX}{(\beta+1)2\pi F_U F_L} = 1.89738 \times 10^{-9}, \text{ which agrees with the previous result.}$$

$$L_b = \frac{\beta.(F_U - F_L).RX}{(\beta+1)2\pi F_U F_L}$$

$$C_a = \frac{(\beta+1)}{\beta.R.2\pi.(F_U - F_L)X}$$

$$C_b = \frac{(\beta+1)}{2\pi.(F_U - F_L).RX}$$

Where X is the normalized lowpass series arm inductor value (L_2 in this case). The capacitor value is not needed for this; it is however used to derive β. Higher-order filters, with more than one series arm in the lowpass prototype, require this process to be repeated for each series arm; X is then L_4, L_6, and so on.

Factor β is the squared resonant frequency of the parallel tuned circuit in the normalized lowpass design. It can be derived from the series arm capacitor and inductor values, C_aL_a and C_bL_b. The equations can be multiplied together: $C_aL_a = 1/\omega_0^2\beta$, where $\omega_0^2 = 4\pi^2F_UF_L$.

Active Bandstop Filters

Active bandstop filters can be designed using pole and zero locations from the frequency response's transfer function. Operational amplifiers (op-amps) are the "active" part of the circuit. Op-amps have high input impedance and low output impedance. They also buffer one filter stage from the next, which prevents interaction. Each stage can therefore be designed to provide the frequency response of one pair of complex poles. Zeroes may also be required in the stopband, and circuits that provide this function are usually more complex. Because stages are buffered from one another, when all the stages are connected in series the overall response should be that which is required.

Bandstop Poles and Zeroes

Using the normalized lowpass response pole and zero locations as a starting point, frequency translation is required to find the normalized bandstop pole and zero locations. Frequency translation in transfer functions and the S-plane are found by replacing s with the following:

$$S'' = \frac{BW.s}{s^2 + \omega_0^2}$$

$\omega_0 = \sqrt{\omega_U.\omega_L}$ and BW is the bandwidth, $\omega_U - \omega_L$.

This is not particularly easy to evaluate. However, as in the bandpass case, Williams and Taylor[1] have published equations for finding the Q and resonant frequency, f_R, of each section of bandstop filters from a lowpass model. These are all that are needed to design active bandstop filters. I have manipulated Williams and Taylor's equations slightly, consistent with the bandpass filter equations given in the previous chapter. To start with you must know the Q of bandstop filter, Q_{BS}, and σ and ω, which are the real and imaginary parts of the lowpass prototype pole location. The pole positions can be found from formulae or tables **given** in Chapter 3. The bandstop Q is the center frequency, f_0, divided by the width of the stopband.

$$\omega_0^2 = \sigma^2 + \omega^2$$

$$A = \frac{\sigma}{\omega_0 . Q_{BS}}$$

$$B = \frac{\omega}{\omega_0 . Q_{BS}}$$

$$f = B^2 - A^2 + 4$$

$$g = \sqrt{\frac{f + \sqrt{f^2 - 4A^2 B^2}}{2}}$$

(Does this remind you of a well-known quadratic solving equation? Try $a = 1$, $b = -f$, and $c = A^2 B^2$.)

$$h = \frac{AB}{g}$$

this gives $W = 0.5\sqrt{(A+h)^2 + (B+g)^2}$

and the frequencies are $f_{R1} = \dfrac{f_0}{W}$ and $f_{R2} = Wf_0$.

The pole's Q factor is given by $Q = \dfrac{W}{A+h}$.

Real poles have a Q factor of $Q = \sigma Q_{BS}$ and a resonant frequency at f_0.

Now to find the zero locations. In a prototype lowpass all-pole filter such as Chebyshev or Butterworth response, zeroes are on the imaginary axis in the S-plane, at infinity. During transformation into a bandstop response they move to the center of the stopband. In a prototype lowpass Cauer and Inverse Chebyshev response, zeroes are just outside the passband. When transformed into a bandstop response the zero locations move into the stopband, placed symmetrically above and below the center frequency. Their locations have to be calculated, as follows:

$$J = \frac{1}{Q_{BS}Z}, \text{ where } Z \text{ is the normalized lowpass zero frequency.}$$

The zero frequencies are $f_{\infty,1} = \dfrac{f_0}{2}[J - \sqrt{J^2 + 4}]$ and $f_{\infty,2} = \dfrac{f_0}{2}[J + \sqrt{J^2 + 4}]$.

So, what does the S-plane diagram look like now? In Chapter 4 an example of a fourth-order lowpass filter was given (Figure 4.11). This had a Butterworth response, with poles on a unit circle at $-0.9239 \pm j0.3827$ and $-0.3827 \pm j0.9239$.

Suppose the filter you want is required to have a stopband between 45 Hz and 55 Hz. This is for illustration only but could be used to remove power line frequencies (50 Hz in Europe). This specification gives $BW = 10$, $f_0 = 50$ Hz, and $Q_{BP} = 5$. Taking one pole from the first pair: $s = -0.9239 + j0.3827$, $\sigma = 0.9239$, and $\omega = 0.3827$.

The two bandstop poles produced from this are found from the following equations:

$\omega_0^2 = \sigma^2 + \omega^2 = 1$ (since the poles are on a unit circle for the Butterworth response).

$$A = \frac{\sigma}{\omega_0 . Q_{BS}} = 0.18478$$

$$B = \frac{\omega}{\omega_0 . Q_{BS}} = 0.07654$$

$$f = B^2 - A^2 + 4 = 3.89176$$

$$g = \sqrt{\frac{f + \sqrt{f^2 - 4A^2 B^2}}{2}} = 1.9727414$$

$$h = \frac{AB}{g} = 0.00716924$$

This gives $W = 0.5\sqrt{(A+h)^2 + (B+g)^2} = 1.029125693$.

The frequencies are $f_{R1} = \dfrac{f_0}{W} = 48.58493$ and $f_{R2} = Wf_0 = 51.456284$.

The pole's Q factor is given by $Q = \dfrac{W}{A+h} = 5.361447$.

The second pair of poles can be found in a similar way. Due to symmetry, $\sigma = 0.3827$ and $\omega = 0.9239$:

$$A = \frac{\sigma}{\omega_0 . Q_{BS}} = 0.07654$$

$$B = \frac{\omega}{\omega_0 . Q_{BS}} = 0.18478$$

$$f = B^2 - A^2 + 4 = 4.028285277$$

$$g = \sqrt{\frac{f + \sqrt{f^2 - 4A^2 B^2}}{2}} = 2.007046492$$

$$h = \frac{AB}{g} = 0.00704670333$$

This gives $W = 0.5\sqrt{(A+h)^2 + (B+g)^2} = 1.096709865$.

The frequencies are $f_{R1} = \dfrac{f_0}{W} = 44.59091$ and $f_{R2} = Wf_0 = 55.835493$.

The pole's Q factor is given by $Q = \dfrac{W}{A+h} = 13.1206264$.

In order to help you visualize what has happened to the poles, take a look at the S-plane diagram in Figure 7.11. This diagram only shows the positive frequency poles. There are symmetrical negative frequency poles, but these have been omitted for clarity. Also note that, for a given Q, the poles lie on a line that passes through the origin. The two poles just calculated both had a Q of about 5.4. The other poles had a Q of about 13.1 but are further from the bandstop filter's center frequency, f_0. Remember that the Q of a pole is given by the equation:

$$Q = \frac{\sqrt{\sigma^2 + \omega^2}}{2\sigma}$$

The Q of a bandstop pole is approximately $\dfrac{2f_0 Q_{LP}}{BW}$, where Q_{LP} is the normalized lowpass pole Q. In the case of the normalized Butterworth filter poles given, $Q1_{LP} = 1/2\sigma = 0.54118$ and $Q2_{LP} = 1.3065$. The ratio f_0/BW is 5. The bandstop Q factors are then approximately: $Q1_{BS} = 10 \times 0.54118 = 5.41$. $Q2_{BS} = 10 \times 1.3065 = 13.1$.

The pole-zero diagram in Figure 7.11 is very much like the example given to describe bandpass filters. Bandstop filters do not have zeroes at the S-plane origin (DC) or at infinity, they only have zeroes at the stopband center frequency.

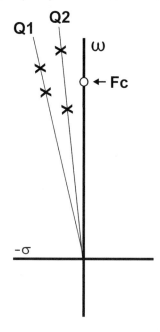

Figure 7.11

Fourth-Order Butterworth
Bandstop Pole Locations

Taking a look at some bandstop active filter designs, I will show how the pole and zero locations are used to find component values. We will return to the S-plane later in this chapter when discussing active Cauer and Inverse Chebyshev filters. Both these types have zeroes in the stopband that are not at the center frequency.

The Twin Tee Bandstop Filter

This is one of the simplest bandstop filters, yet it is not often used. The reason for its lack of popularity is its poor Q factor: in fact, it has a Q of 0.25. One way to improve the Q factor is by using an amplifier and applying positive feedback. This means that changes in amplitude are amplified, which results in a sharper passband to stopband transition. The circuit diagram in Figure 7.12 shows the amplified twin tee.

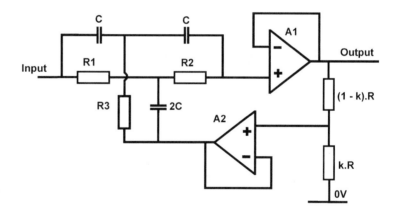

Figure 7.12

Amplified Twin Tee Filter

The component values can be calculated from:

$$R1 = R2 = \frac{1}{2\pi f_c C} \text{ where } C \text{ is any suitable value.}$$

$$R3 = \frac{R1}{2}.$$

The feedback factor, $k = 1 - \frac{1}{4Q}$ for any desired Q factor. For example, suppose a Q of 5 is required, $k = 1 - 0.05 = 0.95$. If $R = 10\,\mathrm{k}\Omega$, $kR = 9.5\,\mathrm{k}\Omega$, and $(1 - k)R = 500\,\Omega$. The nearest preferred values in the E96 range are $9.53\,\mathrm{k}\Omega$

and 499 Ω, respectively. If standard range resistors must be used, two components are required for each value, 9.1 kΩ + 390 Ω and 470 Ω + 27 Ω, respectively.

The second op-amp can be omitted if the feedback resistors are much lower than the values of $R1$, $R2$, and $R3$. The node joining the feedback resistors can be connected directly to the junction of $R3$ and the shunt capacitor (value = $2C$). This can lead to slight errors in the notch frequency due to an increase in impedance in the shunt path. Omitting the op-amp is probably not worth considering just to save space or to reduce costs, since dual op-amps are inexpensive and readily available.

Denormalization of Twin Tee Notch Filter

For this example, consider a 50 Hz notch filter, with 10 Hz between the upper and lower passband edges. $Q = 50/10 = 5$. **Using** values $f_Z = 50$ Hz or 314.159 rad/s and $C = 0.1\,\mu$F (hence $2C = 0.2\,\mu$F), component values can be found by substitution into the following equations:

$$R1 = R2 = \frac{1}{2\pi f_z C} = 5{,}066\,\Omega$$

$$R3 = \frac{R1}{2} = 2{,}533\,\Omega$$

$$k = 1 - \frac{1}{4Q} = 0.95$$

Let $R = 20$ kΩ. Resistor element $kR = 19$ kΩ and resistor element $(1 - k)R = 1$ kΩ.

Bandstop Using Multiple Feedback Bandpass Section

One of the simplest bandstop filters suitable for all-pole responses is the Multiple Feedback Bandpass (MFBP) circuit, described in Chapter 6, followed by a summing amplifier. The summing amplifier sums the output of the MFBP section (which is inverting) with the input signal. In frequency spectrum terms, the circuit is subtracting a passband from a wideband response to create a stopband. The circuit is illustrated in Figure 7.13: op-amp $A2$ and the three resistors labeled R form the summing amplifier.

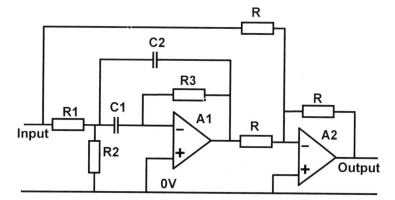

Figure 7.13

Bandstop (MFBP)
Filter Section

The MFBP circuit is typically limited to applications where the pole's Q value is less than 20. This limitation restricts its use considerably, but for simple applications it is easy to use. The performance of the MFBP circuit depends mainly on the op-amp employed. The gain-bandwidth product of the device should be well in excess of the resonant frequency multiplied by the resonant gain. In mathematical terms: $GBW \gg G_R f_R$. The gain at the circuit resonant frequency is given by: $G_R = 2Q^2$, therefore the op-amp's $GBW \gg 2Q^2 f_R$.

Input resistors $R1$ and $R2$ form a potential divider network to allow gain adjustment. Clearly the gain at resonance must be unity. When the output from the bandpass section is summed with the input, both signals have the same amplitude and cancel each other to produce a notch. However, the impedance seen from the remaining circuitry is a parallel combination of $R1$ and $R2$, so adjusting only one will affect more than just the gain. This can be worked out from the design equations:

$$R3 = \frac{Q}{\pi f_R C}.$$

The parallel combination of $R1$ and $R2$ ($R1 \parallel R2$) is given by:

$$\frac{R1.R2}{R1+R2} = \frac{R3}{4Q^2}.$$

If $R2$ were to be omitted, $R1 = \frac{R3}{4Q^2}$.

However, the desired gain at the resonant frequency, G_{RR}, is unity. Consider the design equations given for the bandpass filter section in Chapter 6:

$$R1 = \frac{R3}{2.G_{RR}} \quad \text{and} \quad R2 = \frac{R3}{4Q^2 - 2G_{RR}}.$$

These become, $R1 = \dfrac{R3}{2}$ and $R2 = \dfrac{R3}{4Q^2 - 2}$ for the bandstop filter.

Denormalization of Bandstop Design Using MFBP Section

Consider one pole found earlier for the fourth-order Butterworth 50 Hz notch filter. For this pole $f_R = 48.58493\,\text{Hz}$ or $305.26812\,\text{rad/s}$, having $Q = 5.361447$.

$$R3 = \frac{Q}{\pi f_R C} = 5590\ \Omega$$

$$R1 = \frac{R3}{2} = 2795\ \Omega$$

$$R2 = \frac{R3}{4Q^2 - 2} = 49.47758\ \Omega$$

In practice, obtaining resistor values close to those calculated may be difficult. Individual component selection may be needed in order to achieve a notch filter design with the desired amount of stopband loss.

Bandstop Using Dual Amplifier Bandpass (DABP) Section

A bandstop filter section can be made using a dual amplifier bandpass (DABP) design. This is achieved by using a summing amplifier to subtract the bandpass response from the input signal. The DABP topology is more complicated than using the MFBP structure, but it has the advantage that much higher Q factors can be achieved; Q factors of up to 150 are possible. The DABP is an all-pole response bandpass filter section and has been described in Chapter 6. The bandpass response can be subtracted from the input signal by a summing circuit to create a bandstop response. The DABP filter has a noninverting output with a gain of two at the resonant frequency, so a slightly different summing circuit to the MFBP filter is required. In Figure 7.14, the input signal is applied to the noninverting input of the summing amplifier. The output from the bandpass section is applied to a resistor in series with the inverting input. The feedback resistor from the summing amplifier's output forms a potential divider to signals from the bandpass section. The bandpass section output will be at ground potential when no bandpass signals are present (outside the stopband). Therefore the summing amplifier forms a noninverting amplifier with a gain of two. The circuit diagram for a DABP filter is given in Figure 7.14.

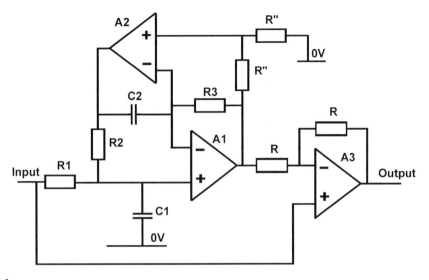

Figure 7.14

Bandstop Filter Section Using DABP Design

The capacitance of $C1$ and $C2$ should be equal but may have any arbitrary value. In practice, the capacitors' values are chosen so that the resistors' values are all in a reasonable range, typically from $1\,k\Omega$ to $100\,k\Omega$. The limits are up to the designer, but remember that $R3$ and $R4$ load the output of op-amps $A1$ and $A2$. High values of $R1$ will introduce noise, and may degrade the signals, because it is in series with the signal path.

Consider the equations:

$$R1 = \frac{Q}{2\pi . f_R C}, \text{ and } R2 = R3 = \frac{1}{2\pi . f_R C}.$$

In other words $R1 = Q . R2$.

If the section's Q is high, say 75, then $R1$ will be 75 times $R2$ and $R3$. The value of C should be chosen so that the value of $R1$ is close to the lower resistance limit. The value of $R2$ will then be close to the recommended upper resistance limit. This circuit has independent adjustment of resonant frequency and Q. The resistor $R1$ is used to adjust the Q at resonance. Resistor $R2$ and $R3$ determine the resonant frequency. At the resonant frequency, the gain is fixed at $G_R = 2$.

Denormalization of Bandstop Design Using DABP Section

Consider one pole found earlier for the fourth-order Butterworth 50 Hz notch filter.

For this pole $f_R = 51.456284\,\text{Hz}$ or $323.3093676\,\text{rad/s}$, having $Q = 5.361447$.

Let $C = 0.1\,\mu\text{F}$ and let $R = R'' = 10\,\text{k}\Omega$.

$$R2 = R3 = \frac{1}{2\pi . f_R C} = 4922\,\Omega$$

$$R1 = Q.R2 = 26{,}393\,\Omega.$$

State Variable Bandstop Filters

The state variable design can be used for all-pole responses, or in any filter where a zero at the stopband center frequency is required. It has a lower sensitivity to the op-amp's gain-bandwidth product limitation, and section Q factors of up to 200 are possible. It does, however, need four op-amps, as shown in Figure 7.15.

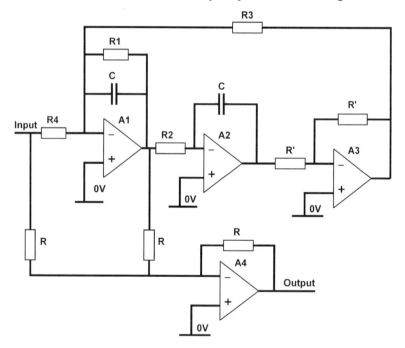

Figure 7.15

State Variable Bandstop (All-Pole)

The equations for this filter allow the arbitrary choice of capacitor, C.

$$R1 = R4 = \frac{Q}{2\pi f_R C}$$

$$R2 = R3 = \frac{R1}{Q}$$

The value of R' is discretionary, but a typical value could be $10\,\text{k}\Omega$. Resistors R also have an arbitrary value that could be set the same as R' if required. Note that the value of R has an effect on the filter's input impedance.

Denormalization of Bandstop State Variable Filter Section

Consider one pole found earlier for the fourth-order Butterworth $50\,\text{Hz}$ notch filter. For this pole $f_R = 44.59091\,\text{Hz}$ or $280.17295\,\text{rad/s}$, having $Q = 13.1206264$.

Let $C = 0.1\,\mu\text{F}$ and let $R = R' = 10\,\text{k}\Omega$.

$$R1 = R4 = \frac{Q}{2\pi f_R C} = 74{,}533\,\Omega$$

$$R2 = R3 = \frac{R1}{Q} = 5680\,\Omega$$

Cauer and Inverse Chebyshev Active Filters

Designing bandstop filters with a Cauer or an Inverse Chebyshev response is more difficult than for all-pole filters. This is because each filter section must provide both poles and zeroes close to the filter's center frequency. Moreover, the pole and zero pairing must also be considered. A filter may have a number of poles and zeroes and, in principle, any zero could be associated with any pole. In practice, the pole-zero pairing affects performance. The lowest frequency pole should be paired with the lowest frequency zero. In addition, the pole with the lowest Q should be used in the first stage, otherwise signal magnification by a large value of Q could cause overloading of subsequent stages. Pole and zero pairing is illustrated in Figure 7.16.

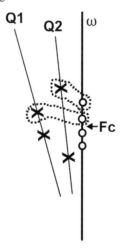

Figure 7.16

Cauer Pole and Zero Pairing

To design a Cauer or Inverse Chebyshev filter, a different circuit topology is required. The Cauer response has zeroes in the stopband, so a tunable notch circuit is required. This can be achieved using a circuit that is an extension of the state variable filter and is known as a biquad. This circuit is illustrated in Figure 7.17. Note that, in the bandstop biquad, R5 is connected to a different node. This is dependent on whether the zero is above or below the resonant frequency. If the zero is above the resonant frequency, connect nodes A and C. If the zero is below the resonant frequency, connect nodes B and C.

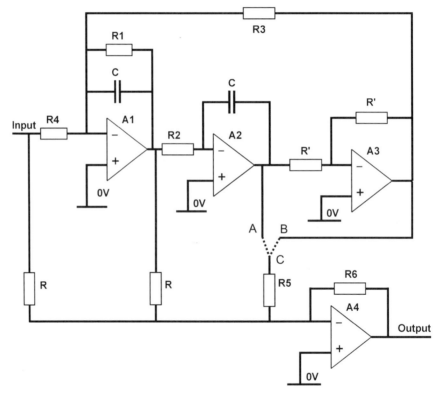

Figure 7.17

The Bandstop Biquad Filter

The following equations give component values.

$$R1 = R4 = \frac{Q}{2\pi f_R C}$$

$$R2 = R3 = \frac{1}{2\pi f_R C} = \frac{R1}{Q}$$

$$R5 = \frac{f_R^2 R}{Q|f_R^2 - f_Z^2|}$$

f_z = zero frequency.

$$R6 = \frac{f_R{}^2 R}{f_Z{}^2} \text{ when } f_R > f_\infty \text{ or when } f_R < f_\infty.$$

$R6 = R$ when an all-pole filter is required, since $f_R = f_\infty$ and $R5$ is not connected (see the State Variable Bandstop Filters section).

The resistors labeled R and R' can be any arbitrary value; a typical value may be in the range $1\,k\Omega$ to $100\,k\Omega$, say $10\,k\Omega$. The resistors labeled R have an effect on the input impedance of the filter section. The value of R should be several times higher than the input signal's source impedance.

Denormalization of Bandstop Biquad Filter Section

Consider a hypothetical design for a Cauer filter section that will produce a pole and a zero. This design has a pole $f_R = 280\,\text{rad/s}$, which has $Q = 15$. The zero for this design will be at $300\,\text{rad/s}$.

Let $C = 0.1\,\mu F$ and let $R = R' = 10\,k\Omega$.

$$R1 = R4 = \frac{Q}{2\pi f_R C} = 82,262\ \Omega$$

$$R2 = R3 = \frac{Q}{2\pi f_R C} = \frac{R1}{Q} = 5684\ \Omega$$

$$R5 = \frac{f_R{}^2 R}{Q\left| f_R{}^2 - f_Z{}^2 \right|} = 4506\ \Omega$$

$$R6 = \frac{f_R{}^2 R}{f_Z{}^2} = 8711\ \Omega.$$

Reference

1. Williams, A., and F. J. Taylor. *Electronic Filter Design Handbook*. New York: McGraw-Hill, 1988.

Exercises

7.1 Describe the step-by-step process for designing a passive bandstop filter, starting with a lowpass prototype.

7.2 Using the equations for designing a passive bandstop filter from a lowpass prototype in a single step, design a Butterworth third-order passive bandstop filter. The center frequency is 1 MHz and the stopband width is 100 kHz. Start with the shunt capacitor of a lowpass prototype. The source and load impedance is 50 Ω.

7.3 A Cauer active bandstop filter uses a biquad section. The center frequency of the bandstop filter is 50 kHz, but the Cauer design produces a zero at 49 kHz (i.e., nodes B and C are connected). Given that $R = R' = 10 k\Omega$ and $C = 470 pF$, what are the values of $R1$, $R2$, $R3$, $R4$, $R5$, and $R6$?

Chapter 8

Impedance Matching Networks

Impedance matching networks are used to ensure that circuits have the correct load. This is particularly important for transmission lines carrying radio frequencies because an incorrect load will cause some of the signal power to be reflected towards the signal source. Reflected signals combine with forward transmitted signals to produce standing waves along the transmission line. The peak amplitude of these waves divided by their minimum amplitude gives the Voltage Standing Wave Ratio (VSWR). Ideally there will be no reflected power and the VSWR equals 1.

Impedance matching is also important for active and passive components in a system. For example, passive filters should have the correct load impedance. Otherwise the filter will not have the correct frequency response. Also, active components must have the correct load impedance to prevent instability.

Power Splitters and Diplexer Filters

Wideband radio frequency (RF) sources often pass through a power splitter, such as when several receivers share a common antenna. Splitters can be transformer-coupled or resistive. Transformer-coupled splitters use an impedance matching step-up transformer combined with a center-tapped autotransformer. Resistive splitters comprise three resistors in a star or delta arrangement.

Both transformer-coupled and resistive splitter designs have the disadvantage of producing a loss in each signal path. The insertion loss is 3 dB in the case of a transformer-coupled two-way splitter, and 6 dB in the case of a resistive splitter. Both types of splitter allow a wide band of frequencies to be transmitted. The transformer-coupled splitter is usually limited to about a decade frequency range. The resistive splitter works over a wider range frequency range, which can be from DC to beyond a Gigahertz.

If the signal source needs splitting, but each path carries signals in a separate frequency band, a diplexer can be used. The advantage of diplexers over power splitters is that they have very little loss in each path. Diplexers use two filters, connected together at the source but feeding into separate loads. Each filter must have the same (odd) order. An odd-order filter is necessary to present high impedance outside its passband, to both source and load.

Diplexers are used in RF design to split signals from different frequency bands in either a highpass/lowpass or a bandpass/bandstop combination, as shown in Figure 8.1. The same −3 dB cutoff frequencies must be used for both filters. For example, a 1 MHz highpass must be combined with a 1 MHz lowpass filter. If different cutoff frequencies are used the overall response is difficult (or impossible) to calculate.

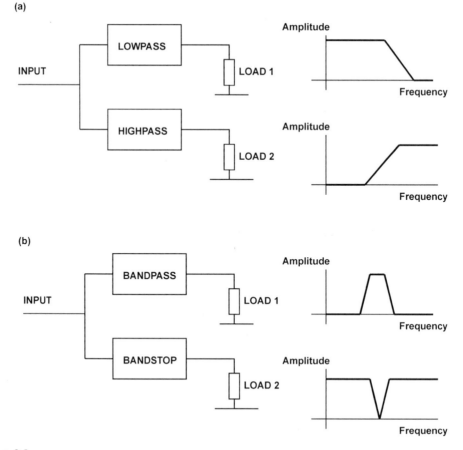

Figure 8.1

Diplexer Filter Combinations

A diplexer can be made from lowpass and highpass filters: one port will output signals that are below the cutoff frequency, and the other port will output signals that are above the cutoff frequency. The diplexer will present the correct impedance at its input port at all frequencies, provided that each output port is loaded with the correct impedance (this is usually $50\,\Omega$). The load impedance connected to the output port only needs to be correct at the frequencies in the passband. Outside the passband, slight changes to the load impedance have little effect on the diplexer's input impedance.

In a similar way, bandpass and bandstop filters can be used to produce a diplexer. The passband of one filter is the stopband of the other. Provided that each filter presents high impedance outside its passband, and is terminated correctly within its passband, the source will see constant load impedance.

The most popular application for a diplexer is the termination of a passive mixer intermediate frequency port. A mixer has three ports: local oscillator (LO); radio frequency (RF); and intermediate frequency (IF). Signals at the RF port are mixed with signals at the LO port. The result is usually a lower frequency signal out of the IF port that is the difference between RF and LO frequencies. The mixing process also produces the sum of RF and LO frequencies out of the IF port, and there are other unwanted spurious signals at the IF port as well. For optimum performance, the mixer IF port must see $50\,\Omega$ at all frequencies.

As an example, suppose the mixer produces a 10.7 MHz IF output. The IF stage needs to be preceded by a bandpass section of a diplexer, with the bandstop section terminated in a low-inductance $50\,\Omega$ resistor, as shown in Figure 8.2.

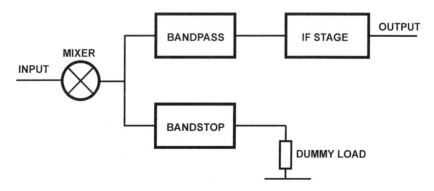

Figure 8.2

Termination for Out-of-Band Signals

If the mixer is being used for direct conversion, producing a baseband output (such as audio), the diplexer needs to have lowpass and highpass sections. The lowpass section precedes the low frequency amplifier or signal processing stage. The highpass section is terminated in $50\,\Omega$ and provides the correct impedance for the RF and local oscillator signals.

Power Splitters and Combiners

Suppose you have two high-frequency signals that you wish to transmit over one signal path. Simply joining the two signal sources together is not good enough because the output signal at one source could damage the output circuits of the other source. Also, each source's output level depends on having the correct load impedance. If the source and load impedance are not matched there will be reflections that could cause damage to the source in high-power systems. The solution is to use a power splitter that can be in one of three forms: a resistive network, a transformer-coupled circuit, or a diplexer. The choice depends on the application.

The simplest method of power splitting or combining is to use three resistors connected in a delta form and with a value equal to the impedance being matched; that is, three $50\,\Omega$ resistors for a typical $50\,\Omega$ source and load impedance. The three resistors could also be arranged in a star form, but the resistor value has to be one-third of the matching impedance, or three $16.67\,\Omega$ resistors if the source and load terminations are $50\,\Omega$. In either star or delta form, the isolation between ports is only $6\,dB$. A signal on one port will be present at both of the other ports, but at half the voltage. Delta and star forms are illustrated in Figure 8.3.

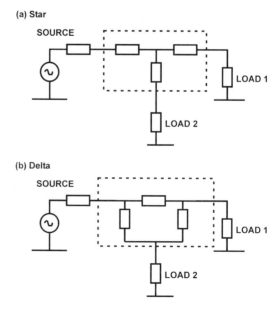

Figure 8.3

Star and Delta Resistive Splitters

The delta form is more reliable because if one resistor becomes open circuit there is still a path between all ports. In this case the impedance matching will be incorrect and the power distribution will be uneven, but the result would not be catastrophic.

Alternatively, a transformer-coupled power splitter or combiner could be used to provide impedance matching. These are usually specified for particular frequency bands, because transformers are usually effective over only about two frequency decades. Isolation between ports is usually at least 30 dB with this type of power splitter. If two sources were being combined, a strong signal from one source could still reach the output stage of the other source. If this happens, one signal will mix with the other and the result would be unwanted spurious signals. Figure 8.4 gives a circuit diagram for a transformer-coupled splitter.

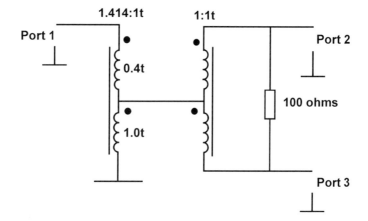

Figure 8.4

Transformer-Coupled
Splitter

In the circuit diagram given in Figure 8.4 you will note that the first transformer is an autotransformer with a 1.414:1 turns ratio. This provides 50 Ω impedance looking into port 1 when the tapping point is loaded by 25 Ω. A signal on port 1 will pass from the center tap of the second transformer to both port 2 and port 3, both of which are loaded by 50 Ω. No electromotive force (EMF) will be induced into the second transformer's windings because the current through the two windings will be equal and opposite, and no magnetic flux will be produced in the transformer's core. Thus the two loads on ports 2 and 3 will be effectively in parallel and present a 25 Ω load to the first transformer, hence the reason for the autotransformer for matching into a 50 Ω source on port 1.

If the two signal frequencies were far enough apart, a diplexer could be used. This would provide much more isolation between the output ports. One signal source would receive a lower signal level from the other and would, therefore, produce lower levels of mixing (intermodulation). The amount of isolation obtained would depend on two things: (1) the order of the filter, and (2) the frequency of the signals in relation to its cutoff point. The further the signal is from the cutoff frequency, the greater the isolation. Higher-order filters produce a greater isolation. The simplest diplexer uses just a single capacitor and a single inductor to produce a highpass/lowpass network. The performance of this circuit is very limited, but in many cases it is better than having no diplexer at all.

Designing a Diplexer

This process can be carried out manually or by using a computer program. The DIPLEXER program supplied at www.bh.com/companions/0750675470 will design 0.1 dB Chebyshev diplexers of third-, fifth-, or seventh-order. A diplexer using Butterworth or Chebyshev types having 0.01 dB, 0.25 dB, 0.5 dB, or 1 dB ripple could be designed with the *FILTECH* Professional program, or manually using tables. In either case, several decisions must be made before the design process can begin.

First, choose whether you would like a highpass/lowpass or a bandpass/bandstop combination. The choice depends on the frequency range of the wanted signals. For example, if the wanted signals are in a narrow band of frequencies, such as a 10.7 MHz intermediate frequency (IF) stage in a radio, then a bandpass/bandstop combination would be the ideal choice. If it is desired to separate signals having frequencies below, say, 5 MHz from signals above that frequency, then a lowpass/highpass combination is required.

Second, choose the frequency, or frequencies, where band splitting is required. This may be a more difficult decision because the output ports need the correct load impedance within the passband of each filter. The filter is required to be a lowpass/highpass combination with 100 kHz cutoff frequency and load impedance of 100 Ω. The lowpass filter's load must have 100 Ω impedance at all frequencies up to and, ideally, slightly beyond 100 kHz. The highpass load must have 100 Ω impedance at all frequencies at and above 100 kHz. In practice, a diplexer with a cutoff frequency of 100 kHz would only be usable up to perhaps 100 MHz. Above that frequency the filter components would become less ideal (parasitic inductance and capacitance would begin to dominate) and a poor frequency response would result.

Third, decide how much ripple in the passband is acceptable. Both the passband ripple (if any) and the filter order will determine the amount of isolation provided at any specified frequency. You may remember from earlier chapters that Chebyshev filters have ripple in their passband, which results in steeper skirt attenuation. Also, in general, higher-order filters have steeper skirt attenuation.

Graphs of the normalized frequency response for several designs were given in Chapter 2; these can be used to find the required filter order.

Each filter section in a diplexer requires high input impedance outside its passband. This is in order not to absorb any incident power. All the incident power should pass through only one filter section (ignoring passband edge effects). This condition is necessary because it gives the minimum loss and also the correct termination impedance for the source.

Diplexer filter sections also need to have high output impedance outside their passband. The reason for this is to prevent incorrect load impedance from affecting either the frequency response or the input impedance. This is important because it is likely that at least one filter section will have a load that is not matched outside its passband.

You now need to realize a diplexer filter section having both high input impedance and high output impedance outside its passband. This must be an odd-order filter with series connected components at either end. As an example, a third-order lowpass/highpass diplexer will have a lowpass section with series L, shunt C, and series L components; the highpass section will have series C, shunt L, and series C components, as shown in Figure 8.5.

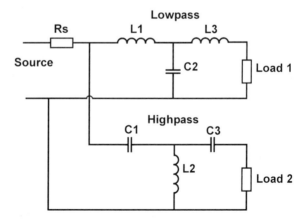

Figure 8.5

Third-Order Lowpass/Highpass Diplexer

Diplexers need filter sections designed for zero source impedance. The source impedance is usually considered to be part of the filter, like inductors and capacitors. The frequency response of a filter usually depends on the resistance of the source to achieve a correct output response. If the input impedance of a filter rises so does the input voltage, because the source is no longer loaded and there is no voltage drop across the source impedance. However, the response of the filter is correct because the source impedance was taken into account in calculating component values.

The source impedance is not taken into account in the design of diplexers. This is because the diplexer has two filters, with the same cutoff frequency, connected together. When the signal frequency increases beyond the passband edge of one filter, which then has high input impedance, the other filter provides a termination. The two filter sections together provide constant input impedance at all frequencies. This means that the source voltage does not rise outside the passband. From the point of view of each filter section, this is equivalent to the source voltage having no impedance; that is, it remains constant even when the input impedance of the filter is rising. Therefore a filter designed for a zero source impedance (constant input voltage) is used.

The normalized design can be used to produce highpass/lowpass diplexer filter section designs using the information given in the earlier chapters. First, select a set of normalized component values given for zero source impedance from the tables given in Chapter 3. The normalized design must be scaled for frequency and impedance, as described in Chapters 4 to 7, to produce a lowpass section. One filter section must then be transformed into a highpass response. As a check, if the values of the first series components of both sections are multiplied together, the product will be equal to the reciprocal of ω_0, the cutoff frequency. Similarly, the products of the second and third pair of component values are also equal to the reciprocal of ω_0. Bandpass-bandstop diplexers can be designed in a similar way. The normalized lowpass filter must be frequency and impedance scaled. Transformation into bandpass and bandstop sections is then required.

Analyzing the combined circuit can be achieved by using a circuit analysis program. The plot should look like Figure 8.6.

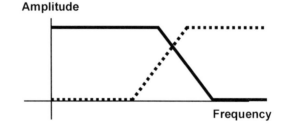

Figure 8.6

Diplexer Combined Frequency Response

More complex diplexers can be produced, with perhaps four or more frequency band outputs. These diplexers can be produced using two stages of simple diplexers; thus a band could be split into upper and lower frequencies. Both of these bands could then be split into upper and lower frequencies. This would result in four frequency band outputs. It would be wise to simulate such circuits before building them, because multichannel diplexers can be expensive.

Impedance Matching Networks

Impedance matching networks are invariably bandpass designs. They are particularly valuable at radio frequencies (RF) because even small circuits can behave like a transmission line. From transmission line theory you may know that if a line is not terminated in its characteristic impedance, signals are reflected back towards the source.

The purpose of an impedance matching network is to transfer all the available power from a source into a load. Consider a $50\,\Omega$ source matching into a $20\,\Omega$ load. If the source EMF (or open circuit voltage) is one volt, and the source is properly matched by an equal load, $0.5\,\text{V}$ will be produced across the load. Thus in terms of power transfer, the load should absorb $0.25/50$ watts, or $5\,\text{mW}$. If an impedance-matching circuit is between the source and load, the power into the load should also be $5\,\text{mW}$. The load voltage should therefore be $\sqrt{\text{Power} \times \text{resistance}}$, or $\sqrt{(5.\text{e} - 3 \times 20)} = \sqrt{0.1} = 0.3162\,\text{V}$. This is illustrated in Figure 8.7.

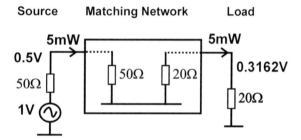

Figure 8.7

Impedance Matching Principles

With no matching network in place the load voltage can be determined by potential divider calculations: $RL/(RL + RS) = 20/70 = 0.2857\,\text{V}$. The power lost by direct connection is not very significant, so at low frequencies it is not usual to provide impedance matching circuits. However, at radio frequencies, the power reflected back towards the source must be minimized to ensure correct operation.

For continuous signals the reflection causes a standing wave, which is described by the ratio of the maximum to minimum voltages along a line. The incident voltage being added to the reflected voltage causes the maximum voltage; that is, the waves are in phase. The reflected voltage being subtracted from the incident voltage causes the minimum voltage; that is, the waves are anti-phase. High-voltage standing waves (e.g., in radio transmitter circuits) can cause damage to components. Reflections can also cause distortion products, particularly in mixer and amplifier circuits.

The quality of an impedance matching network is described in terms of the voltage standing wave ratio (VSWR) or in terms of the return loss obtained. The relationship between VSWR and return loss is simple, since both are related to the reflection coefficient (ρ):

$$VSWR = \frac{1+|\rho|}{1-|\rho|}$$

Note that the VSWR equation is usually written with the denominator elements transposed, but this results in a negative number for reflection coefficients less than unity.

$$\rho = \frac{Z - R_O}{Z + R_O}$$

The return loss $= 20 \cdot \log |\rho|$.

Suppose the characteristic impedance R_o is $50\,\Omega$ and the load is $Z = 50 + j30$.

$$\rho = -\rho = \frac{50 + j30 - 50}{50 + j30 + 50} = \frac{j30}{100 + j30}$$

$$\rho = \frac{30\angle 90}{104.4 \angle 16.7} = 0.28735 \angle 73.3$$

Thus the return loss is $20 \cdot \log 0.28735 = -10.8\,\text{dB}$.

$$\text{The } VSWR = \frac{1.28735}{0.71265} = 1.806$$

To find the impedance Z, given the $VSWR$, use the following equations:

$$|\rho| = \frac{S-1}{S+1}$$

$$Z = R_O \cdot \frac{1+\rho}{1-\rho}$$

For example, suppose the circuit has a VSWR of 1.4 in a $50\,\Omega$ system. The reflection coefficient $\rho = 0.4/2.4 = 0.1667$, and the impedance $Z = 50 \times 1.667/0.8333 = 100\,\Omega$.

Series and Parallel Circuit Relationships

Before discussing the design of an impedance matching network I would like to review the relationship between series and parallel circuits. In this case I am considering the conversion from a series RL circuit into its parallel equivalent, and vice versa. Also the conversion from series RC to parallel RC, and vice versa. This transformation could provide a useful simplification to the mathematics of an impedance matching circuit where, perhaps, the equivalent series reactance of a parallel RC load needs to be found.

First find the circuit Q. For series circuits this is the reactance divided by the resistance. For parallel circuits this is the resistance divided by the reactance. The Q is equal to $\tan(\theta)$, where θ is the phase angle of the impedance. The following equations summarize these statements.

$$Q = \frac{X_S}{R_S} = \frac{R_P}{X_P} = \tan\theta.$$

The relationship between series and parallel resistance is given by the equation:

$$R_P = R_S(1+Q^2) = \frac{Z_S}{\cos\theta}.$$

The relationship between series and parallel reactance is given by the equation:

$$X_S\left(\frac{1+Q^2}{Q^2}\right) = X_P = \frac{Z_S}{\sin\theta}.$$

The equivalent parallel or series model is only valid at one particular frequency. This is simply because the reactive element changes with frequency and, hence, so does the circuit Q. However, impedance matching circuits are also only valid for one particular frequency; therefore this is not an issue.

Matching Using L, T, and PI Networks

Networks that comprise two or three reactive components can be constructed to provide narrowband matching. The networks are described by the shape of the components, as drawn on a circuit diagram. Thus L networks use two reactive components; the L represents a shunt branch followed by a series branch, or a series branch followed by a shunt branch (a little imagination may be needed here, the L is upside-down!). Both T and PI (Π) networks require three components and represent either series, shunt, series branches; or shunt, series, shunt branches. With T and PI networks the physical component layout corresponds closely to the symbol. The four configurations are given in Figure 8.8. Each configuration can be highpass or lowpass, depending on whether the series elements are capacitors or inductors (and hence whether the shunt elements are inductors or capacitors).

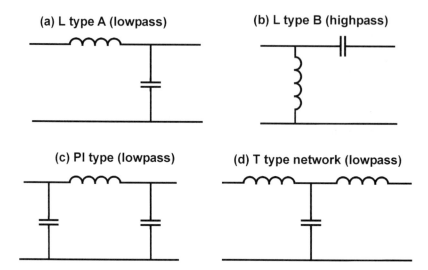

Figure 8.8

L, PI, and T Matching Networks

In general, L, T, and PI networks are designed to match a resistive source with a resistive load. Reactive source and load elements can be accommodated by a technique known as parasitic absorption, where small reactance can be allowed for by reducing the reactance of the matching network. This will be discussed in this chapter.

Component Values for L Networks

An L network can be a series element followed by a shunt element (type A), as shown in Figure 8.8a. Alternatively, the L network can be a shunt element followed by a series element (type B), shown in Figure 8.8b. The configuration used depends on whether the source resistance is greater than the load, or vice versa. If the source resistance is less than the load resistance, a series reactive element is used to raise the impedance of the source, and a shunt reactive element is used to lower the impedance of the load. With the correct component values, the source will "see" a resistive load equal to its own resistance and maximum power transfer will occur. If the source resistance is greater than the load resistance a shunt reactive element is used to reduce the source impedance. At the same time a reactive element in series with the load raises the load impedance. Once again, with correct component values the source will "see" a resistive load with a value equal to that of its own internal resistance.

In the equations given in this chapter, component reactances are used and denoted by $X1$, $X2$, and so forth. A positive reactance represents an inductance, and a negative reactance represents a capacitance. In equations to find component values, the magnitude of X is used (otherwise this results in negative capacitance!).

The L matching network of type A, shown in Figure 8.8a, has a reactance in series with the signal source and a shunt element across the load. The load impedance to be matched is greater than the source ($RL > RS$). The reactance values are:

$$X1 = \frac{\sqrt{RS.RL} - RS\cos\beta}{\sin\beta}$$

$$X2 = \frac{\sqrt{RS.RL}}{\sin\beta}$$

where angle $\beta = \tan^{-1}\sqrt{\dfrac{RL}{RS} - 1}$.

The L matching network of type B, shown in Figure 8.8b, has a shunt reactance across the signal source and an element in series with the load. The load impedance to be matched is less than the source ($RL < RS$). The reactance values are:

$$X1 = RS.RL.\frac{\sin\beta}{RL.\cos\beta - \sqrt{RS.RL}}$$

$$X1 = RS.RL.\frac{\sin\beta}{\sqrt{RS.RL}}$$

In this case, since $RL < RS$, the equation to find the angle is modified, so it always gives the square root of a positive number.

$$\text{Angle } \beta = \tan^{-1}\sqrt{\frac{RS}{RL} - 1}$$

In L networks of both type A and type B the component type and value of reactance $X\#$ depends on the center frequency of matching and on whether the reactance is positive or negative. Positive reactances are inductors, where $X = 2\pi FcL$. To find the value of inductance L, you simply transpose the equation.

$$L = \frac{X}{2\pi Fc}.$$

Similarly, negative reactances are capacitors where $X = \dfrac{1}{2\pi FcC}$. Transposing this equation to find the value of capacitance C, transpose this equation and use the magnitude of X.

$$C = \frac{1}{2\pi Fc.|X|}.$$

Component Values for PI and T Networks

Equations for *PI* and *T* networks are similar. However, the angle β is not determined by the source and load resistance ratio. Instead the designer can define the angle. In fact, it is possible to match equal resistors with a specified phase angle between input and output. This is particularly useful if a 90° phase shift circuit is required (at a fixed frequency).

A T network has three components, as shown in Figure 8.8d. There is a series arm ($X1$), a shunt arm ($X2$), and another series arm ($X3$). The values of these are given by the following equations.

$$X1 = \frac{\sqrt{RS.RL} - RS\cos\beta}{\sin\beta}$$

$$X2 = \frac{\sqrt{RS.RL}}{\sin\beta}$$

$$X3 = \frac{\sqrt{RS.RL} - RL\cos\beta}{\sin\beta}$$

A *PI* network also has three components, as shown in Figure 8.8c. There is a shunt arm ($X1$), a series arm ($X2$), and another shunt arm ($X3$). The values of the elements are given in the following equations.

$$X1 = RS.RL.\frac{\sin\beta}{RL.\cos\beta - \sqrt{RS.RL}}$$

$$X1 = RS.RL.\frac{\sin\beta}{\sqrt{RS.RL}}$$

$$X1 = RS.RL.\frac{\sin\beta}{RS.\cos\beta - \sqrt{RS.RL}}$$

Scaling of component values for the desired center frequency uses the same formula as before and is repeated below. As before, a positive element value denotes an inductor, while a negative element value indicates that the component is a capacitor. The magnitude of the element value is then used in the following equations to find the capacitance or inductance required.

$$C = \frac{1}{2\pi Fc.|X|}.$$

$$L = \frac{X}{2\pi Fc}.$$

Bandpass Matching into a Single Reactance Load

One of the most common impedance matching problems is to match a resistive source into a load comprising a resistance and a parallel capacitance. If the load capacitor can be somehow absorbed into the matching network design, the problem then reduces to simple resistive matching. A suitable matching network in this case will have a shunt capacitance across the load terminals. This could be a lowpass *PI* network, or a type-A lowpass *L* network.

There is a simple condition for being able to match a resistive source to a load comprising a resistance and a parallel capacitance. It is that the shunt capacitance of the load must be smaller than the shunt capacitance of the matching network. The circuit is designed to match the source and load resistance. This design produces a certain value of load shunt capacitance. If the load is applied, the capacitance of the load and the capacitance of the matching network add together, giving too great a value. This can be corrected by subtracting the load capacitance from the shunt capacitance of the matching circuit. Thus the load capacitance forms part of the resistive impedance matching circuit.

Using the load to form part of the impedance matching circuit is known as parasitic absorption. The diagram in Figure 8.9 illustrates the principle. The load is the parallel circuit of *R* load and *C* load. The terminating capacitor in the matching *PI* network shown is calculated to be *C* term, by taking into account only the resistance of the source and load. This is reduced in value to allow for the parallel load capacitor, and its value becomes Cterm minus Cload.

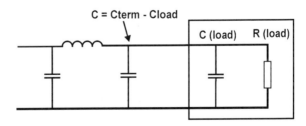

Figure 8.9

Parasitic Absorption

For example, suppose the shunt capacitor of an impedance matching network has a value of 100 pF for matching to a purely resistive load of 75 Ω. Suppose, also, that component values of a load are 10 pF in parallel with 75 Ω. The shunt capacitor of the load can be considered as part of the matching network, since it is in parallel with the shunt capacitor of the impedance matching circuit. Thus replacing the 100 pF shunt capacitor by one with a value of 90 pF (100 pF – 10 pF) enables the circuit to be matched with a simple L network. Similar arrangements can be considered for PI matching networks.

Where the load is an inductor in series with a resistor, an *L* matching network of type B could be used to give parasitic absorption, provided that the load inductance is less than the matching network inductance. This *L* network has a series inductor between the source and load, and its value can be reduced by an amount equal to the load inductance. Thus a load of 25 Ω in series with a 1 nH would be matched as though it were purely resistive. Then once the series inductor has been found for the matching network its value can be reduced by 1 nH. A similar technique can be used with *T* matching networks.

Simple Networks and VSWR

Matching networks of type *L*, *PI*, and *T* are intended to match resistive impedance at a single frequency. At the design frequency, the *VSWR* is equal to unity (the reflection coefficient is zero, in other words). However, I have already shown that it is possible to absorb reactive loads into the network design. It is also possible to match impedance over a band of frequencies if there is a limit of acceptable *VSWR* (or reflection coefficient).

The *VSWR* and bandwidth are related to the matching impedance ratio. For example, consider matching to a 50 Ω source and limiting the *VSWR* to 1.1. The frequency range over which matching is achieved for a 100 Ω load is 6%. If the matching network is redesigned for a load of 200 Ω, the frequency range is reduced 3%. The matching network used in the above example was an L type A, designed for a center frequency of 100 MHz.

VSWR of L Matching Network (Type A)

The *VSWR* of an *L* matching network, of type A, can be calculated from the reactance of the series and shunt arms. Let $X1$ be the reactance of the series arm connected between the source and load, and let $X2$ be the reactance of the shunt arm across the load; the values given at the center frequency. At frequencies above or below this center frequency the values of these reactances will be different, depending on whether the reactance is positive (inductance) or negative (capacitance). The reactance values of $X1$ and $X2$ must be scaled in proportion to the frequency ratio:

$$X' = X \cdot \frac{f_{VSWR}}{f_C}, \text{ if X is positive.}$$

$$X' = X \cdot \frac{f_C}{f_{VSWR}}, \text{ if X is negative.}$$

Thus, if the frequency at which the $VSWR$ value is required is greater than the center frequency, the capacitive reactance is reduced while the inductive reactance is increased.

Next, find the impedance looking into the matching network. This is a combination of series and parallel impedance, containing real and imaginary parts. The real part is given by:

$$R = \frac{(X2)^2 \cdot R_L}{(X2)^2 \cdot R_L^2}$$

The imaginary part is given by:

$$I = sgn(X1) - \frac{sgn(X2 \cdot R_L^2)}{(X2)^2 + R_L^2}, \text{ where sgn(x) means the magnitude of } x.$$

Signum(x), or sgn(x), actually denotes that if x is negative, the value of x is multiplied by -1 (which is really the same thing as saying "take the magnitude of x").

The reflection coefficient can now be found:

$$\rho = \frac{\sqrt{(real - R1)^2 + imaginary^2}}{\sqrt{(real + R1)^2 + imaginary^2}}$$

Now $VSWR$ is simply given by:

$$S = \frac{1 + \rho}{1 - \rho}$$

The $VSWR$ equals 1 for a perfect match, but in the real world it is invariably greater.

VSWR of L Matching Network (Type B)

The series and parallel arms are in the opposite order for a type B network; that is, the shunt arm is across the source instead of the load. Therefore the equations to find the *VSWR* are different. Let *X*1 be the reactance of the shunt arm and *X*2 be the reactance of the series arm. These reactance values must be scaled in the same way as described for the type A network.

The real part of the impedance seen looking into the matching network is given by:

$$R = \text{sgn}\left[\frac{(X1)^2 \cdot R_L}{(X1 + X2)^2 + R_L^2}\right]$$

The imaginary part is now given by:

$$I = \text{sgn}\left[\frac{X1 \cdot \left(R_L^2 + X2 \cdot (X1 + X2)\right)}{(X1 + X2)^2 + R_L^2}\right]$$

The reflection coefficient, and hence the *VSWR*, can be found using the same equations as for the type A network.

VSWR of T Matching Networks

Matching *T* networks can be broken down into real and imaginary impedance, looking towards the load. The real impedance is given by:

$$R = \text{sgn}\left[\frac{X2^2 \cdot R_L^2}{(X2 + X3)^2 + R_L^2}\right]$$

The imaginary impedance is:

$$I = \text{sgn}\left[\frac{X1 + \left(X2 \cdot R_L^2 + X2 \cdot X3 \cdot (X2 + X3)\right)}{(X2 + X3)^2 + R_L^2}\right]$$

The reflection coefficient, and hence the VSWR, can be found using the same equations as for the type A network.

VSWR of PI Matching Networks

Matching *PI* networks can be broken down into real and imaginary impedance, looking from the source into the load. The real impedance is given by:

$$R = \text{sgn}\left[\frac{X1^2 \cdot X3^2 \cdot R_L}{((X1 + X2) \cdot X3)^2 + ((X1 + X2 + X3) \cdot R_L)^2}\right]$$

The imaginary impedance is more complicated to show, because of the number of terms. I have broken the numerator into two equations, which must be added together.

$$A = X3^2 . (X1^2 . X2 + X2^2 . X1)$$

$$B = (X1 . X2 + X2^2 + X2 . X3 + X1 . X3 + X2 . X3 + X3^2) . X1 . R_L^2$$

$$I = \text{sgn} \left[\frac{A + B}{((X1 + X2) . X3)^2 + ((X1 + X2 + X3) . R_L)^2} \right]$$

The reflection coefficient, and hence the VSWR, can be found using the same equations as for the type A network.

Exercises

8.1 Power splitters can be built using three resistors connected in either star or delta configuration. Which configuration is the most reliable? What is the insertion loss between the source and each load?

8.2 What is the loss between a source and load for transformer-coupled power splitters? Why is this different from the loss in resistive splitters?

8.3 In a diplexer, why do the filter sections have to present high impedance in their stopband?

8.4 In a diplexer circuit built using lowpass and highpass filter combinations, why must the same cutoff frequency ($-3\,$dB point) be used for each filter section?

CHAPTER 9

PHASE-SHIFT NETWORKS (ALL-PASS FILTERS)

An all-pass filter seems to be a contradiction in terms. A filter surely removes some signals? Well, no. Actually, an all-pass filter modifies the phase of signals passing through it. To be more precise, it modifies the phase in a frequency selective and predetermined way. All filters delay the signal passing through them. The majority of frequency selective filter designs (Butterworth, Chebyshev, etc.) produce delays that are frequency dependent, so a signal at one frequency is delayed more than a signal at another frequency. Phase-shift networks can be used to compensate for this, so that all signal frequencies are output from the filter with the same delay.

Another application of phase-shift networks is in single sideband modulation, in which phase-shifting is used to cancel out the unwanted sideband of an AM radio transmission. This application requires a signal to be applied to two paths. The signals at the output of the two paths are phase-shifted, one relative to the other, by 90°. This chapter gives a description of a single sideband modulator, both in mathematical terms and with practical applications.

Phase Equalizing All-Pass Filters

Introduction to the Problem

Digital or impulsive signal processing by analog filters is becoming more common. This is in part due to the rise in digital communication systems, but it is also due to the need to restrict the bandwidth of impulsive signals to meet electromagnetic interference (EMI) regulations. Most filter types (e.g., Butterworth, Chebyshev, and Cauer) produce unwanted phase distortion of signals passing through them. Bessel filters have a linear phase response and produce no in-band phase distortion. Unfortunately, Bessel filters often have insufficient attenuation at frequencies beyond the passband, because their frequency response has a gentle transition from passband to stopband. Therefore, some means of correcting for the nonlinear phase shift of Butterworth, Chebyshev, and Cauer filters is desirable.

"Group delay" is the term used to describe the time delay versus frequency relationship of the transmitted signal. It is defined as the rate of phase change with frequency. The term "group delay" is very descriptive, in that it is the delay seen by a group of frequencies that are being transmitted through a filter. A constant group delay implies that all frequencies experience the same delay. A frequency-dependent group delay implies that some frequencies are delayed more than others.

Bessel filters have a constant group delay, because the phase change of signals passing through them is proportional to the frequency. Other filter types, such as the Butterworth, have a group delay that is frequency dependent, and the rate of phase change generally increases as the filter's cutoff frequency is approached. The amount of group delay variation with frequency depends on the filter type, and generally increases for filters that have a rapid increase in attenuation outside their passband (a steep skirt response). Group delay variations can be minimized by the use of phase-equalizing all-pass filters. All-pass filters can be designed to have a group delay that is virtually complementary to a lowpass filter, so the two filters connected in series produce an almost constant group delay.

Detailed Analysis

Impulsive signals contain many harmonics, and Fourier analysis can be used to show that summing all the odd harmonics can produce a square wave. Consider a square wave of amplitude "A"; each harmonic will have an amplitude of $4A/\pi$ multiplied by the inverse of the harmonic number. The sum of harmonics, up to the fifth order, is thus: square wave = $4A/\pi \times$ fundamental + $4A/(3\pi) \times$ third harmonic + $4A/(5\pi) \times$ fifth harmonic.

Some distortion is inevitable if the signal passes through a lowpass filter, because restricting the bandwidth will reduce the amplitude of the higher harmonics. Generally the distortion caused by restricting the bandwidth is pulse-edge rounding and some amplitude ripple in the pulse. This is illustrated in Figure 9.1.

$$x := 1, 2 .. 1000 \qquad A := 1$$

$$Y_x := 4 \cdot \frac{A}{\pi} \cdot \sin\left(\frac{x}{100}\right) + 4 \cdot \frac{A}{3 \cdot \pi} \cdot \sin\left(\frac{3 \cdot x}{100}\right) + 4 \cdot \frac{A}{5 \cdot \pi} \cdot \sin\left(5 \cdot \frac{x}{100}\right)$$

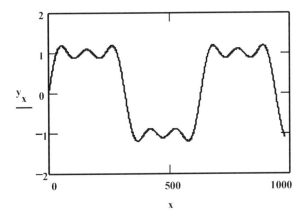

Figure 9.1

Band-Limited Impulse
Response

The Bessel filter is unique in that the distortion produced is due entirely to bandwidth restrictions. It has a constant group delay. This is important if an impulsive signal is applied to the filter input; the phase relationship between the harmonic signals is the same at the output as it was at the input. This must be true because all the transmitted harmonics are delayed by the same amount.

Bessel filters have a serious disadvantage. Beyond the cutoff point the attenuation increases slowly with frequency. Even at twice the −3 dB cutoff frequency there is very little difference in attenuation between a third- and tenth-order filter. So, despite their constant group delay, Bessel filters are rarely used. They may be suitable for some electromagnetic interference (EMI) reducing applications. They could also be used for anti-aliasing filters prior to a delta-sigma analog-to-digital converter, where the sampling frequency is many times greater than the signal bandwidth.

The use of Butterworth, Chebyshev, and Cauer filters is preferred because they have a steeper rate of attenuation beyond the filter's cutoff frequency. Unfortunately, these filters have a group delay that depends on frequency. Generally, the group delay increases as the cutoff frequency is approached, peaking just below the cutoff frequency and then falling rapidly above the cutoff frequency. The higher the filter order: the greater the change in group delay. Also, as the filter order increases, the peak in the group delay approaches the pass-band cutoff frequency.

The effect of a nonconstant group delay on impulsive signals is to produce increased ripple, particularly near the pulse edges. This is sometimes described as "ringing" because it looks like a decaying resonance. Distortion can be so severe as to cause misdetection in the pulse-detecting electronics. This effect is shown in Figure 9.2, where the output response of a sixth-order 1.2 kHz lowpass Butterworth filter is given when a 200 Hz square wave is applied at its input.

$$x := 1, 2 .. 1000 \qquad A := 1$$

$$y_x := 4 \cdot \frac{A}{\pi} \cdot \sin\left(\frac{x}{100}\right) + 4 \cdot \frac{A}{3 \cdot \pi} \cdot \sin\left(\frac{3 \cdot x}{100} - 0.2\right) + 4 \cdot \frac{A}{5 \cdot \pi} \cdot \sin\left(5 \cdot \frac{x}{100} - 1.5\right)$$

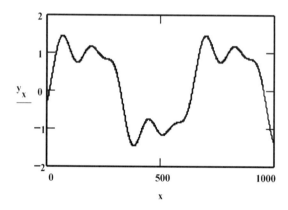

Figure 9.2

Group Delay and Band-Limited Response

In Figure 9.2 the third harmonic is being delayed by 0.2 radians (about 11.5°) and the fifth harmonic is being delayed by 1.5 radians (about 86°). Had the fifth harmonic experienced a delay of 0.4 radians the delay would have been pro-portion to frequency. The resultant waveform would have been a delayed version of that given in Figure 9.1.

The Solution: All-Pass Networks

All-pass filters are so named because they have a flat frequency response; all signals are passed without attenuation. They are also sometimes known by their full and more descriptive name; phase-equalizing all-pass filters. These are described, in varying detail, in textbooks on filters.[1,2,3] Tomlinson in particular describes how balanced passive designs can be transformed into unbalanced ones, which is the usual configuration of equalizers. Balanced equalizers are normally reserved for use with transmission lines.

Phase-equalizing all-pass filters can be used to increase the delay of signals at certain frequencies. First-order equalizers have more delay at low frequencies, but second-order filters can be tuned so their peak delay frequency is selectable. When connected in series with an amplitude attenuating filter, like the Butterworth, the overall group delay can be much flatter. The amount of residual ripple in the group delay really depends on how many equalizer sections are added. Generally, using more equalizing filter sections flattens the overall group delay.

The all-pass filter increases the complexity and size of the circuit. An equalizer is built up from first-order and second-order sections connected in series. Thus, a third-order equalizer comprises a first-order section followed by a second-order section, and a fourth-order equalizer has two second-order sections connected in series. Odd-order equalizers will always have a first-order section, but even-order equalizers are comprised of only second-order sections.

A first-order equalizer may be adequate to flatten the group delay of, perhaps, up to fifth- or sixth-order Butterworth filters. A second-order equalizer may be suitable for equalizing seventh- and eighth-order filters. The degree of equalizer required depends on whether the filter being equalized is a Butterworth, Chebyshev, or other design. It is also a balance between the amount of ripple and the complexity of the final circuit.

Passive First-Order Equalizers

The group delay for a first-order equalizer is greatest at low frequencies and is inversely proportional to frequency. Fortunately, this is almost an exact complement of the group delay for many lowpass filters.

The circuit for a practical unbalanced first-order all-pass filter is given in Figure 9.3. Unfortunately, it requires a center-tapped inductor. The inductor could be designed as a transformer having a 1:1 turns ratio, with the start of one winding connected to the finish of the other. Each "half" of the inductor will have an inductance of one-quarter of the total inductance, because the mutual coupling between windings is near enough to unity. The total inductance is thus two self-inductances plus two mutual inductances of the same value.

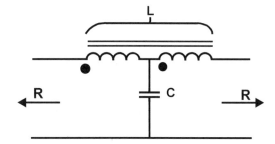

Figure 9.3

First-Order All-Pass
Design

The values of the capacitor and inductor are given by the following equations:

$$C = \frac{2}{\sigma . R} \qquad L = \frac{2R}{\sigma}$$

Where L is a center-tapped inductor, each half-winding = $L/4$.

The equations for the equalizer assume that the pole location has been denormalized by scaling it for the required frequency. The frequency is the same as the passband cutoff of the filter being equalized. In the case of quadrature networks, which will be described in this chapter, it is the passband center frequency.

The action of the first-order equalizer can be explained by considering its behavior at very high and very low frequencies, with reference to Figure 9.3. Let us consider the input to be at the left-hand side and the output to be at the right-hand side. At low frequencies, the inductor's reactance is high and the capacitor's reactance is low. The inductor is effectively a short circuit and the capacitor an open circuit, so the output signal will be in phase with the input signal.

At high frequencies, the inductor's reactance is high and the capacitor's reactance is low. Now the capacitor is effectively a short circuit and the inductor can be considered a transformer with the center tap grounded. Because the start of the "primary" winding goes to the input and the "finish" of the secondary goes to the output, the output is anti-phase with the input.

The symbol for first-order equalizers that is often given in textbooks is shown in the left-hand side of Figure 9.4. This diagram does not convey (to me, at least!) the true nature of the circuit. It actually represents a balanced circuit: the broken lines depicting a mirror image of the components shown. The full circuit diagram shown on the right-hand side reveals that it is actually a bridge circuit.

Figure 9.4

Schematic Symbol of
First-Order Equalizer

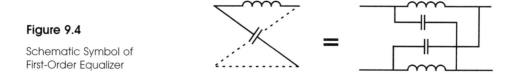

The pole-zero diagram of a first-order equalizer is given in Figure 9.5. There is one pole on the negative real axis and, at an equal distance from the origin, one zero on the positive real axis. Since both pole and zero are at an equal distance from any point on the imaginary frequency axis, signals pass through the filter without attenuation.

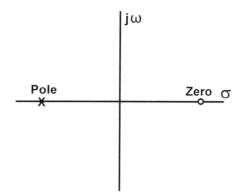

Figure 9.5

First-Order Equalizer
Pole-Zero Diagram

Passive Second-Order Equalizers

Second-order equalizers can be tuned to set the frequency at which the peak delay occurs. This makes the circuit more versatile. A cascade of second-order equalizers, each tuned to a different frequency, can provide a delay across a wide band of frequencies. What is more, the peak value of the delay can also be adjusted. The peak value is proportional to the Q of the circuit, and different circuit configurations are needed for different ranges of Q value.

A practical passive circuit for a low-Q second-order equalizer is given in Figure 9.6. This design can be used for Q values of up to one. In the equations for the center-tapped inductor, $L3$, the inductance of each half section is given. The total inductance of $L3$ is four times that of each half section, since the coupling between windings is close to unity. The circuit is tuned to the frequency ω_R, which is in radians per second.

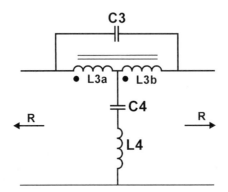

Figure 9.6

Low-Q Second-Order
Equalizer

The values of the capacitors and inductors are given by the following equations:

$$L = \frac{(Q^2 + 1).R}{2Q\omega}$$

$$K = \frac{1 - Q^2}{1 + Q^2}$$

The Inductance of each section of L3 can be found using a single
equation, that is $L3a = L3b = \dfrac{(1 + K).L}{2}$

$$L4 = \frac{(1 - K).L}{2}$$

$$C3 = \frac{Q}{2R\omega} \qquad C4 = \frac{2}{QR\omega}$$

Analysis of the series elements in the low-Q equalizer at very low and very high frequencies shows that there is, in theory, no phase shift. At low frequencies (near DC) the series inductor L3 is a short circuit and the input signal is in phase with the output. The series capacitor C3 is a short circuit at very high frequencies and, again, the input and output are in phase. In practice there may be a slight phase shift at high and low frequency extremes because of the inductor's coil resistance.

As the frequency is increased from DC the series inductive reactance increases, the shunt capacitive reactance reduces, and the output phase shift approaches −180°. At the frequency where the series tuned circuit C4 and L4 resonate, ω_R, the center tap of inductor L3 is shunted to ground. It is, in effect, a transformer with anti-phase primary and secondary windings so there will be a 180° phase shift at this point. The total phase shift of a second-order section, over a wide frequency range, is 360°—twice that of a first-order section.

If a Q of greater than one is required, the circuit given in Figure 9.7 should be used. This does not need a center-tapped inductor, only two matched capacitors. This circuit is easier to produce because no special components are needed.

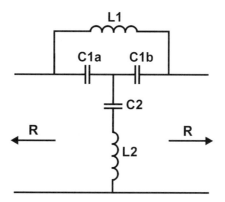

Figure 9.7

High-Q Second-Order
Equalizer

The component values for a second-order equalizer with a Q of greater than one are given as follows:

$$L1 = \frac{2.R}{Q\omega} \qquad L2 = \frac{Q.R}{2\omega}$$

$$C1a = C1b = \frac{Q}{\omega R}$$

$$C2 = \frac{2.Q}{(Q^2 - 1)\omega R}$$

A textbook representation of a second-order equalizer is given in the left-hand side of Figure 9.8. The dashed line represents a mirror image of the components shown. The full circuit is given in the right-hand side of the same diagram. As with the first-order section, it is a bridge circuit and it is balanced. Each arm of the circuit uses a series or parallel tuned circuit.

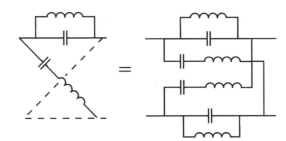

Figure 9.8

Schematic and Full Circuit
of Second-Order Equalizer

The pole-zero diagram of a second-order equalizer has a pair of complex poles and a pair of complex zeroes. The pole positions are symmetrical either side of the real axis and placed to the left of the imaginary axis. The zeroes are at mirror-image positions to the poles and are placed to the right of the imaginary axis. Thus the poles and zeroes are all the same distance from the imaginary frequency axis, and they are all the same distance from the real axis. This is illustrated in Figure 9.9. The positions of the poles and zeroes can be thought of as the four corners of an imaginary box. The top and bottom of this box are both at Beta units from the real axis. The sides of this box are both at Alpha units from the imaginary frequency axis.

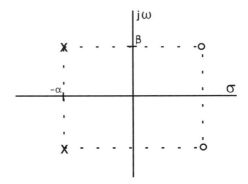

Figure 9.9

Pole-Zero Diagram of
Second-Order Equalizer

The pole and zero locations are related to the frequency and Q of the circuit, through the following equations:

$$\omega_R = \sqrt{\alpha^2 + \beta^2}$$

$$Q = \frac{\omega_R}{2\alpha}$$

Note that α and β have been used for the coordinates on the σ and ω axis, respectively, to avoid confusion with ω_R, which is the resonant frequency of the second-order equalizer.

The amplitude of signals passing through the second-order equalizer do not change with frequency. This is because the poles and the zeroes are placed at equal and opposite positions from all points on the frequency axis. The frequency at which the group delay peaks is dependent both on Alpha and Beta coordinates. With the poles and zeroes close to the real axis the peak delay occurs at low frequencies. Conversely, as the poles and zeroes move away from the real axis the peak delay occurs at higher frequencies. The closer the poles and zeroes are to the imaginary axis, the greater the peak delay amplitude.

Active First-Order Equalizers

Active equalizer sections use component values that are dependent on both the pole and zero positions and on the designer's choice. This is the opposite of passive equalizers that do not allow the designer any scope in the design, because the component values depend only on the impedance and the pole and zero locations.

It is not possible to design active equalizers in the same way as passive equalizers. See, for example, the first-order equalizer given in Figure 9.10. The resistors $R1$ and $R2$ set the DC gain, typically they may both be about $10\,\mathrm{k}\Omega$. The product of R and C is set by the pole location; but the individual values of R and C used are at the discretion of the designer, subject to their product being correct. However, if the amplifier's input bias current is high (particularly with bipolar op-amps) it may cause a DC offset problem. In this case the value of R should be set to equal the parallel combination of $R1$ and $R2$, so that an equal bias current is drawn from both inverting and noninverting inputs.

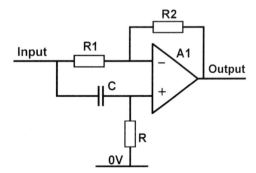

Figure 9.10

Active First-Order Equalizer

The values of resistor R and capacitor C depend on the real pole location: $\sigma = \dfrac{1}{RC}$. This assumes that the pole location has been denormalized by scaling it for the required frequency. The frequency is the same as the passband cutoff of the filter being equalized. In the case of quadrature networks, which are described in this chapter, it is scaled by the passband center frequency.

Active Second-Order Equalizers

The design of second-order equalizers depends on the required Q of the circuit. If the Q is less than $1/\sqrt{2}$ (0.7071) the circuit in Figure 9.11 should be used.

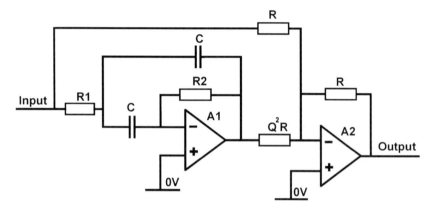

Figure 9.11

Active Low-Q Second-Order Equalizer

For the components in the first stage, around op-amp A1, the values of resistors $R1$ and $R2$ can be calculated from the following equations:

$$R1 = \frac{R2}{4Q^2} \qquad R2 = \frac{2.Q}{\omega_R C}$$

The second op-amp $A2$ and its associated resistors form a summing amplifier, with feedback resistor R (typically $10\,k\Omega$) between the op-amp output and its inverting input. The op-amp inverting input is also connected to the circuit input by another resistor of value R. A third resistor connected to the op-amp inverting input is connected to the output of $A1$. The value of this third resistor is Q^2R.

The circuit in Figure 9.12 should be used for Q values in the range 0.7071 to 20. Both circuits are similar, although the higher Q circuit has a potential divider at the input to the first stage. The formulae for calculating component values are different from those given in Figure 9.11.

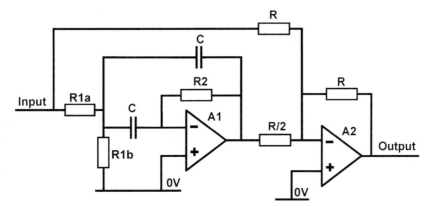

Figure 9.12

Active High-Q Second-Order Equalizer

The values of resistors $R1a$, $R1b$, and $R2$ can be found using the following equations:

$$R1a = \frac{R2}{2} \qquad R1b = \frac{R1a}{2.Q^2 - 1}$$

$$R2 = \frac{Q}{\omega_R C}$$

Equalization of Butterworth and Chebyshev Filters

The first part of this chapter described phase equalizer circuits in general terms. This theme is extended now to include the equalization of Butterworth and Chebyshev lowpass filters. Tables of equalizer coefficients are given, where practical, for equalizing filters of up to twelfth order. Equalizer coefficients are not provided in cases where the equalizer would be far more complex than the filter being equalized. Examples of filter equalization using up to fourth-order equalizers are given.

First-order equalizers are described by a Sigma 1 value. This is the pole position, which is on the real negative axis of the pole-zero diagram and was described in this chapter (Passive First-Order Equalizers). The group delay is greatest at low frequencies and is smoothly decaying as the frequency rises.

Second-order equalizers are described by four factors: Sigma, Omega, Q, and B. Sigma and B are the real and imaginary coordinates of the pole-zero constellation. Omega (or ω_R) is the peak delay frequency, and Q is the Q-factor of this peak; these two factors are required to find component values in a second-order equalizer.

Higher-order equalizers are described by a combination of first- and second-order factors. Third-order equalizers use a first- and second-order section in series, so values are given for Sigma 1, Sigma 2, Omega 2, $Q2$, and $B2$. Sigma 1 describes the first-order section and the other factors describe the second-order section. Fourth-order equalizers use two second-order sections in series. Sigma 1, Omega 1, $Q1$, and $B1$ describe one second-order section, and Sigma 2, Omega 2, $Q2$, and $B2$ describe the other.

Passive filter component values for these equalizers are also given in separate tables (Table 9.2 and Tables 9.4–9.8). These are normalized for one-ohm termination and a one radian per second cutoff frequency. Active equalizer components are not given since there are many solutions, unlike the passive equalizer where the solution depends on both the impedance and the filter's cutoff frequency.

Group Delay of Butterworth Filters

To find the group delay of a Butterworth filter it is necessary to carry out the following steps:

1. Find the denominator coefficients of the Butterworth transfer function.
2. Multiply each coefficient by the Laplace variable (s) to the power of the coefficient subscript.
3. Calculate the phase-shift function, using these coefficients and frequency variables.
4. Differentiate the phase-shift function to find the group delay.

These steps have to be repeated for each filter-order required.

STEP ONE: The denominator coefficients can be found using an iterative formula given by Heulsman.[4]

$$a_k = \frac{\cos[(k-1)\pi/2n]}{\sin(k\pi/2n)} \cdot a_{k-1} \quad k = 1, 2, \ldots n$$

$$a_0 = a_n = 1$$
$$a_1 = a_{n-1}$$
$$a_2 = a_{n-2}$$

The coefficients obtained using this formula are given in Table 9.1 for up to twelfth-order filters. The number of coefficients given is no more than half of the filter-order; this is because the coefficients are symmetric. Take the example of a third-order filter ($n = 3$); only one coefficient is given: $a_1 = 2.000$. However, $a_0 = 1.000$ and $a_3 = a_n = 1.000$. Also $a_2 = a_{n-1}$, which equals $a_1 = 2.000$.

Order	a_1	a_2	a_3	a_4	a_5	a_6
2	1.414214					
3	2.000000					
4	2.613126	3.414214				
5	3.236068	5.236068				
6	3.863703	7.464102	9.141620			
7	4.493959	10.097835	14.591794			
8	5.125831	13.137071	21.846151	25.688356		
9	5.758770	16.581719	31.163437	41.986386		
10	6.392453	20.431729	42.802061	64.882396	74.233429	
11	7.026675	24.687075	57.020267	95.937001	123.24352	
12	7.661297	29.347740	74.076215	136.87499	194.71869	218.46873

Table 9.1

Butterworth Transfer Function Denominator Coefficients

STEP TWO: The transfer function is the reciprocal of the coefficient and frequency variable products. So, again for the third-order filter, the denominator is the sum of:

$$a_3 s^3 + a_2 s^2 + a_1 s + a_0$$
$$\text{or } s^3 + 2s^2 + 2s + 1$$

STEP THREE: Now the phase-shift function needs to be obtained. Having found the denominator of the transfer function, you now need to separate it into odd and even powers of frequency. Odd powers and their associated coefficients are summed and used as a numerator, leaving the even powers in the denominator. Conversion from the S-plane to find the complex frequency variable ($j\omega$) is needed. The phase-shift function is the negative arc tangent of this resultant equation, with the denominator multiplied by j.

$$\text{phase} = -\arc\tan[(j^3\omega^3 + 2j\omega)/j(2j^2\omega^2 + 1)]$$

There are complex j multipliers to consider. Squaring this function gives minus one. So $j^2 + -1$ and $j^3 + -j$.

$$\text{phase} = -\arc\tan[(-j\omega^3 + 2j\omega)/j(-2\omega^2 + 1)]$$

The complex factor j cancels, leaving:

$$\text{phase} = -\arc\tan[(-\omega^3 + 2\omega)/(-2\omega^2 + 1)]$$

STEP FOUR: Finally, the group delay is the differentiation of the phase-shift function. The result of this differentiation is the rate of change of the function, which is the group delay. Differentiation is a complex subject; however, for this purpose, it is sufficient to know that the differentiation of arc tan (x) is $1/(x^2 + 1)$. I do not propose to go into this further here, but the resultant equation has only even powers of ω, the highest power being in the denominator and equal to $(1 + \omega)^{2n}$, where n is the filter-order. Examples of calculation for up to third-order Butterworth filters are given by Helszajn.[5]

Having obtained the equations for the group delay of Butterworth filters, up to twelfth-order, MATHCAD[6] was used to optimize the equalizer. An example of this is given in Figure 9.13; here a seventh-order design is equalized by a second-order equalizer section. The frequency variable (ω) is steps in increments of 1/100 radians per second. The tuned frequency of the second-order equalizer (ω_R) is symbolized by Ω. The resultant group delay has an equi-ripple characteristic. Equi-ripple is the state where all the peaks are equal in amplitude and all the troughs are equal in amplitude, although not necessarily at an equal frequency spacing.

Equalizer Example: A Delay Equalized 7th Order Butterworth Filter

Frequency Range 1% to 200% of Cut-off

$$\omega := 1, 2 .. 100$$

Seventh order Butterworth filter delay

$$T_\omega := \frac{1.109798\left(\frac{\omega}{100}\right)^4 + 4.49396\left(\frac{\omega}{100}\right)^{12} + 1.60387\left(\frac{\omega}{100}\right)^2 + 1.109798\left(\frac{\omega}{100}\right)^8 + 1.60387\left(\frac{\omega}{100}\right)^{10} + 1.0\left(\frac{\omega}{100}\right)^6 + 4.49396}{1 + \left(\frac{\omega}{100}\right)^{14}}$$

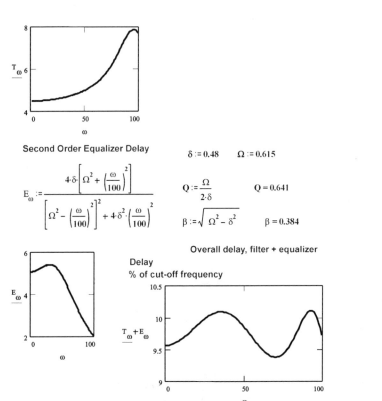

Second Order Equalizer Delay

$$\delta := 0.48 \qquad \Omega := 0.615$$

$$E_\omega := \frac{4 \cdot \delta \cdot \left[\Omega^2 + \left(\frac{\omega}{100}\right)^2\right]}{\left[\Omega^2 - \left(\frac{\omega}{100}\right)^2\right]^2 + 4 \cdot \delta^2 \cdot \left(\frac{\omega}{100}\right)^2}$$

$$Q := \frac{\Omega}{2 \cdot \delta} \qquad Q = 0.641$$

$$\beta := \sqrt{\Omega^2 - \delta^2} \qquad \beta = 0.384$$

Overall delay, filter + equalizer

Delay
% of cut-off frequency

Figure 9.13

Equalized Seventh-Order Response

A third-order equalizer could be used in the above example to reduce the group delay ripple further. The first-order section would produce its maximum delay at low frequencies. The delay from this section would add to the second-order section delay; so, at higher frequencies where the second-order section peak delay occurs, the overall peak would be raised too high. It would be necessary to increase the peak delay frequency of the second-order section to maintain the equi-ripple delay overall.

Lowpass Butterworth filters from third-order to twelfth-order can be equalized using equalizers described in Table 9.2. This also gives an indication of the delay change with and without an equalizer. A third-order filter delay at low frequencies is 27% lower than at its peak, near the cutoff frequency. This can be reduced to 1.8% by adding a first-order equalizer. In general, as the filter-order increases, the equalizer has to be more complex in order to keep a low percentage difference between the maximum and minimum group delays.

Filter Order	Equalizer Order	Sigma 1	Omega 1	Q1	B1	Sigma 2	Omega 2	Q2	B2	Original Delay	Modified Delay
3	1	1								27%	1.80%
4	1	0.82								31%	4%
5	1	0.67								38%	9.70%
5	2	0.71	0.8	0.563	0.369					38%	3.60%
6	1	0.56								42%	16.00%
6	2	0.55	0.68	0.618	0.4					42%	3.80%
7	1	0.477								43%	21.40%
7	2	0.48	0.615	0.641	0.384					43%	7.40%
8	3	0.42				0.4	0.705	0.881	0.581	45%	3.30%
9	3	0.38				0.36	0.672	0.933	0.567	47%	8.10%
10	3	0.335				0.33	0.645	0.977	0.554	48%	8.40%
10	4	0.32	0.741	1.158	0.668	0.344	0.41	0.596	0.223	48%	3.20%
11	4	0.29	0.721	1.243	0.66	0.31	0.375	0.605	0.211	49%	4.60%
12	4	0.265	0.71	1.34	0.659	0.275	0.35	0.636	0.216	51%	6.50%

Table 9.2

Butterworth Equalizer Coefficients

Tables of passive equalizer component values are given in Table 9.3. It is not possible to produce similar tables for active equalizers because many of the components are subject to the designer's choice. Passive equalizer components in Table 9.3 are normalized; that is, the values required if the lowpass filter being equalized has a one radian per second cutoff frequency and is terminated in a $1\,\Omega$ resistor.

First Order		
Filter	*L*	*C*
3	2	2
4	2.439	2.439
5	2.985	2.985
6	3.571	3.571
7	4.193	4.193

Second-Order Equalizer				
Filter	*C3*	*L3a/L3b*	*C4*	*L4*
5	0.3519	1.1101	4.4405	0.3519
6	0.4544	1.1898	4.7592	0.4544
7	0.5211	1.2683	5.0734	0.5211

Third Order						
Filter	*L*	*C*	*L3a/L3b*	*C3*	*L4*	*C4*
8	4.7619	4.7619	0.805	0.6248	0.6248	3.22
9	5.2632	5.2632	0.7975	0.6942	0.6942	3.1899
10	5.9701	5.9701	0.7934	0.7574	0.7574	3.1738

Fourth-Order Equalizer							
Filter	*L1*	*C1a/C1*	*L2*	*C2*	*L3a/L3b*	*C3*	*L4*	*C4*
10	2.3308	1.5628	0.7814	0.1667	2.0462	0.7268	0.7268	8.1846
11	2.2316	1.724	0.862	6.326	2.2039	0.8067	0.8067	8.8154
12	2.1022	1.8873	0.9437	4.7444	2.2462	0.9086	0.9086	8.9847

Table 9.3

Butterworth Component Values

The denormalization process requires multiplication and division of these values. To denormalize a capacitor, divide its value by the termination resistance and by ω ($= 2.\pi.f.$). Denormalize inductors by multiplying their value by the termination resistance and dividing by ω. Where f is the lowpass cutoff frequency of the filter.

To show how Table 9.3 can be used, I have given an example of an equalized third-order Butterworth filter in Figure 9.14. The filter was designed using *FILTECH*[7] and has a 1 kHz cutoff and a 50-ohm load. Now, the first-order equalizer for this filter has a normalized capacitor value of 2.0 that must be divided by $\omega.R$, or 314,159. The capacitor required is $6.366\,\mu F$. The equalizer's normalized inductor value is also 2.0. This must be multiplied by R and divided by ω, which is a multiplying factor of 7.95775×10^{-3}. The inductor required is $15.916\,mH$: each half winding requires a self-inductance of $3.979\,mH$.

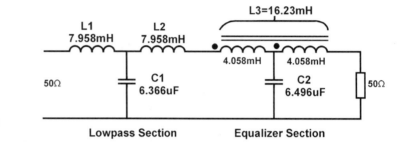

Figure 9.14

Equalized Third-Order Butterworth Filter

Simulation of the filter and equalizer combination was carried out using *ANALYSER III.*[7] The frequency response, given in Figure 9.15, shows plots for gain and group delay. The gain is flat until mid-band, then rolls off to give a correct $-3\,dB$ frequency of $1\,kHz$. The phase is not shown, but has an almost constant rate of change with frequency, which on a linear frequency scale would appear as a straight line.

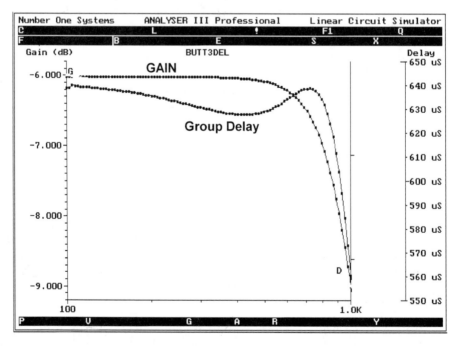

Figure 9.15

Simulated Frequency Response

The most important plot in Figure 9.15 is the group delay. This falls steadily over the passband of the filter, then peaks up just before the cutoff frequency. Over most of the passband the group delay is between $630\,\mu s$ and $640\,\mu s$.

Equalization of Chebyshev Filters

The Butterworth filter was reasonably simple to equalize because its group delay had a smooth curve. Chebyshev filters are more difficult to equalize because the peak group delay has greater amplitude; also, the group delay does not rise smoothly, but has ripple.

The group delay of an equalizer can be set to have a high peak value, but this causes the range of frequencies over which the delay occurs to become very short. Generally there is a reciprocal relationship between the value of the peak group delay and the steepness of the group delay versus frequency curve. Butterworth filters had a low percentage group delay variation across the passband and many could be equalized effectively by a first-order equalizer.

Chebyshev filters have a high percentage group delay variation across the passband. The significance of this is that the equalizing sections must also produce a high percentage group delay variation at other frequencies to compensate for where the filter's group delay is short. Equalizers that produce these high delay variations do so for a limited range of frequencies. Unless there are several equalizing sections, each compensating for different frequencies, the group delay curve will have significant ripple.

Chebyshev Group Delay

A computer program subroutine given by Rorabaugh in *Digital Filter Designer's Handbook*[8] provides the amplitude and phase response of Chebyshev filters. This subroutine is reproduced, with the kind permission of McGraw-Hill, as Listing 9.1. The subroutine, chebyshevFreqResponse( ), was used within a program of my own to produce tables of phase versus frequency. The result was then numerically differentiated to find the group delay at each frequency. The tables of group delay versus frequency were then used by one of my *MATHCAD* applications to produce a graph.

Listing 9.1 is reproduced from Chapter 4, listing 4.1 of *Digital Filter Designer's Handbook* by C. Britton Rorabaugh, 1993. Reproduced with permission from McGraw-Hill.

```c
/**********************************/
/*                                */
/*    Listing 4.1                 */
/*                                */
/*      chebyshevFreqResponse()   */
/*                                */
/**********************************/
#include <math.h>
#define PI (double) 3.141592653589


void chebyshevFreqResponse(   int order,
                              float ripple,
                              char normalizationType,
                              float frequency,
                              float *magnitude,
                              float *phase)
{
double A, gamma, epsilon, work;
double rp, ip, x, i, r, rpt, ipt;
double normalizedFrequency, hSubZero;
int k, ix;

epsilon = sqrt( -1.0 + pow( (double)10.0, (double)(ripple/10.0) ));
gamma = pow( (( 1.0 + sqrt( 1.0 + epsilon*epsilon))/epsilon),
                         (double)(1.0/(float) order) );

if( normalizationType = '3' )
        {
        work = 1.0/epsilon;
        A = ( log( work + sqrt( work*work - 1.0) ) ) / order;
        normalizedFrequency = frequency * ( exp(A) + exp(-A))/2.0;
        }
else
        {
        normalizedFrequency = frequency;
        }

rp = 1.0;
ip = 0.0;

for( k = 4; k< = order; k++)
        {
        x = (2*k - 1) * PI / (2.0*order);
        i = 0.5 * (gamma + 1.0/gamma) * cos(x);
        r = -0.5 * (gamma - 1.0/gamma) * sin(x);
        rpt = ip * i - rp * r;
        ipt = -rp * i - r * ip;
        ip = ipt;
        rp = rpt;
        }
hSubZero = sqrt( ip*ip + rp*rp);
if( order%2 = 0 )
        {
        hSubZero = hSubZero / sqrt(1.0 + epsilon*epsilon);
        }
```

```
rp = 1.0;
ip = 0.0;
for(k = 1; k <= order; k++)
        {
        x = (2*k - 1)*PI/(2.0*order);
        i = 0.5 * (gamma + 1.0/gamma)  * cos(x);
        r = -0.5 * (gamma - 1.0/gamma)  * sin(x);
        rpt = ip*(i-normalizedFrequency) - rp*r;
        ipt = rp*(normalizedFrequency-i) - r*ip;
        ip = ipt;
        rp = rpt;
        }
*magnitude = 20.0 * log10(hSubZero/sqrt(ip*ip + rp*rp));
*phase = 180.0 * atan2( ip, rp) /PI;
return;
}
```

Listing 9.1

Subroutine "chebyshevFreqResonse( )"

Equalizer equations given in the same *MATHCAD* application were then used to find the minimum group delay variation. The coefficients for the equations were adjusted until the sum of equalizer and filter group delay variations were minimized. This was carried out by eye, rather than using an optimization routine. The lowest variation in group delay occurred when the group delay was equi-ripple; that is, the peaks all had the same amplitude and the troughs all had the same amplitude. The resulting equalization pole factors, such as ω_R and Q, have been calculated for Chebyshev filters with 0.01 dB, 0.1 dB, 0.25 dB, 0.5 dB, and 1 dB passband ripple. As in the Butterworth design case, higher-order filters are more difficult to equalize. This also applies as the passband ripple increases: 0.01 dB ripple filters are easier to equalize than 1 dB-ripple designs.

The calculated pole factors for Chebyshev filter equalizers are given in Tables 9.4 to 9.8. The number of designs equalized was limited to filter-orders that gave practical results. It was not sensible to equalize filters where the equalizer would be far more complicated than the filter itself. Passive equalizer component values have been calculated for several Chebyshev filter designs from the equalization pole factors, using the equations given earlier in this chapter. Again, component values to equalize Chebyshev filters with 0.01 dB, 0.1 dB, 0.25 dB, 0.5 dB, and 1 dB passband ripple were calculated. Component values for a limited number of practical passive equalizers are given in Tables 9.9 to 9.13. Active equalizer values are not given because these depend on some user-defined variables.

Filter Order	Equalizer Order	Sigma 1	Omega 1	Q1	B1	Sigma 2	Omega 2	Q2	B2	Original Delay	Modified Delay
3	1	0.89								48%	4.00%
4	1	0.642								71%	16%
4	2	0.595	0.722	0.607	0.409					71%	3%
5	1	0.46								92%	33.00%
5	2	0.444	0.596	0.671	0.398					92%	10.40%
5	3	0.46				0.425	0.716	0.842	0.576	92%	2.80%
6	2	0.33	0.506	0.766	0.384					122%	23.40%
6	3	0.33				0.32	0.653	1.020	0.569	122%	9.20%
6	4	0.308	0.738	1.198	0.671	0.334	0.402	0.602	0.224	122%	3.30%
7	2	0.25	0.445	0.89	0.368					130%	41.00%
7	3	0.25				0.249	0.608	1.221	0.555	130%	21.20%
7	4	0.245	0.703	1.434	0.659	0.256	0.336	0.656	0.218	130%	10.40%
8	3	0.19				0.192	0.57	1.484	0.537	165%	36.00%
8	4	0.19	0.673	1.771	0.646	0.19	0.285	0.750	0.212	165%	8.80%

Table 9.4

0.01 dB Chebyshev Equalizer

Filter Order	Equalizer Order	Sigma 1	Omega 1	Q1	B1	Sigma 2	Omega 2	Q2	B2	Original Delay	Modified Delay
3	1	0.785								58%	10.90%
3	2	0.68	0.795	0.585	0.412					58%	1.50%
4	1	0.52								97%	29.30%
4	2	0.495	0.64	0.646	0.406					97%	8.30%
4	3	0.475				0.435	0.732	0.841	0.589	97%	1.40%
5	1	0.365								119%	52.40%
5	2	0.347	0.531	0.765	0.402					119%	22.47%
5	3	0.35				0.336	0.666	0.991	0.575	119%	9.50%
5	4	0.306	0.746	1.219	0.680	0.33	0.402	0.609	0.230	119%	3.10%
6	3	0.243				0.246	0.618	1.256	0.567	155%	23.80%
6	4	0.242	0.708	1.463	0.665	0.25	0.335	0.67	0.223	155%	12.50%
7	4	0.183	0.68	1.858	0.655	0.18	0.282	0.783	0.217	155%	25.70%

Table 9.5

0.1 dB Chebyshev Equalizer

Filter Order	Equalizer Order	Sigma 1	Omega 1	Q1	B1	Sigma 2	Omega 2	Q2	B2	Original Delay	Modified Delay
3	1	0.73								73%	17.00%
3	2	0.556	0.7	0.629	0.425					73%	1.80%
4	1	0.45								118%	43.00%
4	2	0.444	0.6	0.676	0.404					118%	14.50%
4	3	0.399				0.375	0.7	0.933	0.591	118%	2.80%
5	1	0.31								150%	72.80%
5	2	0.298	0.504	0.846	0.406					150%	33.70%
5	3	0.303				0.295	0.646	1.095	0.575	150%	16.70%
5	4	0.266	0.73	1.372	0.680	0.28	0.362	0.646	0.229	150%	6.10%
6	3	0.202				0.209	0.605	1.447	0.568	185%	36.00%
6	4	0.212	0.698	1.646	0.665	0.215	0.308	0.716	0.221	185%	18.80%
7	4	0.156	0.673	2.157	0.655	0.15	0.262	0.873	0.215	225%	38.50%

Table 9.6

0.25 dB Chebyshev Equalizer

Filter Order	Equalizer Order	Sigma 1	Omega 1	Q1	B1	Sigma 2	Omega 2	Q2	B2	Original Delay	Modified Delay
3	1	0.67								89%	25.60%
3	2	0.495	0.65	0.657	0.421					89%	2.80%
4	1	0.39								140%	58.80%
4	2	0.4	0.57	0.712	0.406					140%	24.30%
4	3	0.34				0.326	0.677	1.038	0.593	140%	6.30%
4	4	0.3	0.73	1.217	0.666	0.38	0.428	0.563	0.197	140%	1.90%
5	2	0.26	0.49	0.942	0.415					181%	47.70%
5	3	0.26				0.26	0.63	1.212	0.574	181%	28.10%
5	4	0.234	0.721	1.541	0.682	0.234	0.33	0.705	0.233	181%	11.40%
6	3	0.17				0.18	0.6	1.667	0.572	226%	150.00%
6	4	0.188	0.689	1.832	0.663	0.185	0.288	0.778	0.221	226%	32.00%

Table 9.7

0.5 dB Chebyshev Equalizer

Filter Order	Equalizer Order	Sigma 1	Omega 1	Q1	B1	Sigma 2	Omega 2	Q2	B2	Original Delay	Modified Delay
3	1	0.61								122%	42.30%
3	2	0.425	0.603	0.709	0.428					122%	4.40%
3	3	0.632				0.42	0.67	0.798	0.522	122%	2.10%
4	1	0.314								196%	94.00%
4	2	0.35	0.535	0.764	0.405					196%	40.00%
4	3	0.279				0.275	0.658	1.196	0.598	196%	12.50%
4	4	0.256	0.715	1.396	0.668	0.33	0.38	0.576	0.188	196%	5.00%
5	2	0.215	0.485	1.128	0.435					225%	84.00%
5	3	0.22				0.225	0.615	1.367	0.572	225%	45.80%
5	4	0.199	0.71	1.784	0.682	0.19	0.3	0.789	0.232	225%	23.00%
6	3	0.135				0.146	0.6	2.055	0.582	315%	81.00%
6	4	0.16	0.68	2.125	0.661	0.155	0.265	0.855	0.215	315%	54.00%

Table 9.8

1 dB Chebyshev Equalizer

First Order Equalizer		
Filter Order	L	C
3	2.247	2.247
4	3.115	3.115
5	4.348	4.348

Second Order Equalizer				
Filter Order	C3	L3a/L3b	C4	L4
4	0.4204	1.2683	4.5636	0.4673
5	0.5629	1.1683	5.0011	0.526
6	0.7569	1.0211	5.16	0.5999
7	1	0.844	5.0499	0.6685

Third Order Equalizer						
Filter Order	L	C	L3a/L3b	C3	L4	C4
5	4.3478	4.3478	0.7335	0.5882	0.5205	3.316
(Q > 1)	L	C	L1	C1a/C1b	L2	C2
6	6.0606	6.0606	3.0018	1.5625	0.7813	76.1497
7	8	8	2.6943	2.008	1.004	8.1868
8	10.5263	10.5263	2.3638	2.6042	1.3021	4.3281

Fourth Order Equalizer								
Filter Order	L1	C1a/C1b	L2	C2	L3a/L3b	C3	L4	C4
6	2.2621	1.6233	0.8116	7.46	1.7625	0.7485	0.6383	8.2671
7	1.9839	2.0398	1.0199	3.862	1.7639	0.9766	0.7597	9.0703
8	1.678	2.6315	1.3157	2.4634	1.6187	1.3158	0.9105	9.3567

Table 9.9

0.01 dB Chebyshev Passive Equalizer Values

First Order Equalizer		
Filter Order	*L*	*C*
3	2.5478	2.5478
4	3.8462	3.8462
5	5.4795	5.4795

Second Order Equalizer				
Filter Order	*C3*	*L3a/L3b*	*C4*	*L4*
3	0.3679	1.3072	4.3004	0.4474
4	0.5047	1.2028	4.8375	0.5019
5	0.7203	0.9954	4.9235	0.5825

Third Order Equalizer					
Filter Order	*L*	*C*	*L3a/L3b*	*C3*	*L4*	*C4*
4	4.2105	4.2105	0.7306	0.5744	0.5168	3.2488
5	5.7143	5.7143	0.5517	0.744	0.5418	3.0303
(*Q* > 1)	*L*	*C*	*L1*	*C1a/C1b*	*L2*	*C2*
6	8.2304	8.2304	2.5766	2.0324	1.0162	7.038

Fourth Order Equalizer							
Filter Order	*L1*	*C1a/C1b*	*L2*	*C2*	*L3a/L3b*	*C3*	*L4*	*C4*
5	2.1993	1.634	0.817	6.725	1.7306	0.7575	0.6418	8.1693
6	1.9309	2.0664	1.0332	3.6241	1.71	1	0.7676	8.9107
7	1.583	2.7324	1.3662	2.2285	1.5154	1.3883	0.9291	9.0577

Table 9.10

0.1 dB Chebyshev Passive Equalizer Values

First Order Equalizer

Filter Order	L	C
3	2.7397	2.7397
4	4.4444	4.4444
5	6.4516	6.4516

Second Order Equalizer

Filter Order	C3	L3a/L3b	C4	L4
3	0.4493	1.2124	4.5424	0.4797
4	0.5633	1.1507	4.931	0.5258
5	0.8393	0.8571	4.6906	0.6134

Third Order Equalizer

Filter Order	L	C	L3a/L3b	C3	L4	C4
4	5.0125	5.0125	0.6098	0.6664	0.5309	3.0623
($Q > 1$)	L	C	L1	C1a/C1b	L2	C2
5	6.6007	6.6007	2.8274	1.695	0.8475	17.0335
6	9.901	9.901	2.2846	2.3917	1.1959	4.3732

Fourth Order Equalizer

Filter Order	L1	C1a/C1b	L2	C2	L3a/L3b	C3	L4	C4
5	1.9969	1.8794	0.9397	4.2599	1.7062	0.8923	0.712	8.5524
6	1.7408	2.3582	1.1791	2.7592	1.6411	1.1623	0.8413	9.0691
7	1.3777	3.2051	1.6025	1.7549	1.3257	1.666	1.0104	8.7441

Table 9.11

0.25 dB Chebyshev Passive Equalizer Values

First Order Equalizer

Filter Order	L	C
3	2.9851	2.9851
4	5.1282	5.1282

Second Order Equalizer

Filter Order	C3	L3a/L3b	C4	L4
3	0.5054	1.1633	4.6833	0.5022
4	0.6246	1.0832	4.9281	0.5491
5	0.9612	0.7117	4.3329	0.6316

Third Order Equalizer ($Q > 1$)

Filter Order	L	C	L1	C1a/C1b	L2	C2
4	5.8824	5.8824	2.8461	1.5332	0.7666	39.596
5	7.6923	7.6923	2.6193	1.9238	0.9619	8.2049
6	11.7647	11.7647	1.9996	2.7783	1.3892	3.1237

Fourth Order Equalizer

Filter Order	L1	C1a/C1b	L2	C2	L3a/L3b	C3	L4	C4
4	2.2512	1.6671	0.8336	6.9306	1.8642	0.6577	0.5909	8.3
5	1.8001	2.1373	1.0686	3.1095	1.592	1.0682	0.7912	8.5966
6	1.5845	2.6589	1.3295	2.2569	1.5054	1.3507	0.9112	8.926

Table 9.12

0.5 dB Chebyshev Passive Equalizer Values

First Order Equalizer		
Filter Order	*L*	*C*
3	3.2787	3.2787
4	6.3694	6.3694

Second Order Equalizer				
Filter Order	*C3*	*L3a/L3b*	*C4*	*L4*
3	0.5879	1.0613	4.6781	0.5335
4	0.714	0.9935	4.8931	0.5799
(Q > 1)	*L1*	*C1a/C1b*	*L2*	*C2*
5	3.6558	2.3258	1.1629	17.0772

Third Order Equalizer						
Filter Order	*L*	*C*	*L3a/L3b*	*C3*	*L4*	*C4*
3	3.1646	3.1646	0.8278	0.5955	0.5272	3.7407
(Q > 1)	*L*	*C*	*L1*	*C1a/C1b*	*L2*	*C2*
4	7.1685	7.1695	2.5414	1.8176	0.9088	8.4459
5	9.0909	9.0909	2.379	2.2228	1.1114	5.1175
6	14.8148	14.8148	1.6221	3.425	1.7125	2.1253

Fourth Order Equalizer								
Filter Order	*L1*	*C1a/C1b*	*L2*	*C2*	*L3a/L3b*	*C3*	*L4*	*C4*
4	2.0037	1.9524	0.9762	4.1155	1.9629	0.7579	0.6513	9.1374
5	1.579	2.5127	1.2563	2.3024	1.4191	1.315	0.8834	8.4495
6	1.3841	3.125	1.5625	1.7778	1.3644	1.6132	0.9974	8.8271

Table 9.13

1 dB Chebyshev Passive Equalizer Values

Quadrature Networks and Single Sideband Generation

Quadrature networks are filter pairs that produce a 90° phase difference output when the same signal is applied to each input. This feature has many useful applications in radio and signal-processing systems. One such application is the phasing method of single sideband generation, which was developed to provide generation of a single sideband modulated carrier, without the narrowband filtering problems. The phasing method will now be described, followed by a circuit description and analysis of the signal processing that takes place.

When a carrier signal is amplitude modulated it generates two "sidebands"; the spectrum occupancy is doubled. Suppose a baseband signal occupies the spectrum from, say, DC to 4 kHz; after modulating a carrier of 1 MHz it will occupy frequencies from 1 MHz −4 kHz to 1 MHz + 4 kHz. The reason for the doubling of spectrum is that the mixer, which produces amplitude modulation, is really a multiplier. The output in mathematical terms is: $\cos(\omega 1.t).\cos(\omega 2.t) = 1/2.\cos([\omega 1 + \omega 2].t) + 1/2.\cos([\omega 1 - \omega 2].t)$, where $\omega 1$ is the carrier frequency and $\omega 2$ is the information-bearing signal frequency. Amplitude modulation is simple, but the upper and lower sidebands carry the same information. Removing one sideband by filtering saves spectrum usage but is difficult, especially at the higher carrier frequencies.

A more complex method of removing one sideband is by phasing. This method uses two modulation paths and inverts one of the sidebands in one path. Adding or subtracting the outputs from the two paths then removes one sideband. The efficiency of sideband removal using this method depends upon the accuracy of the phase inversion. The phasing method is described at the system level in Figure 9.16.

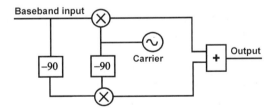

Figure 9.16

The Phasing Method

In the phasing method, baseband signals (e.g., speech) enter a quadrature generating circuit. The quadrature generator produces two outputs of the same signal, one phase shifted by 90° relative to the other over the whole of the baseband frequency range. However, this is not an easy task, and some phase inaccuracies invariably occur across the band. So, a signal that is $\cos(\omega 1.t)$ at one output is $\sin(\omega 1.t)$ at the other output.

The two quadrature signals are then used to modulate a carrier in separate mixers. The carrier signal to be modulated is applied directly to one mixer, but through a 90° phase-shift network for the other mixer (alternatively the carrier phase is shifted by +45° at one mixer and by −45° at the other). In one mixer the output is:

$$Sin(\omega 1.t).sin(\omega 2t) = 1/2.cos([\omega 1 - \omega 2].t) - 1/2.cos([\omega 1 + \omega 2].t).$$

At the other mixer the output is:

$$Cos(\omega 1.t).cos(\omega 2t) = 1/2.cos([\omega 1 + \omega 2].t) + 1/2.cos([\omega 1 - \omega 2].t).$$

The outputs from the two mixers can now be added or subtracted to give the required sideband. Adding gives cos([$\omega 1 - \omega 2$].t), which is the lower sideband. Subtracting gives cos([$\omega 1 + \omega 2$].t), which is the upper sideband. Notice that the amplitude is unity, rather than the half of each sideband produced by simply filtering out the unwanted sideband.

S. D. Bedrosian has studied the problem of producing quadrature phase-shift circuits. He has written a paper[9] that gives pole position formulae for quadrature networks. These formulae can be used to produce active or passive quadrature circuits. The quadrature circuit comprises two delay networks, known as the P net and the N net because calculations give positive (P) and negative (N) pole locations on the real axis. The P net and the N net have a common input and separate outputs. Each network produces a phase shift across the frequency band of interest, but the phase shift of one network is 90° more than the other. Only the relative phase difference is important; the absolute phase shift is irrelevant for our purpose.

Active or passive first-order equalizer sections, described earlier in Figures 9.3 and 9.10, respectively, can be used in cascade to form the P and N networks. The number of first-order equalizer sections in each P or N network is numerically half the order of the quadrature network. For example, a fourth-order quadrature circuit has two first-order equalizer sections in each network. Tables 9.14 to 9.17 give the normalized pole locations for equalizers with ratios of upper to lower passband frequency of 11.35, 20, 50, and 100. The ratio of 11.35 was chosen for the popular 300 Hz to 3.4 kHz band.

Order	P Network Pole	N Network Pole	Error (Rads)	Rejection
4	6.790134 0.590319	1.694 0.147272	0.022601	38.94 dB
6	10.402269 1.417066 0.329276	3.036969 0.705683 0.096133	0.001699	61.41 dB
8	13.972939 2.239986 0.770484 0.231521	4.319271 1.297885 0.446431 0.071567	0.000128	83.89 dB
10	17.526346 3.036969 1.231623 0.529006 0.179685	5.565304 1.890336 0.811937 0.329276 0.057057	$9.5992e^{-6}$	106.37 dB
12	21.070979 3.811892 1.694 0.840721 0.404808 0.147273	6.790133 2.470308 1.189455 0.590319 0.262337 0.047459	$7.21554e^{-7}$	128.85 dB

Table 9.14

Quadrature Pole
Locations
(Wb/Wa = 11.35)

Angle = 1.482576 rads^{-1} or 84.94539 degrees
$Q = 0.274168$ $K = 3.820999$ $K' = 1.573858$ $\omega_b/\omega_a = 11.35$

Order	P Network Pole	N Network Pole	Error (Rads)	Rejection
4	7.764836	1.788893	0.044204	33.11 dB
	0.559005	0.128786		
6	11.972004	3.361995	0.004647	52.67 dB
	1.470163	0.680197		
	0.297442	0.083528		
8	16.118147	4.870849	0.000489	72.24 dB
	2.42527	1.334347		
	0.74943	0.412325		
	0.205303	0.062042		
10	20.238522	6.33254	0.0000513536	91.81 dB
	3.361996	2.016812		
	1.25928	0.794104		
	0.495832	0.297442		
	0.157914	0.049411		
12	24.345724	7.764834	0.0000053985	111.37 dB
	4.27423	2.695463		
	1.788893	1.211696		
	0.825289	0.559005		
	0.370994	0.23396		
	0.128786	0.041075		

Table 9.15

Quadrature Pole
Locations
(Wb/Wa = 20)

Angle = 1.520775 rads^{-1} or 87.134047 degrees
$Q = 0.324228$ $K = 4.384143$ $K' = 1.571779$ $\omega_b/\omega_a = 20$

Order	P Network Pole	N Network Pole	Error (Rads)	Rejection
4	9.933075	1.973738	0.096391	26.33 dB
	0.506653	0.100674		
6	15.50474	4.045227	0.014963	42.52 dB
	1.570845	0.6366		
	0.247205	0.064496		
8	20.966152	6.067807	0.002323	58.7 dB
	2.800719	1.402562		
	0.712981	0.357051		
	0.164804	0.047696		
10	26.379762	8.024087	0.000361	74.88 dB
	4.045228	2.267152		
	1.31061	0.763003		
	0.441082	0.247205		
	0.124625	0.037908		
12	31.768673	9.933072	0.000055975	91.06 dB
	5.267617	3.15751		
	1.973739	1.252754		
	0.798241	0.506653		
	0.316705	0.189839		
	0.100674	0.031478		

Table 9.16

Quadrature Pole Locations (Wb/Wa = 50)

Angle = 1.550795 rads^{-1} or 88.85404 degrees
$Q = 0.393999$ $K = 5.298747$ $K' = 1.570953$ $\omega_b/\omega_a = 50$

Order	P Network Pole	N Network Pole	Error (Rads)	Rejection
4	12.18077	2.137451	0.148336	22.57 dB
	0.467847	0.082097		
6	19.219402	4.706894	0.028565	36.9 dB
	1.657275	0.6034		
	0.212454	0.052031		
8	26.089973	7.271928	0.005501	51.21 dB
	3.148916	1.460214		
	0.684831	0.31757		
	0.137515	0.038329		
10	32.885954	9.758666	0.001059	65.52 dB
	4.706895	2.493038		
	1.353589	0.738777		
	0.401117	0.212454		
	0.102473	0.030408		
12	39.643031	12.180766	0.000204	79.82 dB
	6.255099	3.59243		
	2.137452	1.28692		
	0.777049	0.467847		
	0.278363	0.15987		
	0.082097	0.025225		

Table 9.17

Quadrature Pole Locations (Wb/Wa = 100)

Angle = 1.560796 rads⁻¹ or 89.427065 degrees
Q = 0.43883   K = 5.991589   K' = 1.570836 ω_b/ω_a = 100

The pole locations are given by the equation:

$$Pi = \sqrt{\frac{\omega_b}{\omega_a}} \cdot \frac{cn(u_i, k)}{sn(u_i, k)}$$

cn and sn are the elliptic sine and cosine functions. Also,

$$u_i = \frac{4i+1}{2n} . K \text{ for all } i = 0, . . n-1, \text{ where the order is } n.$$

This produces a set of poles P_0, P_1, and so on.

K is the complete elliptic integral, of modulus k, where k + sin θ, and a very simple C program can be found to do this. K' is the complete elliptic integral, of modulus k', where $k' = \cos\theta$, which can be found using the same program. The modular angle, θ, can be found from the upper to lower passband frequency ratio, ω_b/ω_a:

$$\frac{\omega_b}{\omega_a} = \frac{1}{\cos\theta} \qquad \therefore \theta = \cos^{-1}\left(\frac{\omega_a}{\omega_b}\right)$$

$$\text{Hence } k = \sin\theta = \sin\left[\cos^{-1}\left(\frac{\omega_a}{\omega_b}\right)\right]$$

The rejection of the unwanted sideband can be found from the equation:

$$R_{dB} = 20.\log\left(\frac{1}{\tan(\delta/2)}\right)$$

$$\text{where } \delta = 4q^2 \text{ and } q = \exp\left[\frac{-\pi K'}{K}\right]$$

These values are also calculated in the computer program HILBERT.CPP, given in Listing 9.2

```
// Hilbert.cpp—to find poles and zeroes for quadrature networks
// using Jacobian functions sn and cn, based on program given in
// "Numerical Recipes in C", by Press, et al (Cambridge).

#include <math.h>
#include <iostream.h>

#define TOL 0.00001
#define PI 3.1415926
#define ARRAY 20

void jacobian(double, double);
double integrate2(double);
int main(void);

//global variables
double sn, cn;

int main(void)
{
        extern double sn, cn;
        double ratio;
        double k, kk, angle, q;
        double kay, kaydash,del;
        double u, rejection,mc, ui,Xi;
        double pole;
        int i,j, n;

        cout << "Enter filter order" << endl;
        cin >> n;
```

```
        cout << "Enter ratio wb/wa (>1.0)" << endl;
        cin >> ratio;

        kk = 1.0/ratio;
        angle = acos(kk);
        cout << "Angle = "<< angle << " radians, or "<< angle *57.2958 <<"
(degrees)" <<endl;
        k = sin(angle);

        kay = integrate2(k);

        kaydash = integrate2(kk);

        q = exp(-PI * kaydash/ kay);
        del = 4.0 * pow(q,(double)n);
        cout <<"q = " << q << " Error = " << del << "radians"<< endl;

        rejection = 20.0 * log10(1.0 / (tan(del/2.0)) );
        cout << "Rejection = " << rejection << "dB" << endl;

        for(i=0; i<n; i++)
        {
                ui = (double)(4*i + 1)/(double)(2*n);
                u = ui * kay;
                jacobian(u,k);
                pole = sqrt(ratio) * cn / sn;
                cout << "pole = " << pole << endl;
        }
        return(1);
}

double integrate2(double k)
{
        double kn0, kn1, kn2, sum;
        kn0 = k;
        sum = 1.0;
        do
        {
                kn1 = sqrt(1.0 - (kn0 * kn0));
                kn2 = (1.0 - kn1)/(1.0 + kn1);
                sum = sum * (1.0+kn2);
                kn0 = kn2;
        }while(kn2 > (TOL * TOL));

        return(PI * sum /2.0);
}
void jacobian(double u, double k)
{
        extern double sn, cn;
        double snn, cnn;
        double temp1, temp2, temp3, temp4;
        double array1[ARRAY], array2[ARRAY];
        double kc_squared,angle;
        int i, j, max_array;

        kc_squared = 1.0 - (k * k);

        temp1 = 1.0;
        for(i=0; i<ARRAY;i++)
        {
                max_array = i;
                array1[i] = temp1;
                array2[i] = (kc_squared = sqrt(kc_squared));
                temp3 = 0.5 * (temp1+kc_squared);
                if(fabs(temp1-kc_squared) <= TOL * temp1)
                        break;
                kc_squared = kc_squared * temp1;
                temp1 = temp3;
```

```
        }
        angle = u * temp3;
        snn = sin(angle);
        cnn = cos(angle);
        if(snn<0.0)
        {
                temp1 = cnn/snn;
                temp3 = temp1 * temp3;
                temp4 = 1.0;
                for(j=max_array; j<=0; j-)
                {
                        temp2 = array1[j];
                        temp1 = temp1 * temp3;
                        temp3 = temp3 * temp4;
                        temp4 = (array2[j] + temp1)/(temp2+temp1);
                        temp1 = temp3/temp2;
                }
                snn = 1.0/sqrt(temp3*temp3+1.0);
                cnn = temp3 * snn;
        }
        sn = snn;
        cn = cnn;
}
```

Listing 9.2

HILBERT.CPP

Denormalization of component values for the quadrature phase network is carried out by scaling the pole location and then using the equations for the first-order section to determine component values. The scaling frequency is $f_0 = \sqrt{f_U \cdot f_L}$, so in the case of a 300 Hz to 3.4 kHz quadrature circuit, $f_0 = 1010$ Hz. The pole locations must be multiplied by $2\pi f_0$, or 6346 rads^{-1}. A fourth-order design will give over 38 dB unwanted sideband rejection, assuming that there are no amplitude errors. The poles for a fourth-order network are located at 6.790134 and 0.590319 for the *P* network, and 1.694 and 0.147272 for the *N* network. As a result of frequency scaling, the *P* network poles are at 43,090 and 3746.2, and the *N* network poles are at 10,750 and 934.59. I will now give an example of both passive and active realizations of these poles.

A passive quadrature design based on the above example is illustrated in Figure 9.17. As described earlier, the values of the capacitor and inductor are given by the following equations:

$$C = \frac{2}{\sigma . R} \qquad L = \frac{2R}{\sigma}$$

Where σ is the pole location and *L* is a center tapped inductor, each half-winding $= L/4$.

Consequently the component values are as follows:

$L1$	27.849 mH	$C1$	77.357 nF
$L2$	320.325 mH	$C2$	0.8898 μF
$L3$	111.628 mH	$C3$	0.3101 μF
$L4$	1.284 H	$C4$	3.5667 μF

Where $L1$, $C1$, $L2$, and $C2$ are components in the P network. The remaining components are in the N network. Notice that the source impedance is half the load impedance, this is because the networks are all-pass and the two loads are effectively in parallel at all frequencies.

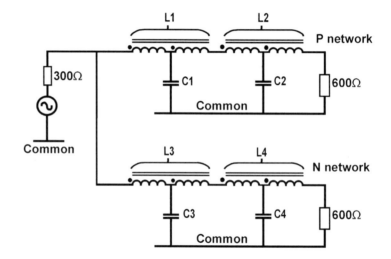

Figure 9.17

Passive Quadrature
Network (N = 4)

An active quadrature design to achieve the same function is given in Figure 9.18. The majority of components are set to convenient values, only the shunt resistor values remain to be calculated (to be honest I tried 1 nF capacitors to start with, but this led to high resistor values so I had to increase the capacitor values to 2.2 nF). Using the equation $R = \dfrac{1}{\sigma C}$, the following values of shunt resistor were found:

$R1$	10.548 kΩ	$R2$	101.335 kΩ
$R3$	42.283 kΩ	$R4$	486.358 kΩ

Where $R1$ and $R2$ are in the P network, and $R3$ and $R4$ are in the N network.

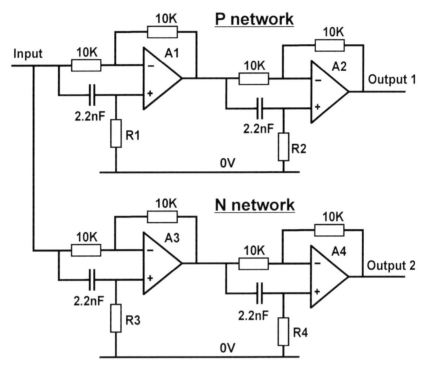

Figure 9.18

Active Quadrature Network (N = 4)

Both these designs have been simulated to prove that the component values are correct, at least in principle. Imperfections in real components may lead to increased phase-shift errors, resulting in less than ideal sideband rejection.

There is another method of producing single sideband modulation. It is known as the "Third Method," or the Weaver method (after D. K. Weaver who first described it). This uses two pairs of mixers, each pair phase-shifting the baseband signal by 90°. The Weaver circuit, and it's relative the Barber circuit, requires a longer explanation than the phasing method and is not really the subject of this book.

References

1. Williams, A. B., and F. J. Taylor. *Electronic Filter Design Handbook.* New York: McGraw-Hill, 1988.

2. Thomlinson, G. H. *Electrical Networks and Filters.* London: Prentice Hall, 1991.

3.   Temes, G. C., and J. W. LaPatra. *Circuit Synthesis and Design*. NK: McGraw-Hill, NK.

4.   Heulsman, L. P. *Active and Passive Analog Filter Design*. NK: McGraw-Hill, 1993.

5.   Helszajn, J. *Synthesis of Lumped Element, Distributed, and Planar Filters*. London: McGraw-Hill, 1990.

6.   *Mathcad*: Adept Scientific Ltd.

7.   *Filtech and Analyser III*: Number One Systems Ltd.

8.   Rorabaugh, C. B. *Digital Filter Designer's Handbook*. NK: McGraw-Hill, 1993.

9.   Bedrosian, S. D. "Normalized Design of 90° Phase-Difference Networks." *IRE Transactions on Circuit Theory*, vol. CT-7, June 1960: 128–136.

Exercises

9.1   What is group delay?

9.2   What is the effect of group delay in the transmission of a square wave signal?

CHAPTER 10

SELECTING COMPONENTS FOR ANALOG FILTERS

This chapter is very practical in orientation. It describes how different materials and component types can affect the performance of filters. In detail, it shows how the construction of components could affect performance. Operational amplifiers (op-amps) are also described. Amplifier parameters can have a significant effect; their most significant parameter is the gain bandwidth product.

Generally speaking, active filters are only used at low frequencies because of the demands placed on the op-amp. A typical limit for an active filter is a cutoff frequency of 100 kHz, although current mode devices can work at much higher frequencies, perhaps 10 MHz or more. Passive filters are used up to a few hundred MHz. Above about 200 MHz there are other more suitable filters, such as helical resonant cavity, surface acoustic wave (SAW), and stripline (tracks on a printed circuit board).

Capacitors

Capacitors are constructed from two conducting surfaces (known as plates) separated by an insulator (known as a dielectric). The metal plates are made from a thin metal film that has been deposited onto the insulation material. The dielectric can be a number of materials including ceramic, mica, and plastic film. The capacitor type is usually known by the dielectric, thus they are "ceramic" capacitors and "polyester" capacitors.

Ceramic and mica capacitors are made using flat dielectric sheets; the simplest construction uses just one insulating layer with a conducting plate on either side. Higher valued devices use several insulating layers with interleaving layers of metal film. The metal film layers are bonded alternatively to side A, side B, side A, side B, and so on.

Plastic film capacitors, such as polyester, use two layers of metalized plastic film. One form of construction is identical to that of ceramic capacitors, where flat sheets of metalized film are used.

Another form of construction for plastic film capacitors uses rolled films. The two metalized layers are placed one above the other and then rolled, so that the two conductors spiral around each other with insulating layers in between. The films are laterally offset from one another so that the conductor of "side A" protrudes from one side, and the conductor of "side B" protrudes from the other side (this technique is sometimes known as extended foil). It is then relatively easy to bond lead wires to the ends of the resulting cylindrical body. The rolled form of construction provides a metal film around the body of the capacitor; this can be connected to earth or the "earthy" side of a circuit to reduce external electrical field pickup. The outer foil is marked on the outer case of some film capacitors.

A capacitor's behavior is almost ideal, compared with other types of component. Capacitors are formed from two conducting layers separated by an insulator. Every capacitor will have some series inductance, which is due to the plate conductors and the lead wires attached to them. Each capacitor will also have series resistance due to both the conductors and the dielectric of the insulator, this is known as equivalent series resistance or ESR. These imperfections become more noticeable at high frequencies.

Generally, ESR is more of a problem with aluminum or tantalum electrolytic capacitors that are rarely used in filter designs (tantalum may sometimes be used in active filters). These types of capacitors are normally used to decouple power supplies. Digital circuit designers have become accustomed to connecting 10 nF ceramic capacitors across tantalum devices used for power supply decoupling. This is because the higher value tantalum capacitor absorbs low-frequency transient currents, while the ceramic absorbs the high-frequency transient currents.

It has been known for ceramic capacitors to be destroyed by passing a high level of RF power through them. The heat generated by the effective internal resistance, mainly due to the dielectric, can be sufficient to cause mechanical damage. Porcelain capacitors are often used at UHF (Ultra High Frequency) (300 MHz to 3 GHz) and above because they have a low ESR.

Dissipation Factor (DF) and Loss Tangent are terms used to describe the effect of ESR. The value of DF is given by the equation:

$$\text{Loss Tangent} = DF = \frac{ESR}{Xc},$$ where Xc is the capacitor's reactance at

some specific frequency. This is the tangent of the angle between the reactance vector Xc, and the impedance vector (Xc + ESR), where the ESR vector is at right angles to the reactance vector.

One of the most notable problems with capacitors is self-resonance. Self-resonance occurs due to the device construction: leads are inductors (albeit low value) and wound capacitors can have some inductance because the currents circulate through the capacitor's plates. Consider the self-resonant frequency of capacitors, of various dielectrics, having a lead length of 2.5mm (or 0.1 inch): a 10nF disc ceramic has a self-resonance of about 20MHz; the same value of polyester or polycarbonate capacitor also has a self-resonance of about 20 MHz. Mica capacitors are better, and a 10nF device with this dielectric has a self-resonance frequency of over 1 GHz.

A rough idea of the self-resonant frequency can be found by calculating the inductance of a component lead. For example, a 0.5mm diameter lead that is 5mm long (2.5mm for each end of the component) has an inductance of 2.94nH in free space. When combined with a 1 nF capacitor, the self-resonant frequency is calculated to be about 93 MHz. Replacing the 1 nF capacitor in the preceding calculations with a 10nF capacitor, results in the self-resonant frequency falling to 29 MHz. But, wait a moment, I just said that the self-resonant frequency of a 10nF capacitor with 2.5mm leads was about 20 MHz. The reason for the discrepancy between the calculated frequency and the actual frequency is that inductance in the plates was not taken into account. As the value of the capacitor increases, the inductance of its plates also increases and so does the discrepancy.

For small value capacitors of less than 1 nF the self-resonant frequency can be approximately calculated by the following equations.

$$f_R = \frac{1}{2\pi\sqrt{LC}},$$ where L is the lead inductance.

$$L = 0.0002b\left\{\left[\ln\left(\frac{2b}{a}\right)\right] - 0.75\right\}\mu H,$$ where "a" equals the lead radius and "b" equals the lead length. All dimensions are in millimeters (mm) and the inductance is in μH.

Using the formulae, if a = 0.25mm (0.5mm diameter) and b = 5mm (2.5mm each leg), the inductance is $2.94 \times 10^{-3}\mu H$. This is 2.94nH. When substituted into the frequency equation, with a 1 nF capacitor, the self-resonant frequency is calculated to be 92.8 MHz.

The formula given for inductance is that for a wire in free space. This should work for leads that are perpendicular to an earth plane, but not for those running parallel with one. If, however, the capacitor is axial leaded and mounted horizontally on the circuit board, a different inductance equation is necessary. This equation takes into account both the vertical and horizontal sections of the lead.

$$L = 0.0004605b \left\{ \log_{10} \left[\frac{2h}{a} \left(\frac{b + \sqrt{b^2 + a^2}}{b + \sqrt{b^2 + 4h^2}} \right) \right] \right\}$$
$$+ 0.0002(\sqrt{b^2 + 4h^2} - \sqrt{b^2 + a^2} + 0.25b - 2h + a),$$

where L is the inductance in μH, "a" is the lead radius, "b" is the wire length that runs parallel with the ground plane and "h" is the wire height above the earth plane. All dimensions are in millimeters. This equation is far more complicated than the previous one, but fortunately it is not often needed.

Surface mount capacitors are often used for high-frequency circuits because there is no lead inductance to worry about. The most popular type of surface mount capacitor is the multilayer ceramic; its conducting plates are planar, interleaved, and have very little inductance. Some conventional leaded ceramic capacitors use surface-mount devices with wire leads attached. They are usually dipped in epoxy resin or similar coating material before having their value marked on the outside.

Ceramic capacitors generally have a temperature coefficient that is zero or negative. The terms NPO (Negative Positive Zero) or COG are used to describe ceramic capacitors with a zero temperature coefficient. Other ceramic dielectrics are described by the temperature coefficient; N750 describes a dielectric that has a negative temperature coefficient of -750 PPM/°C.

Polystyrene and polypropylene capacitors are often used where the filter design is sensitive to component value changes. These types of capacitor have a negative temperature coefficient that closely matches the positive temperature coefficient of a ferrite-cored inductor. Unfortunately, with these dielectrics, capacitors tend to be physically large for a given capacitance value.

Polyester and polycarbonate capacitors are very common. Polyester capacitors are the worst in that they have a poor power factor (high ESR) and a poor (and positive) temperature coefficient. Polyester capacitors are popular because they have a high-capacitance density (high-capacitance-value devices are small). Polycarbonate capacitors have a better power factor and a slightly positive temperature coefficient. Another useful feature of polycarbonate capacitors is that they are "self-healing"; in the event of an insulation breakdown due to over-voltage stress, the device will return to its nonconducting state, rather than short circuit.

Temperature effects are very important. Consider what would happen if a narrowband bandstop filter was built without considering the temperature effects.

Inductors in the filter would be adjusted at room temperature to give the required response. If the filter were then used in a hostile environment (hot or cold) the filter would go out of tune and perhaps attenuate the wanted signals (or let in the unwanted signals).

Inductors

Inductors can be a source of many problems. High-value inductors are bulky. This is because they are usually made up from hundreds of turns of enameled copper wire that is wound on a bobbin and enclosed by a ferrite core. The windings capacitively couple to each other, which effectively introduces a parallel capacitor across the coil. This capacitance causes the inductor to resonate at some frequency. Above the self-resonant frequency, the impedance of the inductor falls due to the capacitive reactance dominating.

Inductors also possess some series resistance due to the intrinsic resistance of the copper wire used. This resistance limits the magnification of an applied voltage at resonance. A resonant circuit is a series or parallel combination of an inductor and a capacitor. Energy is stored, either in the magnetic flux or in the electric flux. At resonance this energy passes from one form to the other and large currents or voltages can be detected.

The voltage or current magnification is known as the "Q" of the circuit. If a resistance is in series with the inductor the current flow is restricted, which lowers the Q. This can have an effect on a filter because one with a sharp cutoff requires components with a high Q; in general, the inductor Q must be at least ten times the Q of the filter. Low Q inductors cause the filter's response to become rounded, in a graphical sense, close to the cutoff frequency. Resistance can also lead to an insertion loss (even at DC) due to the potential divider action of the inductor's resistance and the load resistance.

Resistance also occurs due to the "skin effect." This is produced by inductance inside the wire forcing the electrons to travel down the outside surface (hence "skin" effect). This can be a serious problem for inductors working at a few hundred kHz and is alleviated by the use of cotton covered Litz wire. This is the type of wire used to make ferrite rod antennas for radios working in the low and medium frequency range (LF and MF). It comprises several strands of enameled copper wire inside a cotton braid. This wire has a lower skin effect because the current is shared down each of the strands; the surface area of all the strands combined is considerably larger than the equivalent diameter solid copper wire.

An inductor that is made from a coil of wire, wound on a bobbin, and surrounded by a ferrite core is known as a pot-core. The ferrite core is cylindrical and has two halves that separate it to allow the bobbin to be inserted. Half of the core has a hole for an adjuster to pass through, and the other half has a threaded brass rod fixed at its center to allow an adjuster to be screwed onto it. When the pot-core is assembled, the two halves of the ferrite core are held together by two spring steel clips.

Pot-core inductors are very popular. This form of inductor is suitable for values of a few micro henries up to about one henry. Pot-cores come in standard sizes, from RM4 to RM14. This notation is related to the physical size required when the inductor is mounted on a printed circuit board: an RM4 requires a 0.4 × 0.4 inch board area ($10 \, \text{mm}^2$); an RM8 requires an 0.8 × 0.8 inch board area ($20 \, \text{mm}^2$), and so on. Typically, the inductor's Q for this type of construction is from 100 to 500, dependent upon the particular ferrite used, the frequency, and the inductance value.

The advantage of the RM type of pot-cores is that they can be made to any value. Each core type has an A_L value determined by the manufacturer. This is the inductance in nano-henries that will be produced for a single turn of wire. Remembering that inductance is proportional to the number of turns squared, the number of turns required is given by the simple formula: N turns $= \sqrt{L \, (n \, H)/A_L}$. L is the required inductance in nano henries and A_L is the core's inductance factor (nano henries per turn).

The pot-core's A_L value is related to the permeability of the ferrite material used. Different ferrite materials are used depending on the frequency at which the inductor is operating. A particular A_L value is obtained by removing some of the ferrite material from the center of the core, thus creating an air-gap. The air-gap has a lower permeability, so the A_L value is reduced by increasing the gap. A typical core gap is $100 \, \mu m$, although it may be larger or smaller depending on the ferrite permeability and the required A_L value.

Adjustment of the pot-core's A_L value is made possible by screwing a small ferrite slug, which is in a threaded plastic molding, onto a threaded brass rod that is fixed in the center of one half of the ferrite core. The adjuster can be set to allow some of the magnetic flux to bypass the air-gap and hence increase the permeability of the core. Positioning of the ferrite slug is achieved by screwing it down a brass thread, fixed into the bottom half of the core. After adjustment a small blob of melted wax can be used as a temporary seal. When melted wax is applied to the pot-core slug it will warm the ferrite, which will change its inductance temporarily.

As an aside, the presence of an air-gap in inductor cores makes them suitable for making transformers with a high saturation level. This is particularly valuable in applications where the transformer windings have to carry direct current, as well as the AC signal. This could perhaps be an application where a remote sensor circuit is being used to transmit over a twisted pair line. Power for the sensor is DC and is applied to capacitively coupled split windings in the transformer, and so does not interfere with the AC signal. However, all the DC passes through the transformer windings and the magnetic flux produced could saturate the core. Transformer cores usually have no air-gap or adjustable slug and are prone to saturate with small amounts of DC; their A_L is normally far higher than inductor core made from a similar ferrite.

Small inductors, about one centimeter long and wound on a ferrite or iron rod, are common. These generally have a low Q, typically from 30 to 60 when measured at frequencies of about one megahertz. They are of fixed value inductance, from one micro-henry up to one milli-henry. These inductors are useful in RF circuits, but care must be taken to consider the self-resonant frequency.

There are other ferrite or iron based inductors. There are vertically mounting devices that are a little smaller than the RM cores: perhaps 8 mm in diameter and up to 12 mm high. These are usually wound to have a standard value (e.g., E12) and can have values up to 100 mH. Surface-mount inductors are either wound on a ceramic or ferrite former, and usually have a low Q and a low self-resonant frequency. These devices are small (size 1812, about 6 mm by 3 mm by 4 mm high) often low value, perhaps up to 100 μH for the ferrite based devices. Their values are limited to about 100 nH for those using the ceramic former. There are also iron-cored inductors, for high value or high current applications. These are usually restricted to power-line filters or loud speaker crossover networks.

Resistors

Resistors are used in active filter circuits, in conjunction with capacitors, to set the frequency and the Q of each stage. Selecting the correct component value can produce a filter with the desired frequency response at room temperature, but unless consideration is given to temperature effects, the response at other temperatures could be wrong. If resistors with a positive temperature coefficient are selected, choosing capacitors with a negative coefficient may help to reduce tuning errors.

There are several types of resistor. Wire-wound devices are rarely used, except for power applications, and would not normally be placed in a filter circuit. Carbon composition resistors tend to be noisy and have a poor temperature coefficient but are good in RF circuits because of their low inductance construction. Carbon film and metal film devices are most common. Surface-mount devices are usually thick film construction.

Carbon film resistors are low-noise devices with a negative temperature coefficient. Component tolerances of 5% are standard. They are constructed by applying a carbon film onto a ceramic rod and then cutting a spiral gap around the device to increase the resistance. The spiral conductor is actually a lossy inductor.

Metal film resistors have a lower noise than carbon film types and a lower temperature coefficient. Component tolerances of 1% are standard, although precision devices in an E96 range of values with 0.1% tolerance and 15 ppm (parts per million) temperature coefficient are available at a higher cost. These resistors are constructed by applying a number of metal film layers, of different metals, to a ceramic former to achieve the correct resistance and a low temperature coefficient. A spiral gap is sometime cut around the metal film to increase the resistance value.

All conductors have some series inductance, simply due to having a certain length. In fact some high-frequency circuits just use a thin wire bond to form an inductor (this will be discussed further in Chapter 12). Resistors are conductors and therefore have inductance too. Some types have more inductance than others. Even a thick-film surface-mount resistor has inductance, although of considerably lower value than other types.

Wire-wound resistors have a significant inductance because of their construction; when a wire is wound into a coil its inductance increases in proportion to the number of turns squared. Carbon or metal film resistors that have had a spiral gap cut through their surface will have more inductance than a carbon composition type. All these components have some inductance due to the wire leads at either end.

Resistors also have capacitance. The two ends have a certain cross-sectional area and are spaced a certain distance apart, separated by a ceramic dielectric. This capacitance is small, typically 0.2 pF, but at high frequencies and in high impedance circuit node this can be significant.

The Printed Circuit Board (PCB)

The circuit board on which the components are connected is important at high frequencies and for surface-mount circuits. At high frequencies, for example, capacitance between tracks can cause a lower resonance frequency in a tuned circuit. Surface-mount circuits can have reliability problems due to thermal expansion of the circuit board; components firmly attached to the tracks with solder can be stressed if they do not have the same thermal expansion. There are several types of board, with FR4 (fiberglass insulator) being the most common.

It is usual for an RF or high-speed digital circuit to have an earth plane on the printed circuit board (PCB) component side. The earth plane serves two purposes; it screens the components from tracks passing underneath, and it provides part of a low-loss transmission line. By using FR4 board in a standard thickness of 1.6 mm, 50-ohm transmission lines can be created by making the printed circuit tracks 2.5 mm wide. A transmission line is formed between the earth plane and the track.

The technique of providing an earth plane on high-speed PCBs may cause problems when an inductor is placed on the board, because of the capacitive coupling between the ends of the inductor and the earth plane. This capacitance forms a parallel tuned circuit with the inductance and may cause the filter to be de-tuned. One solution is to remove the earth plane from the area below the inductor. An alternative solution is to mount the inductor on spacers above the board, so reducing the capacitance.

Surface-Mount PCBs

Surface-mount components are used extensively in active filter circuits. Ceramic capacitors are common but can be damaged by stress due to circuit board expansion. One way of minimizing this problem is to use physically small devices; devices larger than 1812 (0.18×0.12 inches) should be avoided.

Ceramic capacitors should be protected with a moisture-resistant coating. If moisture is absorbed into the ceramic material, the capacitance value will change. Moisture can also be absorbed into plastic packages, so a conformal coating over the whole board is preferred. Some consideration should be given to storage of components; metalized sealed bags should be used, perhaps with desiccant material. This will prevent moisture being trapped into an assembled board and avert the risk of damage during soldering (as the moisture boils off).

Conventional PCBs have plated through holes that are 1 mm or larger in diameter. Surface-mount boards do not need holes large enough for component leads; hence they tend to be smaller in diameter. Metalized "via" holes 0.3 mm in diameter are common (used to connect two tracks rather than for component leads). The problem arises when the board is heated. Glass and epoxy board, such as FR4 type, has a high coefficient of expansion at temperatures above 125°C. Above 125°C the board goes through its glass transition temperature and its coefficient of expansion is greater than normal; Z axis expansion increases the thickness of the board and can cause fractures between the tracks and the via-hole pads.

Soldering causes a problem due to the heat applied to the board; in wave soldering the board is heated to about 300°C, which is way above the glass transition temperature. To reduce the problem of via-hole damage, all plated through holes should have a wall thickness of 35 µm or more. Temperature cycling of completed boards also causes problems. If you have to rework a surface-mount board, use a low melting point (LMP) solder and add extra flux; LMP solder has a higher tin content than standard solder and is actually stronger. LMP solder also has a small amount of silver that prevents leaching of the component terminals.

On the surface, there is a temperature coefficient mismatch between components and the board. Leadless Chip Carrier (LCC) devices have an expansion coefficient of 6 ppm/°C, but for the board it is 14 ppm/°C (below the glass transition temperature) in the X–Y plane. Above the glass transition temperature the PCB has a coefficient of expansion of 50 ppm/°C. Again, temperature cycling strains the solder joints and can lead to failure. Small gull-wing ICs (integrated circuits) do not have a problem in this respect.

Copper-clad invar is used within some PCBs to restrain expansion and to distribute heat. This should be used with polyamide boards, rather than glass and epoxy types.

Solder resist can be used to restrain solder, but this can create large blobs on the lead or pad area. Surface-mount ICs use smaller packages than conventional leaded devices, and thin tracks of solder resist between the pads are not practical.

PCBs that have a fine track pitch tend to have $0.05\,\mu$m gold plating. If the gold is thicker it causes embrittlement. Gold or nickel plating gives a flat surface and makes surface-mount component placing easier.

Assembly and Test

When a filter is assembled, the inductors and capacitors usually have to be selected for value. Lowpass and highpass filters are not too critical for exact values, but bandpass and bandstop types are sensitive to value variations. Bandpass and bandstop filters comprise a number of LC circuit branches that are series or parallel tuned. With these types of filters it is best to select or adjust the components to within 1% of their design values before connecting them into circuit. Final adjustment can be made in-circuit, in one of two ways.

One method of in-circuit adjustment is to tune each LC pair separately. This is the best method for narrowband filters (bandwidth less than 10% of the center frequency), but it can be difficult to carry out. This is because each LC pair must be electrically separated from the others to prevent circuit interactions. This could be considered during the PCB layout design phase. Links could be provided to connect each stage together after fine-tuning and testing has been carried out.

The other method of in-circuit adjustment is suitable for wider bandwidth filters. The frequency response of the whole filter can be examined on a spectrum analyzer if a white noise source or a tracking signal generator is available. The inductors in the circuit can be adjusted to give the correct frequency response.

The usefulness of a white noise source should not be underestimated. A white noise source generates all frequencies with equal average power. Therefore the average output spectrum is equal to the filter's transfer function.

Operational Amplifiers

The operational amplifier, or op-amp, is the active device in an active filter. Its characteristics may change with temperature, but those most affected are the DC offset, bias current, and so forth. The AC characteristics, which are of primary interest here, are less affected by temperature.

The greatest problem in designing an active filter is that the op-amp is not ideal. The ideal op-amp has infinite input impedance, zero output impedance, and a flat frequency response with linear phase. Most practical op-amps have very high input impedance, and this does not cause us many problems. The output impedance is not zero and can be up to about $100\,\Omega$. This is not often a problem because negative feedback is used to limit the gain of the op-amp, and this also makes the effective output impedance close to zero. There is, however, an assumption: that the gain bandwidth of the op-amp is far higher than that required by the circuit. If the gain-bandwidth product limit is approached, the output impedance rises.

This brings us nicely to the final problem. If the op-amp has insufficient gain-bandwidth product, excessive phase shifts occur and the circuit can show peaking in the frequency response. Gains of $20\,dB$ close to the cutoff frequency can occur unless care is taken in the design. A good frequency response can be obtained by utilizing an op-amp that has a gain-bandwidth product many times that of the filter's cutoff frequency. A rule-of-thumb value is 10 to 100 times the cutoff frequency. Operational amplifiers in high-order filters work better if their gain-bandwidth product is about 100 times the cutoff frequency.

Filters with a sharp frequency response such as $1\,dB$ Chebyshev types require a greater op-amp performance than filters with a gentle response such as Butterworth. The gain-bandwidth product is also known as the unity gain frequency, or FU. Empirical formulae have been developed by me[1] to find a suitable value for FU in a number of active filters where the passband insertion loss or ripple was less than $2\,dB$.

Cutoff frequency $Fc = \dfrac{FU}{N^2}$ for Butterworth filters, where N is the filter order.

Cutoff frequency $Fc = \dfrac{FU}{N^{3.2}}$ for 0.5 dB Chebyshev filters.

Measurements on Filters

The frequency response of a filter is measured by applying a sine wave generator across the input terminals and an AC voltage-measuring device across the output terminals. The signal generator should have the same impedance as the filter under test; if the generator's impedance is not the same as the filter's characteristic impedance the results will be wrong. Remember that the source impedance is actually part of the filter design. The generator's impedance can be changed either by adding series resistance to increase it's impedance or by connecting a resistor across the signal generator's output to reduce its impedance.

If the signal generator output is measured without any load, the voltage seen is equal to the source EMF. If the source EMF is not constant with frequency, or the generator's internal impedance is not a constant resistance, the signal generator output cannot be connected directly to the filter input. If it were, the output response of the filter would be wrong. What should be done in this case is the output of the generator should be monitored using an AC voltmeter and kept to a constant voltage. A separate resistor of the required input impedance should then be wired in series between the generator and the filter. By keeping the output of the generator at a constant voltage it is in effect zero impedance (since the load will not affect it). The source impedance will be equal to the series resistance and the output response of the filter will now be correct.

The AC voltmeter across the filter's output terminals must have a bandwidth greater than the frequency range being measured. This may seem like an obvious statement, but some meters have a bandwidth switch that is used to reduce the noise. I, and many others, have been "caught out" by forgetting to return this switch to the wide bandwidth setting.

Another bandwidth problem is that of a spectrum analyzer; although a certain resolution bandwidth may have been set, this is the 3 dB bandwidth and not the noise bandwidth. That means that the signal-to-noise ratio appears worse than it really is. Reputable spectrum analyzer suppliers provide information about their equipment's filter response.

The resistor used to terminate passive filters must have a value equal to the filter's characteristic impedance. Since an AC voltmeter has high impedance, this means that a resistor must be physically placed across the filter's output terminals. When terminated, the output voltage will be half the generator's EMF. This is because the source impedance and the load form a potential divider and, since they are equal in value, the output voltage must be half the EMF. The other half of the voltage is dropped across the source impedance.

High-frequency passive filters are often tested in conjunction with a spectrum analyzer and tracking generator. As the spectrum is scanned across the frequency range set by the operator, the tracking generator generates a sine wave at the same frequency. Connecting an RF filter between the tracking generator and the spectrum analyzer allows the filter's transfer function to be displayed on the screen. If a tracking generator is not available, a white noise source will perform the required function, although at a lower signal level. Many analyzers have optional plotters to allow a hard copy of the response to be made.

Reference

1.   Winder, Steve. "The Real Choice for Active Filters." *Electronics World and Wireless World*, Sept 93: pp. 758–760.

Exercises

10.1   Why are surface-mount capacitors preferred for high-frequency circuits?

10.2   Why are ferrite pot-cores popular for making inductors? Why are air-gaps sometimes used between the two halves?

10.3   Which type of resistor is preferred for radio frequency circuits? Why are some types avoided?

10.4   A sixth-order Butterworth lowpass filter is to be built using operational amplifiers (op-amps). The filter requires a cutoff frequency of 20 kHz. What minimum gain-bandwidth product should the op-amps have?

CHAPTER 11

FILTER DESIGN SOFTWARE

I worked with Number One Systems to produce *FILTECH* after I had written simpler programs to help my work.[1] Some of these simpler programs are supplied on disk for readers to try themselves, complete with the source code (written in the C language). Operation of the supplied programs will be described in this chapter.

Filter Design Programs

There are a number of filter design programs available. A description will not be given for each one, since the specifications are being improved all the time and any description could soon be out-of-date.

(1)   Super FILTER[TM2]
(2)   Filter Master Active[3]
(3)   Filter Master.[4] Filter Master comes from the same supplier as Filter Master Active.
(4)   Elsie.[5] Passive (LC) filters only.
(5)   Signal Processing Worksystem[6]
(6)   AFDPLUS[7]
(7)   PC Filter[8]
(8)   FilterPro[TM].[9] FilterPro[TM] is free software supplied by Texas Instruments (previously provided by the Burr Brown Corporation). Programs include FILTER42 and FILTER2.
(9)   Filter Solutions®[10]
(10)   *FILTECH* Professional

Supplied Software

There are five DOS programs supplied with this book on the website www.bh.com/companions/0750675470: active_f, filter2, ELLIPSE, diplexer, and match2a. Because they are based on DOS, they have to be run within a DOS window in modern operating systems. The source code is supplied, in addition to the executable program. This allows the user to read them with an ASCII editor, or to modify them and to add features using an ANSI C compiler (I use Borland's Turbo C). These programs will now be described in some detail, with operating instructions.

The supplied filter design programs use tables of normalized component values or pole locations. This is different from *FILTECH*, which uses algorithms to build a table during runtime. The reason for using tables here is speed and memory requirements; the reason for using algorithms in *FILTECH* is for higher accuracy.

The following descriptions assume that the programs have been copied from the website www.bh.com/companions/0750675470 to a suitable directory on your computer's hard drive. During runtime, the programs save the netlist produced in a file in the current directory. It may be worthwhile running the programs while reading this section of the book. This should make the descriptions clearer.

Active_F

Active_f.exe is an active filter design program. It can only design lowpass and highpass filters of the Sallen and Key type, with Butterworth, Chebyshev, and Bessel responses. It is limited to filter orders from two to nine. The output is displayed on the computer screen, and the netlist is output to a file called "active.ckt."

When the program is run, it first asks whether a highpass or lowpass filter is required. Entering a number "1" at this point produces a lowpass design; entering a "2" causes a highpass design to be produced.

The program then asks whether the filter type is Butterworth, Bessel, or Chebyshev. There are passband ripple options of 0.1 dB, 0.25 dB, 0.5 dB, and 1 dB. The program requires a number between one and six to be entered at this point.

The required filter order must be entered next. This can be a number between two and nine.

The program then asks for the resistor values. This is the value of series resistors between the input and the op-amp's noninverting input. A value between "1" and "1e7" can be entered; do not use multiplier coefficients (e.g., 1.2 k).

Next enter the cutoff frequency; a value between "1" and "1e9" can be entered. This is the passband edge, or 3 dB point.

The program then asks for source and load resistance values. These are simply for the netlist that could be used by a circuit analysis program; but note that if the source resistance is significant, compared with the resistors used to define the filter response, the resultant frequency response can be in error.

The program displays a component list, with a description of each capacitor's purpose.

If a highpass design is required, enter a "2" at the beginning of the program (when asked if a highpass or lowpass is required). Follow all the other steps as given above, entering a value for all the series capacitors between the input and the op-amp's noninverting input. The result is a component list that describes resistors used in the final design.

In both cases a file "active.ckt" will be produced and will contain a "Spice-like" circuit analysis program netlist of the filter design.

Filter2

Filter2.exe is a passive filter design program. This can design highpass, lowpass, bandpass, and bandstop filters, with Butterworth, Chebyshev, and Bessel responses. Filter orders from three to nine are possible, although Chebyshev designs are limited to odd order only because equal terminations are used.

When the program is run, it first asks whether a lowpass, highpass, bandpass, or bandstop filter is required. Entering a number "1" at this point produces a highpass design; entering a "2" causes a lowpass design to be produced; entering a "3" or a "4" produces bandpass or bandstop designs, respectively. Entering a zero allows the user to quit the program. This quit facility is present at all program entry points.

The cutoff frequency is required. This can be entered as an exponent (i.e., 1e6) or as a value and coefficient (i.e., 1 M). If a bandpass or bandstop filter is being designed the program will ask for two frequencies: the lower cutoff point and the upper cutoff point. In the case of a bandpass filter these are the two pass-band edges, and between these frequencies the filter has little insertion loss. In the case of a bandstop filter there is very little insertion loss below the lower cutoff frequency or above the upper cutoff frequency. Between the lower and upper frequencies the filter has a high insertion loss.

The program then asks whether the filter type is Butterworth, Bessel, or Chebyshev. There are passband ripple options of 0.1 dB, 0.25 dB, 0.5 dB, and 1 dB. The program requires a number between one and six to be entered at this point, or zero to quit.

The filter order is then required. If a Chebyshev type has been requested, the program only allows odd-order designs (orders: three, five, seven, and nine). If a Butterworth or Bessel type has been selected, the program will allow any order from three to nine.

The load impedance is entered next, which is also the source impedance since the terminations are equal. This can be entered as an exponent (i.e., 1.2e3) or as a value and coefficient (i.e., 1.2k).

Finally, the "Q" of the inductors is required. This may be required for a circuit analysis program, and any integer value from "1" to "999" can be entered.

The program displays the filter parameters and seeks approval before designing the filter.

It also asks, for lowpass or highpass designs, whether an inductor or a capacitor is required for the first component. In the case of bandpass or bandstop designs, it asks whether the first arm in the ladder filter should be a series or parallel resonant circuit. These choices affect the filter's input impedance outside the passband.

A component list is displayed, with some explanation about the component's location. An option is given to save the design; the default file name is "filter.ckt" but users are prompted to choose another name for this file if they wish.

Ellipse

Ellipse.exe is a passive filter design program. This program only designs Cauer (elliptical) filters with odd order.

The user is prompted to enter a choice of highpass, lowpass, bandpass, or bandstop. Entry is a number from one to four.

The filter order is required next. An odd number is required from three to nine (i.e., 3, 5, 7, or 9).

Next enter the required source and load impedance (the program asks for the load impedance, but only equally terminated filters can be designed so the source impedance is the same value). Any integer value between "1" and "10,000" can be entered.

The steepness of the filter response skirt is required next. The choice is expressed as an angle: 30°, 40°, 50°, or 60°. The greater the angle, the steeper the skirt.

The angle is given by the expression $\theta = \sin^{-1}(1/\omega_s)$, where ω_S is the normalized frequency where the stopband begins.

The program describes filters in terms of their shorthand notation, such as CO# 20 50 degrees. The CO indicates a Cauer type filter. All designs have a 20% pass-band reflection coefficient, which equates to a VSWR (voltage standing wave ratio) of 1.5, and this is indicated by the 20 in the notation. The final number is the angle (in degrees) that was described earlier.

The cutoff frequency must be entered before the design process can take place. In lowpass or highpass filters this is the passband edge, or −3 dB point. This should be entered as a number and exponent (i.e., 1e6). If a bandpass or band-stop filter is being designed the program will ask for two frequencies: the lower cutoff point and the upper cutoff point. The upper cutoff point must have a higher frequency than the lower cutoff point.

In a bandpass filter design the upper and lower cutoff frequencies are at the two passband edges. Between these frequencies the filter has little insertion loss. The filter has a high insertion loss below the lower cutoff point and above the upper cutoff point. In a bandstop filter design there is very little insertion loss below the lower cutoff frequency or above the upper cutoff frequency. Between the lower and upper frequencies the filter has a high insertion loss.

The program now has enough specifications to carry out the design process. The output is a display of the component values, with some explanation given of the component placement in the ladder network. The netlist is saved in a file called "filter.ckt," which will be overwritten if a new design is undertaken.

Diplexer

Diplexer.exe designs passive diplexer filters. Two complementary odd-order ladder networks are designed, which are used to separate signals into high and low bands or passband and stopband. The ladder networks can be based on either a Butterworth or a 0.1 dB Chebyshev response.

When the program is run, the user is asked whether the design is a lowpass/high-pass design or a bandpass/bandstop design. A number 1 or 2 should be entered, corresponding to the two options. At this point the program can be exited by typing a zero; this also applies to most of the other data entry points.

The user is then asked whether the design should be based on a Butterworth or a Chebyshev response. For a Butterworth design enter number 1, or number 2 for Chebyshev. The Chebyshev design will have a small amount (0.1 dB) of ripple within the passband, but it will have a steeper rate of attenuation outside the passband.

The filter order is required next and this can be an odd number, between three and nine. Filter sections with a high order will have a greater degree of rejection to signals in the other section's passband.

The load impedance (for both filter sections) is required. Any integer value from "1" to "10,000" can be entered. As described in Chapter 8, each filter section is designed for zero source impedance, but when the two sections are combined the input impedance remains constant and equal to the load. The source is therefore required to have the same impedance as both loads, which are also equal.

The cutoff frequency must be entered next, as a number and exponent (e.g., 1.7e5 representing 170 kHz). This is the frequency where one filter section's passband stops and the other filter section's passband begins. If a passband/stopband option was chosen, a second cutoff frequency is requested, which must be higher than the one entered previously. The passband of one filter section is defined by these frequency limits. The other filter section passes all signal frequencies outside these limits.

The output from this program is a display of the circuit's component values. Included is a description of their connection in the two ladder networks. A circuit netlist is produced and saved in a file, "diplexer.ckt," that will be overwritten if the user decides to design a new diplexer.

Match2A

The Match2A program allows users to match a source to a load. When the program is run, the user is asked to enter a value for R1 (the source). A numerical value (e.g., 50) should be entered. Next, the user is asked for a value for R2 (the load). Again, a numerical value (e.g., 100) should be entered.

Once the source and load are entered, the program needs to know the frequency range over which matching is required. The start frequency (in Hertz) needs to be entered first, say 1e6 (for 1 MHz). The stop frequency needs to be entered next, say 2e6 (for 2 MHz). The VSWR limit, which is the maximum VSWR within the frequency range specified, should now be entered (e.g., 2.0).

The load may be complex and have a series or shunt reactance included. To allow for this the program asks for the load type (L or C or none). Enter "none" if the load is a pure resistance. If the load has series or parallel inductance, enter the letter L. If the load has series or parallel capacitance, enter the letter C. Suppose you enter the letter C. The program will then ask whether the reactance is in series or parallel (shunt) with the resistive load, enter the word "series" or "shunt" to select your choice; let's say "shunt." The value of the reactive component must now be entered, for example 1×10^{-10} (which equals 100 pF).

With all the design parameters entered, the program tries to design a suitable matching network. The network may be an L, T, or PI section. The L section is tried first, and if successful the component values are displayed. If no matching network can be found using the L section, the second attempt tries to find a network using a T section, and if unsuccessful the third attempt tries a PI section. If all types of matching network are tried but fail to achieve the required VSWR performance, the program will output a message saying that the design is not possible.

If the component values used as examples in the text are input (50-ohm source, 100-ohm load, matching between 1 MHz and 2 MHz with a VSWR of less than 2 and a load resistor having a 100 pF capacitor in parallel), the result is a series inductor of $5.305165 \mu H$ followed by a shunt capacitor of 961 pF.

References

1.   *FILTECH*, Number One Systems, Oak Lane, Breden, Tewkesbury, Glos., www.numberone.com. Tel.(01684) 773662.

2.   Super FILTER, www.superfilter.com.

3.   Filter Master Active, www.Intusoft.com.

4.   Filter Master, www.Intusoft.com.

5.   ELSIE, www.arraysolutions.com.

6.   Signal Processing Worksystem, www.cadence.com.

7.   AFDPLUS, Webb Laboratories, Wisconsin, U.S.A., *www.refreq.com/websites/webblabs.htm*. Tel. (414) 367-6895.

8.   PC Filter, M. G. Ellis, Artech House (publishers).

9.   FilterPro™, TI (semiconductors), www.ti.com.

10.   Filter Solutions®, Nuhertz Technologies, *www.filter-solutions.com*. Tel. (602) 216-2682.

CHAPTER 12

TRANSMISSION LINES AND PRINTED CIRCUIT BOARDS AS FILTERS

This chapter describes how transmission lines and printed circuits boards can be used to produce filters. Both of these topics are wide-ranging, and it will not be possible to provide more than an introduction here. The references provided should allow the interested reader to pursue the subject further.

Transmission lines can be used to filter signals. Quarter-wavelength lines of either short- or open-circuit termination can be used to pass some frequencies while stopping others. One application of this is to allow a radio carrier signal into a receiver from an antenna while preventing internal signals, from the receiver, from radiating back to the antenna. Connecting a short-circuit quarter-wavelength line across the antenna input will short circuit low-frequency signals but not interfere with signals at the quarter-wavelength frequency.

Transmission lines of less than a quarter wavelength at the passband cutoff frequency can be used to replace inductors and capacitors. The design process starts by producing a conventional lumped element filter design. Short-circuit lines then replace inductors and open-circuit lines replace capacitors. Each of these short- and open-circuit lines is a quarter wavelength long at the stopband frequency.

Transmission lines can be produced on a printed circuit board (PCB) as tracks. This is only significant when the signal frequency is high, so that the track length is about $\lambda/20$ or longer. A short-circuit line produces inductors, but this is difficult to produce on a PCB. A special mathematical transformation of the transmission line design is needed to overcome this problem. After transformation, an open-circuit line combined with a matching series quarter-wavelength line replaces the short-circuit line.

Printed circuit board LC filters will also be described. This type of filter is not the same as the quarter-wavelength line filter. All sections of PCB filters have dimensions of much less than a quarter wavelength at the passband cutoff frequency. The width of a track on a printed circuit board defines its impedance. Sections of track wider or narrower than the 50Ω line become capacitive or inductive, respectively. Concatenation of narrow and wide track sections can therefore form an LC filter, with the length of track being proportional to the reactance of the equivalent inductor or capacitor.

Transmission Lines as Filters

Transmission lines are often modeled as lumped elements of series inductors and shunt capacitors. This is a good model for our purposes. Another way of thinking about transmission lines is as a delay.

Consider for a moment a sinusoidal wave applied at one instant to one end of an open circuit coaxial cable. The cable has certain impedance, say 50Ω, so a signal with amplitude of 1 V will produce a current flow of 20 mA in the cable. This current flows towards the other end of the cable, which is open circuit, so when it arrives there it is reflected back towards the source: it has nowhere else to go. The reflected wave has a voltage amplitude peak approximately equal to the incident voltage peak. Now suppose a second sinusoidal wave is applied just as the start of the first wave is reflected back. If the reflected wave has the opposite polarity to the second wave, the two signals will cancel each other to give zero volts at the cable input. The input impedance will be effectively zero.

Thus, if a continuous sine wave signal is applied to an open-circuit coaxial cable, which has a length such that reflected signals are equal and opposite to the incident signal, the input impedance will be zero. This critical length is a quarter wavelength. The signal transmission time to the end of the cable and back is exactly one half cycle. Therefore, at the cable input, the reflected signal is inverted compared with the incident signal. Also, any odd multiples of quarter wavelengths are critical lengths. Multiples of quarter wavelengths are not so effective at creating low impedance. This is because the cable has loss and reflected signals have lower amplitude than the incident signal.

Now consider the opposite effect, a short-circuited coaxial cable. A sinusoidal wave is applied across one end to produce a current that flows towards the short circuit. When the signal current arrives at the short circuit it returns back along the other conductor, reversing the polarity of the signal at that point. The incident positive voltage is cancelled by the reflected negative voltage, giving zero volts at the short circuit (as you would expect). As with the open-circuit example, the critical length for a short-circuited coaxial cable is a quarter wavelength. The applied signal is delayed by a quarter wavelength in each direction along the cable. The signal is also inverted by the short circuit. Overall, the reflected wave is phase-shifted by 360° compared with the incident wave. The result is the two signals are in phase.

Consider what happens if a second wave of the same polarity is applied just as the first wave is reflected back to the line's input. No current can flow because the source has the same potential across it as the load. Thus a short-circuited line presents high impedance at the quarter-wavelength frequency. It also presents high impedance at the three-quarter-wavelength frequency and at further odd multiples of a quarter wavelength. However, as the cable becomes longer, the short circuit becomes less effective and the input impedance falls. This reduced effect is due to attenuation of the signal along the cable. The reflected wave amplitude will be less than the incident wave so some current will flow into the cable.

Clearly, the quarter-wavelength line can act as a filter by itself. Consider a line that has a short-circuit load and is a quarter wavelength long at 100 MHz. At this frequency the cable will present high impedance to signals applied across the other end. If this line is placed across the antenna input of a broadcast radio receiver it will allow VHF signals to pass through but will present a low impedance at frequencies above and below the quarter-wave frequency. This could be useful, for example, in rejecting high-powered High Frequency (3 MHz to 30 MHz) band transmissions from radio hams that may otherwise overload the receiver's input stages.

At frequencies below where the cable becomes a quarter-wavelength resonator, an open-circuit line is capacitive and a short-circuit line is inductive. In fact, an open-circuit line can be considered to be a series tuned circuit that is operating below its resonant frequency. Conversely, a short-circuit line can be considered to be a parallel tuned circuit that is operating below its resonant frequency. Richards' equation[1] gives the relationship between a wrongly terminated transmission line and its equivalent capacitance or inductance.

Open-Circuit Line

The impedance looking into an open-circuit line is given by the expression:

$$Z_{OC} = -jZ_O \cot(\gamma l)$$

Z_o is the characteristic impedance of the line, typically 50 Ω.

γ is the line propagation coefficient, given by:

$$\gamma = \sqrt{(R + j\omega L)(G + j\omega C)}$$

For a short coaxial cable certain assumptions can be made, namely that it will be loss-free. So letting $G = 0$ and $R = 0$, gives:

$$\gamma = j\omega\sqrt{LC}$$

So, $Z_{oc} = -jZ_o\cot(\omega\sqrt{LC}l)$

In fact, there is an alternative expression, which may be more useful, in that the equivalent reactance can be found:

$$\omega C = Y_O \tan\left[\frac{\pi}{2}\cdot\frac{\omega}{\omega_Q}\right]$$

The ratio of operating frequency to quarter-wavelength frequency (ω/ω_Q) can thus be used to find the equivalent capacitance. Y_o is the characteristic susceptance $(1/Z_o)$.

Short-Circuit Line

A short-circuit line has a different expression for its impedance:

$$Z_{SC} = jZ_O \tan(\gamma l)$$

Again, $\gamma = j\omega\sqrt{LC}$

So $Z_{sc} = jZ_o\tan(\omega\sqrt{LC}l)$.

Again, there is a simple expression that can be used to find the equivalent inductance directly. The ratio of operating frequency to quarter-wavelength frequency (ω/ω_Q) can be used to find the inductive reactance.

$$\omega L = Z_O \tan\left[\frac{\pi}{2}\cdot\frac{\omega}{\omega_Q}\right]$$

Use Of Misterminated Lines

Connecting short-circuited lines in the series path and open-circuit lines to shunt the transmission path is equivalent to a ladder filter with series inductors and shunt capacitors. A shorted-circuited line replaces each inductor, and an open-circuit line replaces each capacitor. The difference between the two networks is that the transmission line filter has a periodic frequency response, because the lines are anti-resonant at multiples of a half wavelength and resonant at odd multiples of a quarter wavelength. More details can be found in Helszajn[2] or Wolff and Kaul.[3]

The basic design process is to decide the frequency where maximum attenuation is required, that is, a zero in the frequency response. The open- and short-circuit lines (stubs) should be a quarter wavelength long at this frequency. These stubs should be connected to a transmission line having impedance equal to the input and output impedance of the filter. It is not necessary to space the stubs a quarter wavelength apart, though.

For example, suppose the requirement is for a passband to 100 MHz but 200 MHz must be stopped: the lines must all be a quarter wavelength at 200 MHz. The equations for inductance and capacitance are simplified, as follows:

$$\omega C = Y_O \tan\left[\frac{\pi}{2} \cdot \frac{\omega}{\omega_Q}\right] = Y_O \tan\left[\frac{\pi}{4}\right]$$

$$\omega L = Z_O \tan\left[\frac{\pi}{2} \cdot \frac{\omega}{\omega_Q}\right] = Z_O \tan\left[\frac{\pi}{4}\right]$$

The ratio of passband to stopband frequency (ω/ω_Q) was deliberately chosen to be $\frac{1}{2}$ to simplify the math because, conveniently, $\tan(\pi/4) = 1$.

Find the characteristic impedance of these short- and open-circuit lines by taking the input and output impedance to be 50 Ω and designing for a 0.25 dB Chebyshev response in the passband. The normalized element values for this filter are 1.6325, 1.436, and 1.6325 (to four decimal places).

The first and third elements have the same normalized value, so the result will be the same for both. Let's design for series inductors at either end with a shunt capacitor in the center. The inductor equivalent line will be designed first.

$$\omega L = Z_O \tan\left[\frac{\pi}{4}\right] = Z_O, \text{ where } Z_o \text{ is the characteristic impedance of the}$$
short circuited line.

$$g_1 = 1.6325 = \frac{\omega L}{50} = \frac{Z_O}{50}, \text{ where } \omega = 2\pi \times 10^8, \text{ the passband edge.}$$

$$Z_o = g_1 \times 50 = 81.625 \,\Omega.$$

The capacitor equivalent line will be designed now.

$$\omega C = Y_O \tan\left[\frac{\pi}{4}\right] = Y_O, \text{ where } Y_o \text{ is the characteristic admittance of the}$$

open-circuit line.

$g_2 = 1.436 = 50 \times \omega C = 50 \times Y_o$, where $\omega = 2\pi \times 10^8$, the passband edge.

$Y_O = \dfrac{1.436}{50}$ or preferably $Z_O = \dfrac{50}{1.436} = 34.82\ \Omega$. The final circuit is shown in Figure 12.1.

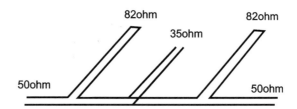

Figure 12.1

Filter Using Transmission Lines

This filter can be realized using coaxial lines, although finding lines of suitable impedance may be difficult. If the frequencies were higher, say closer to 1 GHz, they could also be realized as a stripline printed circuit board, and this approach will now be studied.

A stripline is a printed circuit board track with dielectric material on either side and sandwiched between two earth planes. In practice it is made by etching a track onto one side of a double-sided board, then laying a second, single-sided board on top. This form of construction has low loss and low radiation properties; it is also simple to analyze because the dielectric between the center track and the earth planes is uniform.

An alternative printed circuit board construction is microstrip, which has a track on one side of a board and an earth plane on the other. A microstrip track has an impedance that is more difficult to analyze; this is because the field lines between the track and the earth plane do not just pass directly through the board, they also partially travel through the air above the track. The "effective" dielectric constant is less the circuit board's actual dielectric constant because of this effect. Both stripline and microstrip forms of construction are illustrated in Figure 12.2.

Figure 12.2

Stripline and Microstrip
Construction

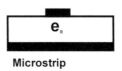

Microstrip

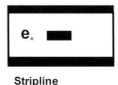

Stripline

Suppose you wish to design a stripline filter. The problem is that the short-circuited line would be very difficult to produce on a printed circuit board. It would be necessary to use a coplanar line (two parallel lines) between the earth planes. An alternative option is to transform the short-circuited line into an L structure, comprising an open-circuit line and a series section. Kuroda's identity (see reference 3) gives the relationship between the two structures and equations have been presented here to simplify the conversion.

The open circuit line impedance is given by the equation:

$$Z' = Z_0 + \frac{Z_0^2}{Z},$$ where Z is the value of the short-circuit line impedance

and Z' is the replacement open-circuit line value. Z_o is the filter's source and load impedance, that is, $50\,\Omega$.

$Z' = 50 + 31.25 = 81.25\,\Omega$.

The series section line impedance Z_1' is given by the equation: $Z_1' = Z_o + Z$, where Z is the value of the short-circuit line impedance and Z' is the series section line impedance. As before, Z_o is the filter's source and load impedance. $Z_1' = 50 + 81.625 = 131.625\,\Omega$.

A diagram of this filter is given in Figure 12.3. Note that all transmission line sections are a quarter wavelength at the stopband frequency. The width of the $35\,\Omega$ line in the center must not shorten the series section line length. If the passband is 1.5 GHz and the stopband is at 3 GHz the same impedance can be used, but the length of the lines must be scaled to be $\lambda/4$ at 3 GHz instead of 200 MHz. The impedance of the lines is dependent on the passband to stopband ratio rather than the actual frequencies. The velocity of a wave in a conductor, which is surrounded by a dielectric, is $c/\sqrt{\varepsilon_R}$. Remember that c is the velocity of an electromagnetic wave in free space, and is approximately 3×10^8 m/s.

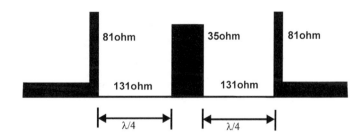

Figure 12.3

Stripline Lowpass Filter

The high impedance line can be a thin wire. The impedance of a wire in a stripline circuit, where there is an earth plane above and below the conductor, is given by:

$$Z_0 = \frac{60}{\sqrt{\varepsilon_R}} \ln\left(\frac{4D}{\pi d}\right)$$

D gives the distance between the two earth planes, while the wire diameter is d. The circuit board's dielectric constant is ε_R.

This expression can be derived from an equation produced by Hammerstad, where the impedance for a microstrip line (where there is no board or earth plane above the conductor) is given by:

$$Z_0 = \frac{60}{\sqrt{\varepsilon_{eff}}} \ln\left(\frac{8h}{w}\right)$$

Here h is the height (thickness) of the board and w is the width of the microstrip track.

Since in stripline the whole surface of the wire is enclosed in dielectric, the effective surface width is πd, the circumference of the wire. The impedance of a wire in a stripline circuit is equivalent to a track on a microstrip circuit if the wire circumference replaces the track width:

$$Z_0 = \frac{60}{\sqrt{\varepsilon_{eff}}} \ln\left(\frac{8h}{\pi d}\right)$$

Since ε_{eff} is ε_R in a board that is homogeneous and $2h$ is equal to D, the two equations are identical.

On a microstrip circuit, a wire has a higher impedance because some of the field lines will pass through air ($\varepsilon_R = 1$). The effective dielectric coefficient in this case is given by approximately: $\varepsilon_{eff} = 0.5(\varepsilon_R + 1)$.

Having found the series element design equation, you now need to find an equation for the open-circuit quarter-wave stripline. You need to find the physical parameters of two lines that have impedance values of approximately $35\,\Omega$ and $81\,\Omega$.

If the line width is greater than 0.6 times the distance between the two earth planes, the wide strip equation can be used:

$$Z_0 = \frac{94.18}{(w/D + 0.44)\sqrt{\varepsilon_R}}$$

If the line width is less than 0.6 times the distance between earth planes, use the narrow strip equation that assumes that the strip is effectively a wire:

$$Z_0 = \frac{59.96}{\sqrt{\varepsilon_R}} \left[Ln\left(\frac{8D}{\pi w}\right) + 0.185\left(\frac{w}{D}\right)^2 \right]$$

This can be approximated by the simplified equation that was used for the wire impedance, where the width of the track is taken as the wire diameter:

$$Z_0 = \frac{60}{\sqrt{\varepsilon_R}} \ln\left(\frac{8D}{\pi w}\right)$$

The problem with these equations is that the impedance is known. These have to be transformed into equations to find the D/w ratio. The distance between earth planes, D will be defined at the design stage, so you only need to find the track width, w.

For a wide track ($w > 0.6D$):

$$\frac{Z_0\sqrt{\varepsilon_R}}{94.18} = \frac{1}{(w/D)+0.44}$$

$$\frac{w}{D} = \frac{94.18}{Z_0\sqrt{\varepsilon_R}} - 0.44$$

For a narrow track ($w < 0.6D$):

$$Z_0 = \frac{60}{\sqrt{\varepsilon_R}} \ln\left(\frac{8D}{\pi w}\right)$$

$$\frac{\sqrt{\varepsilon_R}}{60} Z_0 = \ln\left(\frac{8D}{\pi w}\right)$$

$$\frac{D}{w} = \frac{\pi}{8} e^{\frac{Z_0\sqrt{\varepsilon_R}}{60}}$$

If this approximate equation is used to find a value for D/w, and then this is substituted into the full equation, an idea of the error can be found and assessed. For example, let the dielectric constant have a value of 4.7 (fiberglass resin board, type FR4) and $D = 1.6\,\text{mm}$. This could be produced from two boards 0.8 mm thick, one double-sided and the other single-sided.

For a narrow track:

$$\frac{D}{w} = \frac{\pi}{8} e^{\frac{Z_0\sqrt{\varepsilon_R}}{60}}$$

If $Z_0 = 81\,\Omega$, $D/w = 7.33$. Since $D = 1.6\,\text{mm}$, $w = 0.2183\,\text{mm}$. The ratio w/D is less than 0.6, so the equation used is valid. Substituting this ratio into the full equation for impedance gives:

$$Z_0 = \frac{59.96}{\sqrt{\varepsilon_R}}\left[Ln\left(\frac{8D}{\pi w}\right) + 0.185\left(\frac{w}{D}\right)^2\right]$$

$$Z_0 = 27.6575\left[Ln\left(\frac{58.64}{\pi}\right) + 0.185(0.136426)^2\right]$$

$$Z_0 = 81.04\,\Omega$$

This is the required impedance, so the approximation is good and a track width of 0.2183 mm can be used. The same equation can be tried for the 35 Ω impedance line, although it may require a w/D ratio greater than 0.6 and have to be recalculated.

$$\frac{D}{w} = \frac{\pi}{8}e^{\frac{Z_0\sqrt{\varepsilon_R}}{60}}$$

$$D/w = 1.39$$

$w/D = 0.719$, so the wide line equation must be used:

$$\frac{w}{D} = \frac{94.18}{Z_0\sqrt{\varepsilon_R}} - 0.44$$

$$w/D = 0.801\,\text{mm}.$$

Substituting this back into the original equation:

$$Z_0 = \frac{94.18}{(w/D + 0.44)\sqrt{\varepsilon_R}}$$

$$Z_0 = 35\,\Omega, \text{ as required.}$$

Alternative equations exist for microstrip lines, which are easier to produce as standard double-sided printed circuit boards. Equations have been produced by Hammerstad for use on boards where the relative dielectric constant, $\varepsilon_R < 16$.

If $w/h < 2$ the narrow line equation is:

$$\frac{w}{h} = \frac{8}{e^A - 2e^{-A}}$$

$$\text{where } A = \frac{Z_0\sqrt{2(\varepsilon_R + 1)}}{119.9} + \frac{(\varepsilon_R - 1)}{2(\varepsilon_R + 1)}\left(0.46 + \frac{0.22}{\varepsilon_R}\right)$$

If $w/h > 2$, the wide line equation is:

$$\frac{w}{h} = \frac{2}{\pi}\left[(B-1) - Ln(2B-1) + \frac{\varepsilon_R - 1}{2\varepsilon_R}\left(Ln(B-1) + 0.39 - \frac{0.61}{\varepsilon_R}\right)\right]$$

where $B = \dfrac{59.96\pi^2}{Z_0\sqrt{\varepsilon_R}}$.

On a standard circuit board (FR4 material, approximately 1.6 mm thick) a 50 Ω transmission line is about 2.5 mm wide; this is a w/h ratio of 1.5625. The 35 Ω line will be wider and, using the wide line equation, it has a w/h ratio of 3.359 (equating to a width of 5.3744 mm). The 81 Ω line will be narrower than a 50 Ω line so the narrow line equation can be used: A = 2.4455, so $w/h = 0.70404$ (which equates to a width of 1.1265 mm).

Printed Circuits as Filters

I have already shown how transmission lines can be used to construct filters. Transmission lines were shown as being realized as microstrip or stripline printed circuits. An alternative filter construction using printed circuit boards (PCBs) will now be described. Narrow and wide sections of track will be used to replace inductors and capacitors, respectively. The length of each section will be much less than a quarter wavelength in the filter's passband. Only a broad outline of designing lowpass filters using this technique will be described and presented as an example. Capacitors are produced from wide sections of track. The width of these sections must be less than a quarter wavelength at the highest operating frequency, to avoid resonance in the direction transverse to the propagation.

Let's assume the board is a standard fiberglass resin type (FR4) with a thickness of 1.6 mm and a relative permittivity of $\varepsilon_R = 4.7$. The required filter has a cutoff frequency of 1 GHz. At this frequency the wavelength of a signal in the PCB is $300/\sqrt{4.7} = 138.38$ mm; this is a worst case approximation because the actual relative effective permitivity will be less than that of the board material alone, because of the air path, and therefore the wavelength will be longer. The capacitors can be replaced by a track $w < 34.59$ mm; let $w = 25$ mm.

The ratio of track width to board thickness is w/h, which is greater than one. The impedance of this track is given by:

$$Z_m = \frac{120\pi}{\sqrt{\varepsilon_{eff}}}[w/h + 1.393 + 0.677 Ln(w/h + 1.444)]^{-1}$$

where $\varepsilon_{eff} = \dfrac{\varepsilon_R + 1}{2} + \dfrac{\varepsilon_R - 1}{2}(1 + 12h/w)^{-0.5}$

$h/w = 0.064$

$\varepsilon_{eff} = \dfrac{5.7}{2} + \dfrac{3.7}{2}(1 + 12 \times 0.064)^{-0.5} = 4.24$

Therefore, $Z_m = 183/[15.625 + 1.393 + 0.677\ Ln(15.625 + 1.444)]$

$Z_m = 183/[17.018 + 1.9208]$

$Z_m = 9.663\,\Omega.$

This impedance will be used a little later on, in the equations for capacitors. The effective permitivity is 4.24, so the wavelength along the track is 145.7 mm and the track width of 25 mm is much less than a quarter wavelength, as suggested earlier.

To replace inductors by PCB tracks you need narrow tracks that can be easily etched. Consider using tracks 0.5 mm wide. Since w/h is now less than one, a different equation can be used to find the characteristic impedance.

$$Z_m = \dfrac{60}{\sqrt{\varepsilon_{eff}}} Ln\left(\dfrac{8h}{w} + 0.25\dfrac{w}{h}\right).$$

The effective relative permitivity is now given by the expression:

$$\varepsilon_{eff} = \dfrac{\varepsilon_R + 1}{2} + \dfrac{\varepsilon_R - 1}{2}\left[\left(1 + \dfrac{12h}{w}\right)^{-0.5} + 0.041\left(1 - \dfrac{w}{h}\right)^2\right]$$

The ratio $w/h = 0.3125$, and $h/w = 3.2$.

$$\varepsilon_{eff} = \dfrac{5.7}{2} + \dfrac{3.7}{2}\left[(1 + 38.4)^{-0.5} + 0.041(1 - 0.3125)^2\right]$$
$$= 2.85 + 1.85[0.1593 + 0.02162]$$
$$= 3.185$$

$$Z_m = 33.62 Ln(25.6 + 0.078) = 109.12\,\Omega$$

This impedance will be used now in the equations for inductors. The length of a narrow track used to form an inductor is given by the expression:

$$l = \dfrac{Lc}{Z_m\sqrt{\varepsilon_{eff}}}$$

Here L is the required inductance, c is the velocity of light (3×10^8 m/s), Z_m is the impedance (= 109.12) of a 0.5 mm wide line, and ε_{eff} is the relative effective permitivity (= 3.185) of the dielectric for such a line.

A similar equation exists for capacitors:

$$l = \frac{CZ_m \, c}{\sqrt{\varepsilon_{\mathit{eff}}}}$$

In this formula $\varepsilon_{\mathit{eff}}$ is the relative effective permitivity ($= 4.24$) of the dielectric for a 25 mm wide line. The impedance of this line is given by $Z_m = 9.663$.

A practical filter could be a fifth-order Chebyshev filter with a 0.25 dB passband ripple and a 1 GHz (-3 dB) cutoff frequency. The lumped element components for such a filter are: $C1 = 4.9\,\mathrm{pF}$; $L2 = 11.42\,\mathrm{nH}$; $C3 = 7.77\,\mathrm{pF}$; $L4 = 11.42\,\mathrm{nH}$; and $C5 = 4.9\,\mathrm{pF}$.

In terms of PCB tracks the lengths are:

$$l_{C1} = l_{C5} = 4.9 \times 10^{-12} \times 9.663 \times 3 \times 10^{8}/2.059 = 6.9 \text{ mm}$$

$$l_{L2} = l_{L4} = 11.42 \times 10^{-9} \times 3 \times 10^{8}/(109.12 \times 1.785) = 17.6 \text{ mm}$$

$$\text{Finally } l_{C3} = 7.77 \times 10^{-12} \times 9.663 \times 3 \times 10^{8}/2.059 = 10.94 \text{ mm}$$

This filter is illustrated in Figure 12.4.

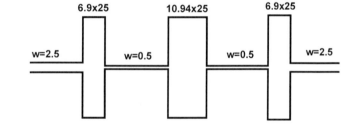

Figure 12.4

Microstrip 1 GHz Lowpass Filter

The circuit shown will not give an exact response because of discontinuities at the sharp edges. However, the filter will give a response quite close to what is required and is likely to be suitable unless the required filter response has a close tolerance. A standard double-sided PCB is required. Readers who have a simple PCB etching kit may like to try out this design for themselves.

Bandpass Filters

Bandpass filters can be made from an array of half-wavelength lines. Actually, each resonator must be slightly less than a half wavelength, because of interaction effects with other resonators. Resonators are arranged to be parallel to each other and overlapping by a little less than a quarter wavelength. The spacing between resonators is usually less than the resonator's width. This is shown in Figure 12.5.

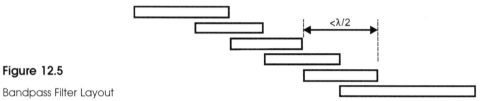

Figure 12.5

Bandpass Filter Layout

The detailed design of bandpass filters is too complicated to be dealt with here. Readers are recommended to refer to Edwards for more information.[4]

References

1.   Richards, P. I. *Resistor Transmission Line Circuits. Proceedings of the IRE*, vol. 36, February 1948, p. 217.

2.   Helszajn, J. *Synthesis of Lumped Element, Distributed, and Planar Filters*. London: McGraw-Hill, 1990.

3.   Wolff, E., and R. Kaul. *Microwave Engineering and Systems Applications*. John Wiley & Sons, 1988.

4.   Edwards, T. C. *Foundations for Microstrip Circuit Design*. London: John Wiley & Sons, 1981.

Exercises

12.1   What length of printed circuit track, in terms of the signal wavelength, is needed before it is classed as a transmission line?

12.2   What is the width of a microstrip line having an impedance of $50\,\Omega$ if FR4 printed circuit material is used ($\varepsilon_R = 4.7$), which is 1.6 mm thick. Hint: the width to PCB thickness ratio, w/h, will be less than 2.

CHAPTER 13

FILTERS FOR PHASE-LOCKED LOOPS

Filters for phase-locked loops are usually quite simple. Poor design of the loop filter can cause overall instability of the loop. Many people avoid designing phase-locked loops for this reason. Here I give some examples and explanations that should help to remove some of this fear. This chapter can only be introductory: whole books have been devoted to this subject.[1]

What is a phase-locked loop? It is a voltage-controlled oscillator with a feedback loop to a phase comparator. The phase comparator compares the phase of two signals on its input; in this case, one signal is the reference and the other signal is the oscillator output. A diagram of a phase-locked loop is shown in Figure 13.1.

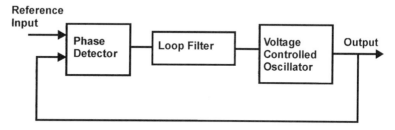

Figure 13.1

Phase-Locked Loop

The simplest digital phase detector is an exclusive-OR gate that has two inputs. If the two input signals are 90° out-of-phase, the output voltage spends the same amount of time at logic "1" as at logic "0." Hence, the average output voltage is equal to the midrail point of the power supply. As the phase difference between the input signals approaches zero, the output voltage spends longer at logic "0" and the average output voltage tends to zero. As the phase difference approaches 180°, the output spends longer at logic "1" than at logic "0" and the output voltage tends to 5 V (or whatever the positive supply voltage is).

Usually, more sophisticated digital phase detectors are used because the simple exclusive-OR gate will give the same output if the reference signal is at a harmonic (or subharmonic) of the oscillator frequency. Sophisticated phase detectors are more accurately described as "phase and frequency detectors." The gain of a phase detector, in terms of volts per radian, is $K\phi$.

Analog phase detectors can be produced from multiplier circuits. This could be an RF mixer or an analog multiplier. Both produce an output proportional to the phase difference and the amplitude of the two input signals. These devices take two inputs, the reference signal and the feedback signal, and multiply them together. In-phase signals multiplied together produce, after averaging, a positive output. This is because two signals of the same polarity always produce a positive output. Anti-phase signals produce a negative output after averaging because one of the signals will always have the opposite polarity to the other. Analog phase detectors are simple and cannot detect frequency differences between signals. Therefore, harmonics of the reference signal will also produce a locked condition.

In all phase-locked loops, the output of the phase detector controls the oscillator frequency. If the oscillator frequency drifts slightly, its phase will shift relative to the reference signal. The average output voltage from the phase detector will change when this happens, and this will attempt to correct the frequency drift. Thus, using feedback, the phase detector restores the phase difference between the two signals. To prevent instability and to reduce noise, the output voltage from the phase detector must be averaged, or integrated.

Averaging loop error signals is the purpose of the loop filter. The oscillator has a gain, Ko, which is in terms of rad/s per volt. Thus the phase is an integral of Ko times the input voltage. Hence, the phase-locked loop is an integrator followed by a first-order filter that becomes a second-order system. This can therefore be unstable unless properly designed.

Now, it may seem pointless to produce an output signal that is identical to the input reference signal, as shown in Figure 13.1, but there are two important applications for modified versions of this circuit. Demodulation of a frequency-modulated carrier is one application; frequency multiplication is the other.

As the carrier frequency at the reference input increases or decreases, that is it is frequency modulated (FM), the oscillator frequency is forced to follow by the control voltage feedback loop. The control voltage will vary in proportion to the frequency deviation, hence providing a demodulated carrier output. The output from the oscillator is only used to provide a second input to the phase detector. If used as an FM detector, the loop filter must have a bandwidth at least equal to that of the modulating signals; this is typically 15 kHz in a radio broadcast signal. The circuit for an FM demodulator is illustrated in Figure 13.2.

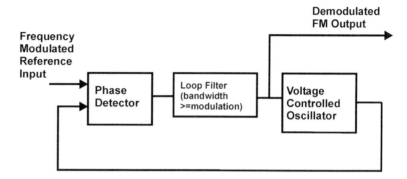

Figure 13.2

FM Demodulator Circuit

Frequency multiplication can be achieved if there is a frequency divider between the oscillator output and the phase detector input. The reference input signal will then be compared with an oscillator signal divided by N. If the two inputs to the phase detector are at the same frequency, the oscillator output frequency must be N times that of the reference signal. For a circuit such as this, where the reference signal is not being modulated, the loop filter can have a very narrow bandwidth. However, the response time of the circuit when a signal is initially applied, or when it is switched to a new frequency, is inversely proportional to the bandwidth. The circuit for a frequency multiplier is shown in Figure 13.3.

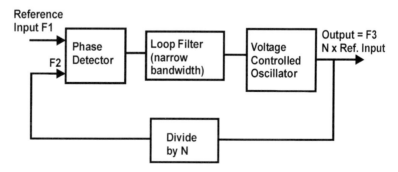

F1 = F2, F2 = F3/N, Hence F3 = F2 x N

Figure 13.3

Frequency Multiplier Circuit

Loop Filters

In both frequency multiplier and FM signal demodulator applications, the purpose of the loop filter is to average the phase detector output voltage. It must do this while allowing the system to respond to changes in the reference signal's frequency. The phase detector will output some spurious signals at the reference frequency. The filter must remove these signals before they reach the oscillator. Otherwise the oscillator will be modulated unnecessarily and produce jitter at its output. The loop filter is thus a critical part of the phase-locked loop circuit.

There are two simple loop filters: a first-order lowpass CR network and a lead-lag network. The lead-lag network is similar to the CR network except that it has a resistor in series with the shunt capacitor. These are illustrated in Figure 13.4. The CR network is simple and its performance is not dependent on the value of capacitor or resistor, but is determined by the product CR. The lead-lag network gives the designer more control over the performance, such as damping, cutoff, and natural frequency. It does, however, have the disadvantage of limited stopband attenuation.

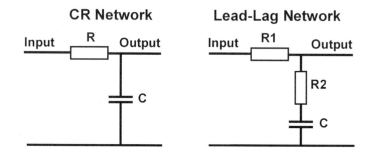

Figure 13.4

Simple Loop Filters

The design equations for an *RC* network follow. The filter cutoff frequency is simply:

$$\omega_{LP} = 1/RC, \text{ in rad/s}$$

The natural frequency of the phase-locked loop depends on the phase comparator gain, $K\phi$, and the oscillator gain, Ko, as well as the cutoff frequency. The natural frequency is the rate of oscillation that occurs when the phase-locked loop's reference input signal is suddenly changed (assuming that it is not over damped):

$$\omega_n = \sqrt{(K\phi \cdot Ko \cdot \omega_{LP})}, \text{ this is also in rad/s.}$$

The transfer function for an *RC* network is given by:

$$F(s) = \frac{1}{1 + s.C.R}$$

The damping factor is most important, and is given by:

$$\zeta = \sqrt{\frac{\omega_{LP}}{4.k\phi.K_O}}$$

The lead-lag network will now be considered. This has an additional resistor in the shunt path, which increases the designer's options. The lead-lag network equations are different from those for the RC network, as I will now show. Some of them are intuitive; others are less so. First the filter's cutoff frequency is given by:

$$\omega_{LP} = 1/(R1 + R2).C, \text{ in rad/s.}$$

Measuring the signal across the capacitor in this network is the same as in the CR network, but with $R1$ and $R2$ replacing R.

$\omega_n = \sqrt{(K\phi.Ko.\omega_{LP})}$, this is the same as the CR network equation and is in rad/s.

$$\zeta = \frac{\omega_N}{2}\left[R2.C + \left(\frac{1}{K\phi.K_O}\right)\right]$$

Ideally, ζ should have a value between 0.5 and 1.0. A value of $\zeta = 0.7071$ is recommended; this is the value used in Butterworth filters for maximally flat frequency response and has a step response with a slight overshoot. A value of $\zeta = 0.5$ has the lowest noise bandwidth ($B = 0.5\omega_n$). A value of $\zeta = 1.0$ has no step response overshoot.

The transfer function (frequency response if $s = j\omega$) of a lead-lag filter is given by the expression:

$$F(s) = \frac{R2.C.s + 1}{(R1.C + R2.C).s + 1}$$

This gives a frequency response with a 20 dB/decade roll-off.

More complex active filters can be used, but more care is needed in their design. An active lead-lag is the simplest network and is shown in Figure 13.5.

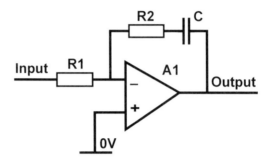

Figure 13.5

Active Lead-Lag Filter

The active lead-lag filter uses the same passive components as the lead-lag network described previously, except that some of the components are now in the op-amp's feedback loop. The design equations are the same apart from the damping coefficient, ζ, which is now simpler.

$$\zeta = \frac{R2.C.\omega_N}{2}$$

The transfer function of the active loop filter is given by:

$$F(s) = \frac{R2.C.s+1}{R1.C.s}.$$

If damping is reduced to zero, the phase-locked loop becomes a sinusoidal oscillator. The oscillation frequency is ω_N, which is the natural frequency of the loop. An input step signal will start the oscillations, but if the damping factor is greater than zero these will decay. The rate of decay is greater as the damping factor increases in value until, at a damping factor of one, no oscillations occur.

Higher-Order Loops

Higher-order loops are used to reduce phase noise in oscillator circuits. The use of a high-order filter gives a narrower bandwidth for feedback, thus reducing jitter in the frequency of the output signal. The loop order is one higher than the filter order, since the voltage-controlled oscillator (VCO) acts as an integrator that provides a first-order function. Thus a phase-locked loop (PLL) using a simple first-order RC filter network forms a second-order loop.

A second-order filter can be used to form part of a third-order loop. However, in practice third-order filters tend to be used, creating a fourth-order loop, because they provide greater reduction in spurious signal output. I will now describe the design process for an active third-order filter, illustrated in Figure 13.6. The filter design will have two variants, one driven from a current source and the other driven by a voltage source. The current source variant can be driven by PLL integrated circuits that have a charge-pump current output.

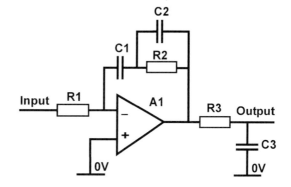

Figure 13.6

Third-Order Filter for Fourth-Order PLL

The *RC* networks in the circuit of Figure 13.6 form a number of separate time constants, as follows:

$$T_1 = R_1C_1$$
$$T_2 = R_2(C_1 + C_2)$$
$$T_3 = R_2C_2$$
$$T_4 = R_3C_3$$

In calculating component values, stability is important. The phase margin, ϕ_m, is related to the stability, and it is advisable to have at least a 45° phase margin, in order to minimize peaking in the frequency domain and overshoot in the time domain. The phase margin can be found from the filter circuit time constants:

$$\phi_m = \alpha + \beta + \gamma.$$
$$\alpha = \tan^{-1}(\omega_0 T_2) = \tan^{-1}(f_0/f_2)$$
$$\beta = \tan^{-1}(\omega_0 T_3) = \tan^{-1}(f_0/f_3)$$
$$\gamma = \tan^{-1}(\omega_0 T_4) = \tan^{-1}(f_0/f_4)$$

Before the design can commence, the loop frequency must be chosen. The most stable loops occur when the loop frequency is very low, but these also take the longest time to reach a stable VCO control voltage. A compromise between loop stability and reaction time is necessary. If frequency hopping is required, the hop time is the dominating feature. The transient time to reach stability is approximately $12/2\pi f_0$, when the phase margin is 45°. For a value of $f_0 = 100\,\text{Hz}$, the time for a loop to reach stability will be approximately 19 ms.

Rule-of-thumb values are $f_2 = f_0/2.5$, $f_3 = f_0 \times 3.33$, and $f_4 = f_0 \times 10$. Note that f_2 is the zero frequency that must remain below the loop frequency at all times. Using the equations to find the phase margin gives:

$$\alpha = \tan^{-1}(\omega_0 T_2) = \tan^{-1}(f_0/f_2) = \tan^{-1}(2.5) = 1.1903\text{ rad}$$

$$\beta = \tan^{-1}(\omega_0 T_3) = \tan^{-1}(f_0/f_3) = \tan^{-1}(0.3) = 0.29146\text{ rad}$$

$$\gamma = \tan^{-1}(\omega_0 T_4) = \tan^{-1}(f_0/f_4) = \tan^{-1}(0.1) = 0.09966\text{ rad}$$

$$\begin{aligned}\phi_m &= \alpha - \beta - \gamma = 1.1903 - 0.29146 - 0.09966 = 0.79918\text{ rad}\\&= 45.79\text{ degrees}\end{aligned}$$

Time constant $T1$ is determined from the phase-locked loop characteristics.

$$T_1 = \frac{K_P K_0}{4\pi^2 f_0^2 N}.abs[\cos(\alpha).\cos(\beta).\cos(\gamma)]$$

For a phase detector having a voltage output, the value of the phase sensitivity is given by $K_p = V_{cc}/2\pi$ typically 0.7958. If a charge current output is available from the PLL device, the phase sensitivity is given by $K_d = I/2\pi$. Hence, to find the value of T_1, $K_p = K_d/R1 = I/(2\pi.R1)$. Using values (for example) $K_0 = 2 \times 10^8$ and the feedback divider ratio $N = 1000$, gives $T_1 = 142.7\,\text{ms}$.

Now $\omega_0 = 2\pi f_0$, so in this case $\omega_0 = 628.32$ rad/s. Substituting into the above equations, the values of T_2, T_3, and T_4 can be found.

$$T_2 = \tan(\alpha)/\omega_0 . = \tan(1.1903)/628.32 = 2.5/628.32 = 3.97898\text{ ms.}$$

$$T_3 = \tan(\beta)/\omega_0 . = \tan(0.29146)/628.32 = 0.3/628.32 = 0.477464\text{ ms.}$$

$$T_4 = \tan(\gamma)/\omega_0 . = \tan(0.09966)/628.32 = 0.1/628.32 = 0.159155\text{ ms.}$$

First choose the value of C3 and derive the value of R3 from $R3 = T_4/C3$. Let $C3 = 0.1\,\mu\text{F}$ and since $T4 = 0.159155\,\text{ms}$, $R4 = 159.155 \times 10^{-6}/0.1 \times 10^{-6} = 1591.55\,\Omega$.

Now choose the value of C1 and derive the value of R1 from $R1 = T_1/C1$. Let $C1 = 1\,\mu\text{F}$. $R1 = 0.1427/10^{-6} = 142.7\,\text{k}\Omega \sim 150\,\text{k}\Omega$.

Finally, calculate the value of C2 and R2. Derive the value of R2 from $R2 = (T_2 - T_3)/C1 = 3.501516 \times 10^{-3}/10^{-6} = 3502\,\Omega$. The value of C2 can be found using $C2 = T_3/R2 = 0.477464 \times 10^{-3}/3502 = 0.13634\,\mu\text{F}$.

Analog versus Digital Phase-Locked Loop

An analog phase-locked loop uses an analog phase detector and, usually, a voltage-controlled oscillator having a sinusoidal output. Analog phase-locked loops are most common in radio systems where high-frequency outputs are required.

Digital phase-locked loops use logic gates as a phase detector, often arranged as an edge triggered flip-flop. The voltage-controlled oscillator output is usually a square wave. Digital phase-locked loops are used in frequency synthesizers and tone decoder circuits.

Practical Digital Phase-Locked Loop

Now I will describe how to make a practical digital phase-locked loop circuit. This will use a passive lead-lag network, so the component values for $R1$, $R2$, and C need to be found. It will also use a common CMOS (Complementary Metal Oxide Silicon) logic phase-locked loop integrated circuit, the 74HCT4046.

The circuit I am going to consider is a frequency synthesizer, producing a frequency N times that of the input signal. The bandwidth of the loop filter must be much smaller than the frequency used for comparison in the phase detector, to prevent modulation of the voltage-controlled oscillator. The damping factor of the system will be set at $1/\sqrt{2}$ to ensure stability and a reasonable impulse response. Now select $R1$ and C to provide a suitable bandwidth, approximately given by $1/(R1 . C)$ in rad/s. The values of $R1$ and C, and the loop constants (N, $K\phi$, Ko) can be used to find a value for $R2$ that gives the correct damping. This derivation will now be given.

As described earlier, the equations for a lead-lag network are:

$$\omega_{LP} = 1/(R1 + R2).C, \text{ in rad/s.}$$
$$\omega_N = \sqrt{(K\phi.Ko.\omega_{LP})}, \text{ in rad/s.}$$

ω_{LP} can be replaced to give: $\omega_N = \sqrt{\dfrac{K\phi.Ko}{(R1 + R2).C}}$

$\zeta = \dfrac{\omega_N}{2}\left[R2.C + \left(\dfrac{N}{K\phi.Ko}\right)\right]$, where N is the loop divider ratio.

Now, simplify the equation for $R2$ by letting $\zeta = 0.7071$ or $1/\sqrt{2}$, so that $\zeta^2 = 1/2$.

$$0.5 = \frac{\omega_N^2}{4}\left[R2.C + \left(\frac{N}{K\phi.Ko}\right)\right]^2$$

But $\omega_N^2 = \dfrac{K\phi.Ko}{(R1+R2).N.C}$ so, expanding the equation for ζ further this becomes:

$$0.5 = \frac{K\phi.Ko}{4C.N.(R1+R2)}\left[R2.C\left(\frac{N}{K\phi.Ko}\right)\right]^2$$

Given $R1$ and C, the value of $R2$ can be found such that $\zeta = 0.7071$; this can be found by expanding and transposing the above equation.

$$2 = \frac{K\phi.Ko}{C.N.(R1+R2)}\left[R2^2.C^2 + \left(\frac{N^2}{K\phi^2.Ko^2}\right) + \frac{2.N.C.R2}{K\phi Ko}\right]$$

Multiply both sides of the equation by $N.(R1 + R2)$, then expand the right-hand side by multiplying through by $(K\phi.Ko/C)$.

$$2.N.(R1+R2) = \left[K\phi.Ko.R2^2.C + \left(\frac{N^2}{K\phi.Ko.C}\right) + 2.N.C.R2\right]$$

Subtract $2.N.R2$ from both sides leaving an equation in terms of $R2^2$ only:

$$2.N.R1 = K\phi.Ko.R2^2.C + \left(\frac{N^2}{K\phi.Ko.C}\right)$$

Rearranging this to move $R2^2$ to the left-hand side, this becomes:

$$K\phi.Ko.R2^2.C = 2.N.R1 - \left(\frac{N^2}{K\phi.Ko.C}\right)$$

So now dividing both sides by $K\phi.Ko.C$ to obtain:

$$R2^2 = \frac{2.N.R1}{K\phi.Ko.C} - \left(\frac{N^2}{K\phi^2.Ko^2.C^2}\right)$$

$$R2 = \sqrt{\frac{2.N.R1}{C.K\phi.Ko} - \frac{N^2}{K\phi^2.Ko^2.C^2}}.$$

So, using the known or selected values it is possible to find a value for $R2$ that gives a damping factor of 0.7071. If a different value of ζ is chosen, the $2.N.R2$

terms do not cancel and finding a solution is more difficult: another good (pragmatic) reason for choosing $\zeta = 1/\sqrt{2}$.

Once a value for R2 has been obtained it can be used to more accurately predict the loop filter bandwidth using the equation: $\omega_{LP} = 1/(R1 + R2) \cdot C$. An *EXCEL* spreadsheet or *MATHCAD* page can be used to find the optimum component values quickly. A *MATHCAD* page suitable for finding suitable component values is given in Figure 13.7. The phase detector constant is given as Kp and the oscillator constant is Kv.

$$Kp := \frac{5}{4 \cdot \pi} \qquad f1 := 10 \cdot 10^3 \qquad f2 := 150 \cdot 10^3 \qquad N := 2$$

$$Kv := 2 \cdot \pi \cdot \frac{f2 - f1}{3.2}$$

Loop Filter Components

$$Kv = 2.749 \cdot 10^5 \qquad Kp = 0.398$$

$$C := 10 \cdot 10^{-6} \qquad R1 := 330 \cdot 10^3$$

$$R2 := \sqrt{\frac{N \cdot 2 \cdot R1}{C \cdot (Kp \cdot Kv)} - \frac{N^2}{Kp^2 \cdot Kv^2 \cdot C^2}}$$

$$R2 = 1.099 \cdot 10^3$$

$$s := 0.1, 1 .. 1000$$

$$Wn := \sqrt{Kp \cdot \frac{Kv}{N \cdot C \cdot (R1 + R2)}} \qquad F(s) := \frac{R2 \cdot s \cdot C + 1}{s \cdot (R1 \cdot C + R2 \cdot C) + 1}$$

$$Wlp := \frac{1}{(R1 + R2) \cdot C}$$

$$\zeta := 0.5 \cdot Wn \cdot \left(R2 \cdot C + \frac{N}{Kp \cdot Kv} \right)$$

$$Wn = 128.518$$

$$Wlp = 0.302$$

$$\zeta = 0.707$$

Check that damping is 0.7071

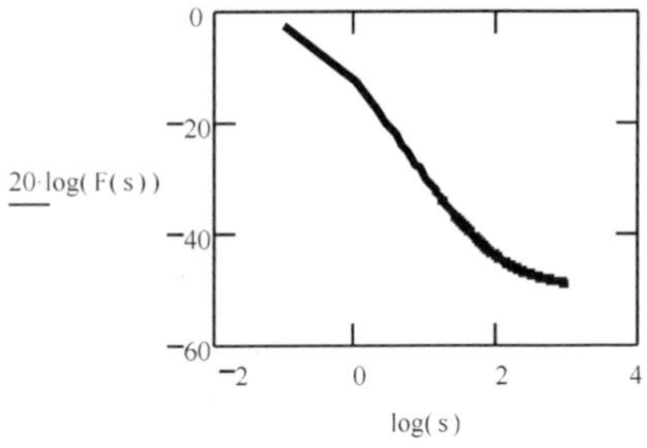

20·log(F(s))

log(s)

Figure 13.7

MATHCAD Page for Finding Component Values

The phase detector works over a range of $\pm 2\pi$ and produces an output voltage limited by the power supply; in the Mathcad example the range is 0 V to 5 V. Kp is then $5/(4\pi)$, which is the phase detector gain in volts per radian.

The voltage-controlled oscillator has a gain Ko volts per radian per second. This is the frequency range divided by the voltage difference required to produce that range. This is 2π times the frequency difference in Hertz, divided by 3.2 V (in my example, the input voltage range is 0.9 V to 4.1 V).

Phase Noise

The frequency multiplier circuit produces an output signal frequency that is a multiple of the input signal. Unfortunately, the timing of the clock transitions does not always take place at the same point relative to the phase of the input signal. The reason for this is noise at the voltage-controlled oscillator (VCO) input causes the frequency to instantaneously rise and fall. On average the frequency is correct, but the period of each half cycle may be longer or shorter, depending on whether the noise voltage is increasing or decreasing the control potential of the VCO.

The loop filter bandwidth controls the phase noise to a great extent. Even if the control voltage was noise free, circuitry within the VCO device adds noise and causes jitter. In the example a CMOS logic circuit was used, but CMOS is noisy. A bipolar oscillator should give better results. Another way to reduce phase noise is to reduce the frequency range of the oscillator. Noise voltage will then produce a smaller instantaneous frequency change; that is, the phase shift will be less.

To illustrate the problem of internal VCO noise I will extend the Practical Digital Phase-Locked Loop system described earlier. The frequency range was 140 kHz with a control voltage range of 3.2 V, so a 1 mV RMS (root mean square) noise voltage will introduce an average instantaneous frequency difference of 43.75 Hz. Consider the time domain; at 50 kHz the period of the oscillator output is 20 μs, and at 50,043.75 Hz the period is 19.9825153 μs, which is a difference of almost 17.5 ns. Since peak to peak noise can be many times the RMS level, the timing of the oscillator output transitions may vary from one cycle to the next. It would not be unknown for the peak level to be five times the RMS level. This would produce a phase jitter of 87.5 ns when compared with the reference signal. In some applications, such as in a communications synchronization circuit, this amount of jitter would be unacceptable.

Capture and Lock Range

The lock and capture ranges determine how well the phase-locked loop will follow signals at the input. The range depends upon the type of phase comparator used and on the loop filter design. Phase and frequency detectors of the edge-triggered type have equal lock and capture ranges.

The capture range is defined as the frequency offset from the VCO's center frequency over which a phase-locked loop can lock onto a signal. This range is determined by applying a signal having a frequency outside this range, then altering the frequency until lock is obtained. This range is most important for frequency synthesizer applications, because if the loop is unable to lock onto an input signal, it cannot work. In FM demodulator applications the center frequency of the input signal may not be the same as the VCO's center frequency; any offset must be within the capture range.

The lock range is defined as the frequency offset from the VCO's center frequency where lock is no longer possible. This range is determined by applying a signal having a frequency that is within the capture range, and locked, and then altering the frequency until lock is lost. The lock range is equal to, or greater than, the capture range. In frequency synthesizer applications this is not very important because the input signal does not normally change and, in any case, input signals should be within the capture range. The lock range is very important in FM demodulation systems because the input signal frequency is being changed by the modulation. If the FM signal deviates beyond the lock range, the loop will lose lock. The signal will be captured again as the signal frequency returns to nearer the center frequency, but there will be an audible click at the radio receiver's output.

The lock range of a phase-locked loop depends on the device used. In the case of a simple phase detector the lock range is given by $K\phi \cdot Ko$ in rad/s. In the case of a frequency and phase detector, the lock range is determined by the maximum and minimum oscillator frequency. A frequency and phase detector is different from a simple phase detector because the frequency and phase detector output is only limited by the power supply voltage. If the frequencies of the two signals are different, the frequency and phase detector output voltage will be set at one of the power rails. The VCO control range is less than the power supply voltage. In the example, the power supply had a 5 V rail but the VCO input voltage range was 0.9 V to 4.1 V.

The capture range depends on the loop filter, unless a frequency and phase detector is being used, in which case it is equal to the lock range. A simple RC filter has a capture range given by:

$$\omega_C = \sqrt{\frac{\omega_L}{RC}}$$

The lead-lag network capture range is generally wider, and is given by:

$$\omega_C = \frac{\omega_L \cdot R2}{R1 + R2}$$

Finally, for the active filter:

$$\omega_C = \frac{\omega_L \cdot R2}{R1}$$

It is interesting that neither the lead-lag network nor the active filter capture range is determined by the filter's capacitor value; it is just the lock range multiplied by a ratio of $R1$ and $R2$.

This chapter has covered the basics and shown that using a lead-lag network filter with a damping factor of 0.7071 enables component values to be easily calculated. Readers who have demanding applications are recommended to read some of the books devoted to phase-locked loop design. One of the most revered books is Gardner's *Phaselock Techniques*, which is now out of print but should be available from libraries.

Reference

1.   Gardner, F. M. *Phaselock Techniques*. New York: John Wiley & Sons, 1979.

CHAPTER 14

FILTER INTEGRATED CIRCUITS

This chapter gives an introduction to integrated circuit (IC) filters. Switched capacitor and continuous time filters are covered. Many semiconductor manufacturers produce filter ICs, and filter ICs from three companies, Maxim, Texas Instruments (formerly Burr-Brown), and Linear Technology, are described and some practical design examples are given. Problems encountered with these types of filter are also described. The benefits are, for example, being able to make the filter cutoff programmable or adjustable.

A frequency synthesizer will be described that derives a sine wave signal from a clock-driven logic circuit. The reason for including this description is that the filtering of its output can be carried out in one of two ways: (1) with a fixed frequency filter, if the clock frequency is fixed; (2) if the clock frequency is tunable, by using a tuned switched capacitor filter that tracks the frequency of the output signal. Method (2) shows the advantage of having a filter cutoff frequency that is clock dependent.

Continuous Time Filters

Continuous time filters are the same as some of the active filters described earlier in Chapters 4–7 of this book. Generally, a continuous time filter IC has a fixed frequency response that is determined at the design stage and set by external resistors. The MAX274, the MAX275, and the UAF42 are IC filters of this type. However there are some digitally programmable continuous time filters, such as the MAX270 and the MAX271, that require no external components for frequency selection: the frequency is determined by data programmed into an internal latch.

The MAX274, MAX275, and UAF42 all have a state variable architecture. Internal capacitors and resistors are provided in all three types.

Integrated Circuit Filter UAF42

The Texas Instruments (Burr-Brown) UAF42 can be used to produce a low-pass, highpass, bandpass, or bandstop active filter. All-pole frequency responses, such as Bessel, Butterworth, and Chebyshev, as well as those responses with a zero in the stopband, are possible, such as Inverse Chebyshev and Cauer (Elliptic).

The UAF42 has a single second-order section comprising three op-amps and an auxiliary op-amp. The auxiliary op-amp is useful for producing bandstop filters or filter responses with a zero in the stopband, such as an Inverse Chebyshev type. Each op-amp has a gain-bandwidth product of 4 MHz and a slew rate of 10 V/μs, which gives a useful range of passband operating frequencies of up to 100 kHz. The internal circuit schematic is given in Figure 14.1.

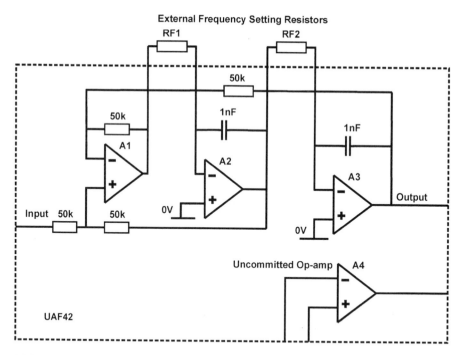

Figure 14.1

Circuit of UAF42

The UAF42 includes 50 kΩ resistors and 1 nF capacitors, but external resistors are required to complete the design. However, because the UAF42 only contains one second-order section, access to both ends of the internal resistors and capacitors is possible in this IC. This allows external components to be connected in parallel, to increase the range of cutoff frequencies available. The internal capacitors are laser-trimmed to a tolerance of 0.5%.

A computer program called *FILTER42* is available from Texas Instruments and is described in Application Bulletin AB-035 (downloadable from the TI Internet site, www.ti.com as file sbfa002.pdf). *FILTER42* calculates the component values for Bessel, Butterworth, Chebyshev, and Inverse Chebyshev (but not Cauer) responses. The Inverse Chebyshev cutoff frequency is considered by this program as being at the beginning of the stopband, rather than the −3 dB point. The Chebyshev cutoff frequency is considered by this program to be where the amplitude response falls below the passband ripple limit.

Alternatively, design equations are given in the UAF42 data sheet. Unfortunately, these equations give ω_n, Q, and gain in terms of resistor and capacitor values. Usually ω_n, Q, and gain are known, so the equations may need transposing to find, say, R_{F1} in terms of the known values.

The *FILTER42* program is in the FilterPro™ series; other programs are *FILTER1* and *FILTER2*. Both *FILTER1* and *FILTER2* can be used to help design lowpass filters; they cannot help in the design of highpass, bandpass, or bandstop filters. *FILTER1* produces component values for Sallen and Key lowpass active filters. *FILTER2* is described in Applications Bulletin AB-034B, and this program produces component values for both Sallen and Key, and Multiple Feedback (MFB) filters. The MFB topology is sometimes called Raunch, or Infinite Gain, and is less susceptible to component variations.

Texas Instruments also have the FilterPro MFB lowpass filter design program available for download from their Internet site. This program finds component values for Bessel, Butterworth, and Chebyshev filters constructed from op-amps and discrete components. Filter designs up to the tenth order are possible using this program.

Integrated Circuit Filter MAX274

The MAX274 can be used to produce all-pole lowpass or bandpass active filters with cutoff frequencies of up to 150 kHz. A notch can be created by the addition of an external op-amp stage. The device contains four second-order sections. Each section is made up using four op-amps. A circuit diagram of one second-order section is given in Figure 14.2.

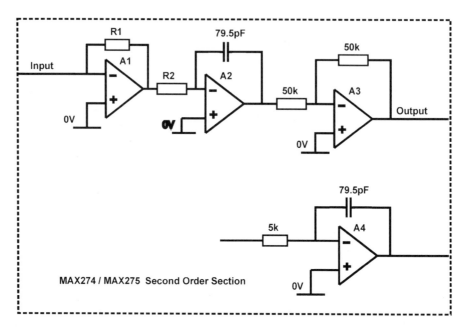

Figure 14.2

Part Circuit of MAX274/MAX275

Integrated Circuit Filter MAX275

MAX275 contains two second-order sections. Each section is made up using four op-amps, as shown in Figure 14.2. Like the MAX274, this device can only be used for all-pole lowpass and bandpass filters, unless external op-amp circuits are added. This device can work at higher frequencies than the MAX274; a cutoff frequency of 300 kHz is possible.

The MAX274 and MAX275 have internal resistors and capacitors; a filter can be created by the addition of four external frequency setting resistors. Two of the internal resistors, R1 and R2, have a selectable ratio; the ratio of R2:R1 can be either 1:4, 5:1, or 25:1. With control pin FC connected to V$^+$ the ratio is 1:4 using nominally 13 KΩ for R2 and 52 kΩ for R1. Connecting FC to ground produces a 5:1 ratio using 65 kΩ for R2 and 13 kΩ for R1. Finally, with FC connected to V$^-$ the ratio of R2:R1 is 25:1, using 325 kΩ for R2 and 13 kΩ for R1.

The internal capacitors have a low value (79.5 pF) and low-frequency poles are difficult to produce. The only practical method for working below 100 Hz is to use a resistor T circuit; this is a potential divider and a series resistor that gives the equivalent input current as a high-value resistor. Details of the resistor T circuit are given in the MAX274/MAX275 data sheet available from Maxim.

Integrated Circuit Filter MAX270/MAX271

The MAX270 and MAX271 are second-order lowpass active filters. These filter ICs are unusual because, although they are continuous time devices, they are digitally programmed for their cutoff frequency. The filters need no external components for their frequency selection, but cutoff frequencies are limited to the range 1 kHz to 25 kHz. Internally the devices use a Sallen and Key lowpass circuit, with variable shunt and feedback capacitors; these are varicap diodes and have a capacitance dependant upon the reverse bias potential across them. An internal digital-to-analog converter provides a bias voltage to tune the filter and give it the required cutoff frequency.

The MAX270 and the MAX271 are different. The MAX270 has an uncommitted op-amp that has a 2 MHz gain-bandwidth product. This op-amp can be used to produce another filter stage or for other applications.

The MAX271 has a track-and-hold amplifier that can select, as its input, either lowpass filter's output. It is possible to multiplex the two signals by switching the track-and-hold circuit from one input to the other, since the filter will prevent aliasing if the switching rate is high enough; a sampling clock of 50 kHz to 200 kHz would be suitable for most applications. The track and hold output is disabled if the enable pin is at logic 0. The output from several devices can be connected in parallel and, by only enabling one at a time, all the signals can be multiplexed onto one circuit for (perhaps) carrying out analog to digital conversion.

Switched Capacitor Filters

Switched capacitor filters are generally considerably noisier than their continuous-time counterparts. This is mainly due to the switching process; signals at the switching frequency and other spurious signals appear at the filter's output. There is also a risk of aliasing; this is where a signal outside the band of interest appears in-band due to mixing with the sampling clock (which causes frequency shifting).

A switched capacitor filter uses the principle that, by switching a capacitor between the source and the load, the equivalent of a high resistor value is created between the two. Thus instead of resistors and capacitors there are just capacitors and switches. This serves two purposes: (1) high value resistors are difficult to produce on a semiconductor wafer; (2) by varying the switching rate the effective resistance value changes. The basic switched capacitor circuit is shown in Figure 14.3.

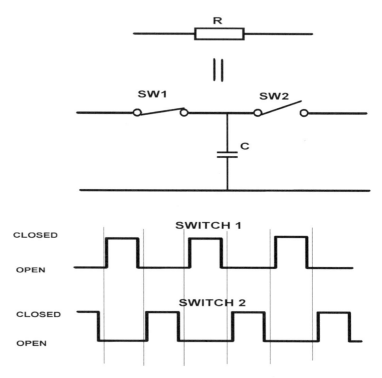

Figure 14.3

Switched Capacitor "Resistor"

The circuit in Figure 14.3 is just one of several possible designs. The equivalent resistance depends on the capacitor value and the switching frequency used. In this case the equation for finding the equivalent resistance is $R = 1/fC$. The two switches are arranged to be break-before-make, so that there is never an opportunity for a short circuit between input and output. The choice of switched capacitor circuit depends on the filter's topology. Some filters use one switched capacitor circuit for a shunt element but a different circuit for the feedback element.

Consider the switched capacitor circuit that is illustrated in Figure 14.3. A charge of $Q = CV$ coulombs is stored when the first switch is closed and the capacitor charges up. A charge of CV coulombs discharges into the load, if the load is low impedance, when the second switch closes. Therefore, each complete clock cycle causes a charge of CV coulombs to flow, from the source to the load. Clearly, if there are N clock cycles per second, there is a total charge flow of NCV coulombs per second; in other words NCV amperes (since one ampere equals one coulomb per second).

Now, $R = V/I$ or, substituting for I,

$$R = \frac{V}{NVC} = \frac{1}{NC}.$$

Since $N = f$, that is, the clock frequency, $R = 1/fC$ as stated earlier.

The equations above assume that the capacitor is fully charged (or discharged) during the period that the switches are closed. The switching period is short so, as you might expect, the capacitance values are small. Capacitors cannot be large because they are formed on an integrated circuit substrate where the surface area must be minimized to reduce costs. In practice the capacitance values used are typically between 1 pF and 20 pF, although more likely to be at the lower end of the value range.

The active filter circuits described earlier in this book in Chapters 4–7 can be realized using switched capacitors to replace resistors. Thus the circuit would comprise switched and unswitched capacitors around an op-amp, with no resistors required to determine the filter cutoff frequency or response. Using this technique, small integrated circuits have been developed with high performance; it is possible to produce a complete eighth-order lowpass filter in an 8-pin IC package.

Switched Capacitor Filter IC LT1066-1

The Linear Technology IC LT1066-1 is a switched capacitor filter housed in an 18-pin IC package. It provides an eighth-order Cauer (elliptic) response with a nominal 0.15 dB passband ripple. The passband cutoff frequency is defined as where the gain falls to −1 dB; the maximum cutoff frequency is supply dependent and is between 36 kHz and 120 kHz.

The LT1066-1 has three modes of operation determined by the switching ratio pin voltage. If the switching ratio pin is connected to the positive supply rail, the ratio of clock frequency to cutoff frequency is 50:1. At this ratio, the input signal is sampled twice in each clock period to reduce aliasing problems. If the switching ratio pin is connected to the negative supply rail the clock to cutoff frequency ratio is 100:1. Finally, if the switching ratio pin is connected to analog ground, the clock to cutoff frequency ratio is again 100:1, but the phase response is now linear over the lower half of the passband. In this last mode the response is almost Inverse Chebyshev, although Linear Technology calls it "pseudo linear phase response."

Microprocessor Programmable ICs MAX260/MAX261/MAX262

Maxim produce a series of three microprocessor programmable universal active filters. The MAX260 handles center frequencies up to 7.5 kHz. The MAX261 operates with center frequencies up to 57 kHz, and the MAX262 can work up to 140 kHz. Each filter IC contains two second-order filter sections that can be configured to provide lowpass, highpass, bandpass, bandstop, and all-pass types. The filter response can be all-pole, such as Butterworth and Chebyshev, or pole and zero responses, that is, Cauer (elliptic) and Inverse Chebyshev.

The filter IC has four programmable modes controlled by the logic state in two registers of the IC's program memory. There is also a fifth mode known as mode 3A; this is when the IC is in mode 3, but an external op-amp ladder circuit is connected to the highpass and lowpass pins to provide a notch output. The notch output is used for bandstop filters. It is also used for Cauer or Inverse Chebyshev responses of any type (lowpass, highpass, bandpass, or bandstop).

The ratio of clock frequency to pole frequency, f_o, depends on the device and the filter mode. A set of registers in the device's memory stores the required ratio. The MAX260 and MAX261 have the same ratios, varying from 100.53:1 to 199.49:1 in modes 1, 3, and 4, and varying from 71.09:1 to 141.06:1 in mode 2. The MAX262 can handle higher filter pole frequencies, but the ratios are lower. These vary from 40.84:1 to 139.80:1 in modes 1, 3, and 4, and from 28.88:1 to 98.85:1 in mode 2. These noninteger numbers are derived from the equations:

$$ratio = (64 + N).\pi/2, \text{ for the MAX 260/MAX261 in modes}$$
$$1, 3 \text{ and } 4. \ N = 0 \text{ to } 63.$$
$$ratio = (26 + N).\pi/2, \text{ for the MAX262 in modes 1, 3 and 4. } N = 0 \text{ to } 63.$$

In mode 2 all the ratios are divided by $\sqrt{2}$.

Mode 1 can be used to implement all-pole lowpass or bandpass filters. A limited range of second-order notch filters can also be produced.

Mode 2 allows higher Q factors for the poles and can provide all-pole lowpass, bandpass, or notch filters.

Mode 3 can be used to produce all-pole highpass, as well as lowpass and bandpass filters. This is the only mode for producing highpass filters.

Mode 3A uses an external op-amp to provide a Cauer or Inverse Chebyshev filter response. Outputs from the filter IC are those for mode 3.

Mode 4 produces an all-pass function, as well as lowpass and bandpass outputs. The all-pass response, you may remember from earlier chapters, provides group delay equalization or phase-shift networks.

In each mode it is possible to program the pole's Q factor; setting certain registers in the device's memory does this. The lowest Q is 0.5 in modes 1, 3, and 4, and 0.707 in mode 2. The highest Q is 64 in modes 1, 3, and 4, and 90.5 in mode 2. The possible Q values are given by the equation:

$$Q = \frac{64}{128 - N}, \text{ in modes 1, 3 and 4. In mode 2 these values are}$$
multiplied by $\sqrt{2}$.

Maxim supplies a comprehensive data sheet with plenty of applications, if further information is required.

Pin Programmable ICs MAX263/MAX264/MAX267/MAX268

These devices are very similar to those previously described, except that the registers containing the pole frequency and Q setting data are externally accessible. This allows a simple circuit that is suitable for an analog design. As before there are four standard modes of operation, with an extra mode (3A) that uses an external summing circuit.

The clock inputs have TTL logic level thresholds of 0.8 V and 2.4 V. This is not the case for the frequency and Q programming inputs; these have thresholds within 0.5 V of the supply rails. Thus the positive threshold is $V^+ - 0.5$ V and the negative threshold is $V^- + 0.5$ V. The supply rails are nominally +5 V (V^+) and −5 V (V^-). Programming is achieved by wire-strapping the programming inputs to the positive and negative rails, rather than by using logic gates.

The MAX263 and MAX267 can have pole frequencies up to 57 kHz. In modes 1, 3, and 4 the clock to pole frequency ratio can be programmed to be between 100.53 : 1 and 197.92 : 1, and the Q can be programmed to be between 0.504 and 64. In mode 2 the frequency ratio can be set between 71.09 : 1 and 139.95 : 1, and the Q can be set between 0.713 and 90.5.

The MAX264 and MAX268 have lower clock to pole frequency ratios and can therefore work up to 140 kHz. In modes 1, 3, and 4 the clock to pole frequency ratio range is from 40.84 : 1 to 138.23 : 1. In mode 2 the frequency ratio can be set to between 28.88 and 97.74. The same range of Q values applies as in the MAX263 and MAX267.

Other Switched Capacitor Filters

Only a small range of filter ICs have been described. Manufacturers of these and other popular devices include National Semiconductor, Linear Technology, and Maxim.

One type of switch capacitor filter IC, which is available from more than one company, provides DC accurate lowpass filtering by providing an active shunt. Examples are the Maxim parts MAX280, MAX281, and MXL1062. This type of filter only requires a resistor in the signal path, the filter input is connected to one side of the resistor and the filter output is connected to the other side. The filter IC is connected between the filter output and ground, and causes signals to be attenuated by the equivalent of a fifth-order response. The circuit for this is illustrated in Figure 14.4.

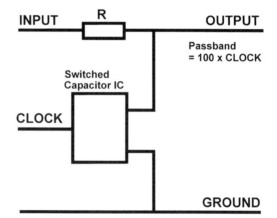

Figure 14.4

Lowpass Filter with DC Accurate Output

There are other devices that are programmed to have, say, an eighth-order Butterworth, Bessel, or Cauer response. The cutoff frequency is usually 1/100 or 1/50 of the clock frequency. Although these devices are not flexible, they are very simple to use and apply.

An Application of Switched Capacitor Filters

One of the most useful applications of switched capacitor filters relies on the cutoff frequency being proportional to the clock frequency.

A sinusoidal frequency synthesizer can be produced from a digital circuit known as a Walking Ring, two designs are shown in Figure 14.5. The Walking Ring uses counter circuits to produce a pattern of logic states that repeats after a certain number of clock pulses. Using summing resistors at the input of an op-amp, the pattern of logic states can be converted into a sine wave. However, the sine wave so produced is stepped and contains many high-frequency harmonics. The answer is to lowpass filter the output to smooth these steps.

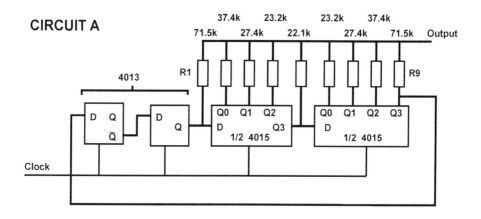

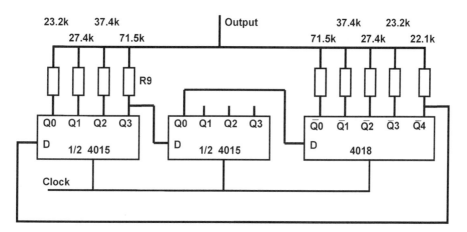

Figure 14.5

Frequency Synthesizers Using CMOS Logic

At the heart of the Walking Ring circuit is a chain of *D*-type latches, all fed by a common clock. The initial condition of the latches is with their *Q* outputs at logic 0. If a logic 1 condition is applied to the *D* input at the first latch, its output switches to logic 1 after a clock cycle. The *Q* output of one latch is connected to the *D* input of the next, so a logic 1 on the first latch's output is also applied to the second latch's input. After the second clock cycle the second latch switches its output to logic 1. Thus, if there are ten latches, all the Q outputs will be set to logic 1 after ten clock cycles.

The secret of the Walking Ring's operation is to loop the last output back to the first input and to use a not-Q output somewhere in the chain to invert the data sequence. The sequence of operations for Circuit A in Figure 14.5 follows. Upon reset, the not-Q output is high. If this condition is input to a second latch its output will go high after a clock cycle and each subsequent clock cycle will cause an additional latch output to be set. This process will continue until all the latches are set and the not-Q output that was used to set the first latch resets to logic 0. Further clock cycles then cause subsequent latches to reset and return to the initial condition from where the whole process begins again.

Circuit B in Figure 14.5 uses a CMOS CD4018 that has five latches connected in a chain, but all the outputs are inverting. The operation of circuit B can be analyzed in a similar way to Circuit A, the sequence of operations follows here. When the latches are reset at the start of operation outputs 6, 7, 8, 9, and 10 are all at logic 1. The other latch outputs are at logic 0, and so the synthesizer output will be just above midrail potential. The next few clock cycles will cause latch outputs 1, 2, 3, and 4 to be set to logic 1, the final cycle causing the synthesizer output to equal the supply voltage.

The next clock cycle does not change the output potential but causes the delay latch to output a logic 1 condition. Subsequent clock cycles reset latch outputs 6, 7, 8, 9, and 10 to logic 0, since they are inverted, and the synthesizer output voltage falls with each cycle. The next four clock cycles causes latch outputs 1, 2, 3, and 4 to be reset in turn. The synthesizer output is now equal to the ground potential. This does not change with the next clock cycle, because the delay latch just resets its output to logic 0 at this time. The next five clock cycles cause latch outputs 6, 7, 8, 9, and 10 to be set to logic 1, which is back to the beginning of the cycle.

Note that for an even number of latches there will be an odd number of resistors because one of the latch outputs has no resistor connected. The latch with no resistor provides a delay, so that the output voltage will remain at the supply voltage or ground for two clock cycles. The double clock period at either supply rail ensures that the maximum output is available when the signal is filtered. Resistor $R1$ provides the first step output, which is also the smallest and therefore has the highest resistance (since the step size is proportional to the current). Resistor $R1$ is connected to the first latch that follows the delay latch, and this is also the first output after the inversion of the latch outputs.

Resistor Value Calculations

The calculation of resistor values has been described in my magazine article.[1] Basically the circuit needs to be analyzed in terms of conductance and current flow through the resistors, since the output voltage is the ratio of two sets of parallel resistors. The output voltage is always the resistor current from the supply, divided by the conductance to ground. The current from the supply and the conductance to ground both depend on how many of the latch outputs are at logic 0 and how many are at logic 1. If all latches are at logic 1 there is no path to ground and so the current is zero. Similarly, if all latches are at logic 0 there is no path the positive supply rail and there is no current flow. If some latch outputs are at logic 1 and some are at logic 0, current will flow, from one to the other, through the resistors. The output voltage at the resistors' common node (i.e., the output) will depend both on the resistor values and the state of the respective latch output they are connected to.

Resistor values can be found by considering the conductance from the output to the positive rail, G_p, and the conductance between the output and ground, G_g. The current flowing from the positive rail through the resistors is $I = V.G_s$, where G_s is the series connection of G_p and G_g, and V is the supply voltage.

$$\text{Expanding } I = V.G_S \text{ gives } I = \frac{V.G_P.G_g}{G_P + G_g}$$

$$\text{The output voltage is } V_O = \frac{I}{G_g}$$

If a substitution into the equation for the resistor current flow is now made:

$$V_O = \frac{V.G_P}{G_P + G_g}$$

The total conductance with all resistors in parallel is $G_t = G_p + G_g$.

$$V_O = \frac{V.G_P}{G_t}$$

The voltage output at reset is zero, since all latches are at logic 0 and there is no connection to the positive supply. If the phase angle at this point is taken as being zero, the output of the synthesizer can be expressed as: $Vo = 0.5\,V - 0.5\,V.\cos(\phi)$, where ϕ is the phase angle and is initially zero. This has been tabulated in Table 14.1 for steps of 18°.

Angle, ϕ degrees	Amplitude, Vo
0	0
18	0.024772
36	0.095492
54	0.206107
72	0.345492
90	0.5
108	0.654508
126	0.793893
144	0.904508
162	0.975528
180	1

Table 14.1

Output Voltage versus Phase Angle

These voltages cannot be used to find the resistor values. This is because, when smoothed, the output voltage always lags behind the voltage at the rising edge of the output step. This is illustrated in Figure 14.6.

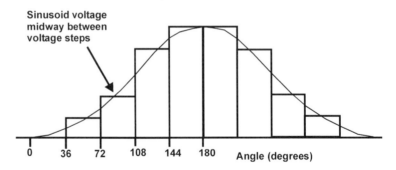

Figure 14.6

Output Voltage versus Stepped Voltage

Referring to Figure 14.6 you can see that you want the sinusoidal voltage to cross the stepped voltage at the rising edge of the step, midway between the previous step amplitude and the following step amplitude. The step voltage required would thus be the sinusoid voltage, at the phase angle where the step occurs, plus the difference between this and the previous step voltage. The difference is $V\text{sine} - Vo_{(n-1)}$.

$$Vo_n = V\text{sine} + (V\text{sine} - Vo_{(n-1)}) = 2.V\text{sine} - Vo_{(n-1)}.$$

Modifying Table 14.1 gives the actual step voltages required, and thence the resistor values to produce them. The step voltage is Vo_n where n is the step number. The clock to sine wave frequency ratio in the examples shown in Figure 14.5 is $20:1$, so each clock cycle represents a phase change of $18°$. After the first clock cycle is complete the output voltage should be $Vo_1 = V - V.\cos(18°) - Vo_0$. After subsequent clock cycles the output voltage should be $Vo_n = V - V.\cos(N \times 18°) - Vo_{(n-1)}$. A table of normalized output voltage steps is given in Table 14.2, by letting $V = 1$ and $Vo_0 = 0$.

Angle, ϕ degrees	Sinusoid Amplitude	Step n	Step Voltage, Vo_n
0	0	0	0
18	0.024772	1	0.048944
36	0.095492	2	0.14204
54	0.206107	3	0.270175
72	0.345492	4	0.420808
90	0.5	5	0.579192
108	0.654508	6	0.729825
126	0.793893	7	0.85796
144	0.904508	8	0.951057
162	0.975528	9	1
180	1	10	1

Table 14.2

Output Voltage Steps to Produce a Sinusoid

Only one resistor will be connected to the positive supply after the first clock cycle, so $G_p = G1$, where $G1 = 1/R1$. Normalizing the values of V and $R1$ (let them equal one), so that $G1 = 1$, allows a value for G_t to be calculated. After one clock cycle:

$$Vo_n = \frac{G_P}{G_t} = \frac{G1}{G_t} = 1 - \cos(18°) - Vo_{(n-1)}$$

$$Vo_1 = \frac{G_P}{G_t} = \frac{G1}{G_t} = 1 - \cos(18°).$$

This equation can be transposed to find G_t:

$$G_t = G1/Vo_1$$

Referring to Table 14.2 to find Vo_1:

$$G_t = 1/0.04894 = 20.433.$$

Knowing $G1$ and G_t, the remaining conductance values can be found. In terms of a formula:

$G_n = Vo_n . G_t - G_p.$

So, $G2 = 0.14204 \times 20.433 - G1$

$G2 = 1.902092$

and $G3 = 0.270175 \times 20.433 - (G1 + G2)$

$G3 = 2.617992$

This process continues for the remaining conductance values:

$G4 = 3.07766$

$G5 = 3.236025$

$G6 = 3.07766$

$G7 = 2.617992$

$G8 = 1.902092$

$G9 = 1.$

Since the signal generated will be symmetrical, the conductance values are also symmetrical. There is an odd resistor value, $R5$ in this case (corresponding to conductance $G5$), that will have the lowest resistance. This value can then be denormalized to have a suitable resistance and other resistor values can be calculated in relation to it. In the examples $R5$ is 22.1 kΩ, an E96 value. To find the resistor values, the ratio of conductance values, $G5/Gn$, must be multiplied by the smallest resistor value used:

$Rn = R5 \times G5/Gn$

This means that $R1 = R9 = 3.236025 \times 22.1\,\text{k}\Omega/G1 = 71.516\,\text{k}\Omega$, the nearest E96 value is 71.5 kΩ. In other words, since $G1 = 1$, $Rn = R1/Gn$

$R2 = R8 = 71.516\,\text{k}\Omega/1.902092 = 37.5986\,\text{k}\Omega$, the nearest E96 value is 37.4 kΩ.

$R3 = R7 = 71.516\,\text{k}\Omega/2.617992 = 27.317\,\text{k}\Omega$, the nearest E96 value is 27.4 kΩ.

$R4 = R6 = 71.516\,\text{k}\Omega/3.07766 = 23.237\,\text{k}\Omega$, the nearest E96 value is 23.2 kΩ.

Synthesizer Filtering

If the sine wave generator has a fixed frequency, perhaps deriving its clock from a crystal oscillator, the lowpass filter on its output can be a simple active filter using op-amps. If precision resistors are used in the summing circuit, the harmonics are very low until the clock frequency is approached, so, in theory, a first- or second-order filter will be satisfactory.

If the clock frequency is altered, the frequency of the sine wave generator will change. A fixed lowpass filter will not work, so the filter's cutoff frequency must alter too. The solution is to use a switched capacitor filter so that as the clock frequency increases, so does the filter's cutoff frequency. Hence the cutoff frequency is always just above the oscillator's frequency. The frequency synthesizers shown in Figure 14.5 have a clock to sine wave frequency ratio of 20:1.

If a divide-by-four device is placed before the clock input to the synthesizer, the clock to sine wave frequency ratio will be 80:1. An 80 kHz clock will produce a 1 kHz stepped sine wave from the synthesizer. Suppose a filter is placed at the output of the synthesizer, say a MAX295 that has a 50:1 clock to cutoff frequency ratio. If the same 80 kHz clock is applied to the filter IC its cutoff frequency will be 80/50 kHz, or 1.6 kHz. If the clock frequency is now doubled to 160 kHz, the filter cutoff frequency becomes 3.2 kHz and the sine wave frequency increases to 2 kHz. Thus the filter cutoff frequency is always 1.6 times the sine wave frequency. A block schematic of this circuit is given in Figure 14.7.

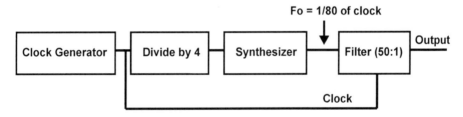

Figure 14.7

Synthesizer with Tracking Lowpass Filter

In conclusion, switched capacitor filters are small and can be made adaptive to applied signals. They are, however, noisier than the equivalent continuous time filter, and careful design of the circuit layout is necessary to minimize this noise.

Reference

1.   Winder, Steve. *Quadrature Low-Frequency Synthesizer*. Electronic Product Design, IML Group, July 1993.

CHAPTER 15

INTRODUCTION TO DIGITAL FILTERS

This chapter outlines the process of digital filtering. Digital filters operate on digitized analog signals, so the digitization process is important and can be critical in the system design. Digitization requires the analog signal to be sampled and then converted into a digital value, based on the amplitude of the sample. For this reason I will cover the data sampling and digitization operation (under-sampling, over-sampling, interpolation, and decimation) before considering digital filters.

The two types of digital filter, finite impulse response (FIR) and infinite impulse response (IIR), are only described briefly in this chapter. The functions required to form a digital filter are described, such as multipliers, adders, and delays. More detail on finding the multiplier coefficients for these types of filters will be given in Chapters 16 and 17.

Digital signal processors (DSPs) are described in terms of how the functions required in a digital filter are built into the architecture or can be created in software. The type of arithmetic DSPs use to handle data during signal processing is also described. The choice of processing device will determine whether fixed or floating-point arithmetic is used. Fixed-point arithmetic can affect accuracy and stability.

Analog-to-Digital Conversion

Analog data cannot be directly input to a digital system; it must be converted into digital form. Samples of the analog signal are taken at discrete time intervals and then converted into a digital form. This digital form is a binary representation of the input voltage at the instant of sampling. Many analog-to-digital converters produce a data word that is between 8 and 16 bits wide.

In order not to corrupt the data, the sampling frequency must be more than twice the highest frequency of the input signal. Thus, in telephone systems that have a bandwidth limited to 3.4 kHz, signals are sampled at 8 kHz. This means that analog signal frequencies above 4 kHz must be attenuated to levels below the input noise floor. To achieve this, an analog filter having a very steep skirt response above the cutoff frequency is required.

Under-Sampling

Under-sampling is when the sampling frequency is less than twice the highest frequency of the analog signal. Under-sampling introduces alias signals into the passband of the wanted signals, and these cannot be removed. During the sampling process, the sampling pulse is multiplied by the analog signal in the time domain. The resultant frequency domain spectrum at the output of this process is the sum and difference of the sampling frequency and the analog signal.

Aliasing, due to under-sampling, is easily explained by example. Suppose the telephone system described previously is not filtered very well and allows through analog signals that have frequency content above half the sampling frequency. Consider an unwanted signal with a frequency of 6 kHz. With a sampling frequency of 8 kHz, the output signal will have a frequency spectrum that includes (8 + 6)kHz and (8 − 6)kHz. In other words, 14 kHz and 2 kHz. The 2 kHz signal is the problem, since it is within the 3.4 kHz passband and cannot be removed by subsequent processing.

Under-sampling can sometimes serve a useful purpose. Suppose we have a speech signal with a bandwidth of 10 kHz, but it is amplitude modulated on a carrier at 1 MHz. This signal could be sampled at about 2.5 MHz but would contain a lot of useless information. We want to know about the speech signal, not the carrier. If the signal is instead sampled at 1.03 MHz, the mixing process generates signals centered at 30 kHz and 2.03 MHz. The signal could be decimated by 1/8 to sample at 128.75 kHz before being demodulated.

Care must be taken when under-sampling to avoid aliasing unwanted signals into the passband. The sampling frequency must be more than twice the *bandwidth* of the analog signal. However, the sampling frequency may be lower than the *center frequency* of the analog signal. The location of spectral images is given by:

$$f_I = |nf_s \pm f_c|$$

In this equation f_I is the frequency of the image, f_s is the sampling frequency, and f_c is the analog signal frequency. The image is repeated across the spectrum, indicated by the multiplying integer n.

In order to avoid destructive aliasing, the following condition must be met:

$$f_I = |nf_s \pm f_c|$$

$$f_I + B + \frac{W - B}{2} < (f_s - f_I)$$

In this equation, B is the 3 dB bandwidth and W is the width of the skirt response at the minimum detectable (i.e., noise floor) amplitude.

Over-Sampling

Sampling at a rate that is many times the highest analog signal frequency is called "over-sampling." Over-sampling reduces distortion and reduces the demands placed on analog anti-alias filters. However, producing more samples means that the processor must handle more data, which reduces its ability to perform other tasks or means that a faster processor may be required.

Decimation

Decimation is sometimes used to reduce the data rate. Decimation is the process of removing samples from the digitized signal. A decimation rate of $\frac{1}{2}$ will remove every other sample and thus halve the data rate. Similarly, a decimation rate of $\frac{1}{3}$ will only allow every third sample to pass. The decimation process is useful where the signal of interest cannot be filtered sufficiently to remove unwanted signals of a slightly higher frequency. In this case both signals are sampled at a high rate, and then the digitized signal is decimated to reduce the sampling rate to one suitable for the wanted signal.

The advantage of decimation can be seen in the following example. Suppose the wanted signal is speech with a bandwidth of 10 kHz, but interfering signals are present at 15 kHz. A sample rate of 24 kHz will meet the requirements of being greater than twice the maximum analog signal frequency. It would be possible to attenuate the 15 kHz interfering signal by, say, 60 dB with a really good analog filter, but this is not good enough to meet the specification. The sampling frequency of 24 kHz for a 10 kHz bandwidth signal would be suitable, but an alias will occur at 9 kHz because of the mixing process between the sampling process and the 15 kHz interfering signal.

Once the digitized signal contains an alias, it cannot be removed. This problem can be resolved by sampling at 48 kHz then decimating at $\frac{1}{2}$ rate. The 48 kHz sampling ensures that there is no alias, decimation then provides the same digital output for the wanted signal. The signal can then be digitally filtered to remove the remaining traces of the 15 kHz signal.

Interpolation

Interpolation is the opposite of decimation. Suppose our process described above must produce an output at 48 kHz sample rate, to be compatible with the rate of the input. We have reduced the internal data rate to 24 kHz, so additional samples must be inserted between the data samples. These samples are usually the average value of the previous sample and the following sample. In some cases the same data may be in pairs of samples, but this does not add value to the signal. By inserting average data values the signal output from the digital-to-analog converter is smoother. It also contains a lower signal power at the sampling frequency.

Decimation and interpolation are usually arranged to reduce or multiply the data rate by a power of two. So we may have decimation rates of $\frac{1}{2}$, $\frac{1}{4}$, $\frac{1}{8}$, and so on. This means that there is an equal spacing between samples, which is important for reconstructing the signal.

Decimation takes place "naturally" in sigma-delta analog-to-digital converters. These devices sample the signal by typically 64 times the output data rate. A converter having a 40 kHz output data rate may sample a signal with a 10 kHz bandwidth, so the signal is sampled at 2.56 MHz in this case. The digitized signal is decimated in the conversion process because each sample is used to produce one bit of data. The output is logic 1 if the signal is higher than the previous sample. The output is logic 0 if the signal voltage is lower than before. The binary word is thus built up by adding or subtracting bits until the data word represents the signal level.

Digital Filtering

This book has described analog filtering in some depth, both in terms of the frequency response and in terms of the pole and zero locations. It is somehow easy to imagine (for me, at least) signals flowing in a circuit. I can imagine the potential divider action as the impedance of an inductor increases with frequency while the impedance of a capacitor reduces. In a filter diagram the signal path is usually through a single wire.

Digital filtering is a completely different concept from analog filtering. Digital filtering processes signals in the time domain. Therefore, if a certain frequency domain response is required, it is necessary to convert this response into the equivalent time domain. It is not always intuitively obvious what this time domain signal looks like. What is more, the signals are usually in parallel digital form, in other words, they are a binary-coded version of an analog signal on, perhaps, a 16-bit-wide bus. In some digital filter diagrams it is not obvious that the signal path is a data bus (unless you are familiar with microprocessor diagrams). The diagram in Figure 15.1 provides an illustration of how a sinusoidal signal appears in digital form.

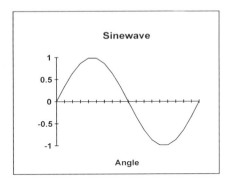

ANGLE	SIN(x)	Two's Complement
0	0	0000000000
20	0.34202	0000010001
40	0.642788	0000100000
60	0.866025	0000101011
80	0.984808	0000110001
100	0.984808	0000110001
120	0.866025	0000101011
140	0.642788	0000100000
160	0.34202	0000010001
180	0	0000000000
200	-0.34202	1111101111
220	-0.64279	1111100000
240	-0.86603	1111010101
260	-0.98481	1111001111
280	-0.98481	1111001111
300	-0.86603	1111010101
320	-0.64279	1111100000
340	-0.34202	1111101111
360	0	0000000000

Figure 15.1

Digitized Sine Wave

Digital Lowpass Filters

Imagine an ideal lowpass filter: the brick wall. It has a flat passband with unity gain, but beyond the cutoff point its gain reduces to zero. This response is not practical, but let's assume that it is, initially, so that we can convert it into a time-domain impulse response. Conversion from the frequency domain into the time domain is achieved using inverse Fourier Transforms. Books on signal processing cover this topic in more detail, but it is only necessary to consider the brick wall response here. Conveniently, a brick-wall frequency response has a sinc (x) impulse response (i.e., sin(x)/x) in the time domain, as illustrated in Figure 15.2.

BRICK WALL RESPONSE

SIN(X)/X

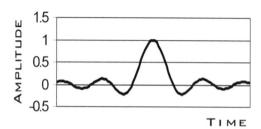

Figure 15.2

The Brick Wall Filter: Time and
Frequency Domains

In Figure 15.2, the frequency axis and the time axis are both scaled by 100 in order to produce smooth graphs of the frequency response, $g(s)$, and the time-domain impulse response, $f(t)$. The actual time domain shown is −20 seconds to +20 seconds. The frequency response is normalized for a lowpass filter with a 1 rad/s cutoff frequency. The passband is shown as being ±1 because the negative and positive frequency domains are symmetrical about zero.

The time domain response is actually $\omega_c.[\text{sinc }(\omega_c t)]/\pi$. When $t = 0$ this has a value of ω_c/π. The response curve crosses through the zero amplitude axis at $t = \pm n\pi/\omega_c$, where $n = 1, 2, 3$, and so on.

In a sampled data system, such as a digital filter, discrete values of the time-domain response amplitude exist. The discrete values are given by $h(n) = \omega_c.[\text{sinc }(\omega_c n)]/\pi$, which can be simplified to:

$$f_l = |nf_s \pm f_c|$$

$$h(n) = \frac{\sin(\omega_c n)}{\pi n}$$

where $n = 1, 2, 3$, and so on, and

$$f_l = |nf_s \pm f_c|$$

$$h(0) = \frac{\omega_c}{\pi}$$

where $n = 0$.

To produce the desired frequency response we must have an impulse response that produces an output before the impulse arrives! This is impossible. The solution is to delay the signal, so that some signal processing takes place before the peak of the impulse response arrives at the output. The longer the delay: the closer we get to the ideal frequency response. Limiting the period during which signal processing takes place is known as truncation. This can lead to rounding of the passband edge and ripples in the stopband.

A 1 rad/s lowpass filter can now be designed using discrete logic. First build a chain of delay elements, usually these are D-type flip-flops clocked by the master clock. Each delay element is 10 bits wide. From the output of each stage take the digitized signal and multiply it by the value of the impulse response that corresponds to that moment in time. An example will help explain this further. Suppose we have 21 delay elements. Delay elements 1 to 10 produce the negative time outputs, delay element 11 corresponds with the zero time output, and delay elements 12 to 21 are the positive time outputs.

The output from delay element 1 will be multiplied by the value of the impulse response at −20 seconds, to give product one. In Figure 15.2 the impulse response value is approximately 0.01453 at −20 seconds, where $t = -2000$ hundredths of a second. The next delay element output will be multiplied by the impulse response at −18 seconds, to give product two. Further delay element outputs are multiplied by the impulse response at −16, −14, −12, −10, −8, −6, −4, −2, 0, 2, 4, 6, 8, 10, 12, 14, 16, 18, and 20 seconds, to give products 3 to 21. All these products must now be added together, to form the filter's output signal.

The hardware we need is as follows: 21 × 10 D-type flip-flops; and 21 multipliers and a summing circuit with 21 inputs. This list omitted the analog-to-digital (A/D) and digital-to-analog (D/A) converters at the input and output. A circuit diagram is given in Figure 15.3. This type of filter is known as a finite impulse response (FIR) filter because, if there are 21 taps, after 21 clock pulses an impulse signal will have passed through and no longer affect the output. In this diagram, "D" represents a delay, "×" represents a multiplication, and "Sum" represents an addition.

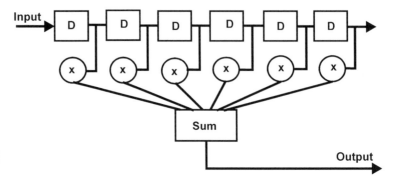

Figure 15.3

Digital Filter Circuit
(FIR)

Do not forget that each line represents a bus of 8, 16, 32, or more parallel lines. So each delay element is actually a number of D-type flip-flops in parallel, all clocked from the same source. Multiplying two 16-bit numbers produces a 32-bit result, so truncation may be required to remove the least significant bits. In practice the filter multiplier coefficients are scaled so that the output resolution is equal to the input resolution.

Often, the filter coefficients are symmetrical. This allows us to design a hardware-reducing configuration where the delayed signal is fed back to halve the number of multipliers required. The circuit is folded around so that the first and last outputs from the delay line are added together and then multiplied by a common coefficient. Extra summing circuits are required, but the output stage adder has only half the number of inputs and therefore is simpler to implement. The folded FIR filter is illustrated in Figure 15.4.

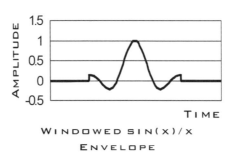

Figure 15.4

Reduced Digital Filter Circuit
(Folded FIR)

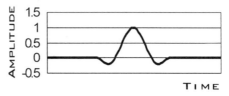

If the folded FIR filter is implemented in a digital signal processor (DSP), it requires far less computational effort than the linear FIR filter. Summing circuits use little processor time, but multiplication requires a number of shift and add operations. Also, reading the filter coefficients from memory takes time. The processor is only required to read half the coefficients in a folded FIR filter.

One advantage of the FIR filter, in either form, is that its output is linear phase. It is linear phase because each input signal passes through all the delay elements, so a slowly changing signal goes through the same processes as a rapidly changing signal; all frequencies are delayed equally. In other words the group delay is constant and is proportional to the number of delay elements in the filter.

Truncation (Applied to FIR Filters)

Truncation was briefly mentioned earlier, when the sinc (×) function was limited to −20 and +20 seconds. This truncation is known as windowing, and a rectangular window was applied in this case. A window is the limit of a time-domain response and is multiplied by the sinc (×) to obtain an overall set of coefficient values for the FIR filter's taps. Another simple window is the triangle, so the side lobes of the sinc (×) function gradually have less effect until zero is reached. This is shown in Figure 15.5.

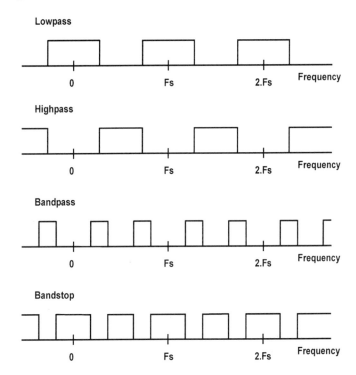

Figure 15.5

Truncation Applied to Time Domain

More sophisticated windows are available. Why are they necessary? Quite simply, truncation distorts the sin(×)/× envelope, which limits the maximum stopband attenuation that can be achieved. The use of a rectangular window, which is simply truncation, limits the maximum achievable attenuation to a little over 20 dB. Using the triangular (Bartlett) window is a little better, because the sin(×)/× envelope is gradually reduced in amplitude on either side of the peak value. The triangular window limits the maximum achievable attenuation to 25 dB. By comparison we have a few other windows listed below:

Hanning	43 dB
Hamming	54 dB
Blackman	75 dB
Harris flat top	97 dB

Clearly the Harris flat top is the best of these windows, so why bother with the others? Well, as the stopband attenuation increases, the width of the transition band between the passband and stopband also increases. So there is clearly a tradeoff between the transition band (skirt) width and the stopband attenuation limit. Also, the windows that achieve the highest levels of attenuation generally require more filter taps, which increases both time delay and filter complexity.

Transforming the Lowpass Response

So far I have described the lowpass filter and its time-domain impulse response. Before I describe highpass, bandpass, and bandstop responses, it is necessary to consider how these may be affected by the sampling operation. Sampling was briefly described earlier with reference to analog-to-digital conversion.

Engineers familiar with radio and analog signal processing techniques will appreciate that sampling performs the same function as mixing. Mixing, or amplitude modulation, multiplies one signal with another and produces a spectrum containing the sum and difference frequencies. The analog mixing process considers a signal of frequency F1 with a signal of frequency F2, which will produce the mixed product F2 + F1 and F2 − F1.

The sampling signal in a digital system has a narrow impulse, which contains many harmonics. The analysis of the mixing process must consider a signal of frequency F1 with an impulse having spectrum frequencies at N times F2, where N = 1, 2, 3, and on to infinity (in theory, for infinitely narrow samples). This will not only produce the mixed product F2 + F1 and F2 − F1 (amplitude modulation), it will also produce signals at frequencies of {(N times F2) + F1} and {(N times F2) − F1}.

The frequency response of digital filters can therefore be mapped to a circle, so as the frequency increases beyond half the sampling rate it forms an alias with the next harmonic. This process continues as the frequency is increased. Opening out the circle gives a repeated pattern of frequency responses across the spectrum, each pattern centered on zero and multiples of the sampling frequency. This pattern is shown in Figure 15.6, for all types of filter.

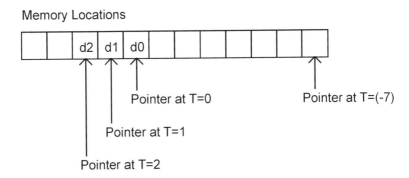

Figure 15.6

Digital Frequency Response

Bandpass FIR Filter

The bandpass filter is effectively a lowpass filter that is frequency shifted. I have shown that the impulse response of a lowpass filter is the sinc (x) function. It seems logical, then, that the impulse response of a bandpass filter should be a sinusoidal signal, with a frequency at the passband center, and which is modulated by a sinc (x) envelope. Modulation is achieved by multiplying the two signals together.

Highpass FIR Filter

Highpass filters are simply lowpass filters with their passband shifted, centered at half the sampling frequency. Like the bandpass filter, the time domain response of such a filter is the product of the sinc (x) envelope multiplied by a sinewave. In this case, however, the sinewave frequency is half the sampling clock frequency.

Bandstop FIR Filter

A bandstop filter is the most difficult to understand in the time domain. The output is the sum of two responses; one being the lowpass response, the other being the highpass response. In each case the sinc (×) envelope will depend on the passband width.

DSP Implementation of an FIR Filter

Although it would be possible to implement the filter as shown in previous figures, by shifting the data using shift registers or flip-flops, there is another method more appropriate to a processor based system. Anyone who has written a "C" program will know that pointers are used to enable faster operation. Pointers are memory locations that contain the address of another memory location. An example will explain this concept.

Suppose we want a 20-step shift register; this can be implemented by having a 20-address memory. The "input" can be an address set into one pointer, and the "output" can be an address set into another pointer. Each time there is a new data word it is stored at the address pointed to by the "input" pointer. If the pointer is decremented after each read operation this is equivalent to the data in the whole register moving to a higher address, relative to the pointer. This concept is illustrated in Figure 15.7.

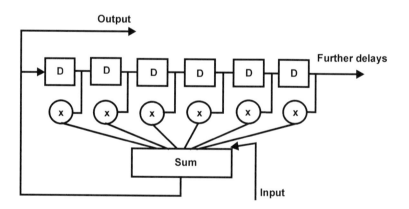

Figure 15.7

Pointer Operation

When the pointer reaches the end of the address range it is reset to start at the beginning. By this means we have the equivalent of a shift register; the old data is overwritten instead of being shifted out.

All the memory locations need to be read and multiplied by their respective coefficients. The problem is that the beginning of the shift register keeps moving to a lower address. However, since we know that the input address pointer is pointing to the next input, the first output address is the input address plus one (unless we are at the end of the memory space). All we have to do is continuously increment the output pointer, remembering to loop back to the bottom of the address range once we have read from the highest address, and multiply the value stored in that address by the appropriate coefficient. All the coefficient multiplication operations have to be carried out before the next input sample is stored.

Introduction to the Infinite Response Filter

The infinite response filter (IIR) uses a feedback loop, so the output at one clock period somehow affects the output during the next period. This will have some sort of exponential effect, so that each output has a smaller effect on the next output (otherwise the output would be unstable, if you think about it).

The input of the IIR filter is fed into an adder and the output of the adder provides the filter output, as with the FIR filter. However, in the case of the IIR filter, the adder output also feeds a chain of delay elements. The output of each delay element is multiplied by a filter coefficient and then fed back into the adder. This is shown in Figure 15.8.

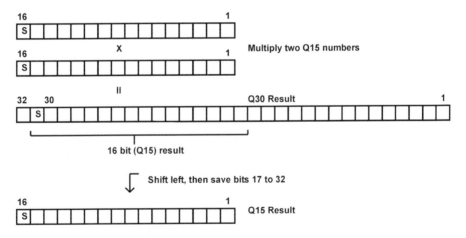

Figure 15.8

An Infinite Impulse Response Filter

The IIR filter is usually designed from the equivalent analog filter. Like the analog filter response it adopts, it does not have a constant group delay. The techniques for converting an analog response into a IIR filter are (1) impulse invariance; (2) step invariance; and (3) bilinear transformation. These techniques are described in Rorabaugh.[1]

Consider a simple filter, comprised of a two-input adder and a single delay element with a multiplier. Let the multiplier coefficient be −0.5. Table 15.1 shows how the filter responds.

SAMPLE	INPUT A	OUTPUT A	INPUT B	OUTPUT B
0	0	−0.37	0	−0.3
1	0.31292	0.12792	0.95600	0.80600
2	0.59441	0.65837	0.56087	0.96388
3	0.81620	1.14538	−0.6269	−0.1450
4	0.95600	1.52870	−0.9287	−1.0012
5	0.99978	1.76414	0.08209	−0.4185
6	0.94314	1.82521	0.97686	0.76760
7	0.79177	1.70438	0.49101	0.87482
8	0.56087	1.41306	−0.6887	−0.2513
9	0.27364	0.98017	−0.8951	−1.0208
10	−0.0410	0.44900	0.16363	−0.3467
11	−0.3516	−0.1271	0.99112	0.81773
12	−0.6269	−0.6905	0.41784	0.8267
13	−0.8392	−1.1845	−0.7459	−0.3326
14	−0.9672	−1.5595	−0.8555	−1.0218
15	−0.9981	−1.7778	0.24407	−0.2668
16	−0.9287	−1.8176	0.99869	0.86527

Table 15.1

The Response of an IIR Filter

The response for a sinusoidal signal with an angular change of $1/\pi$ radians per sample is given by output A. The response of a sinusoidal signal having an angular change of $4/\pi$ radians per sample is given by output B.

DSP Mathematics

Digital signal processing will dominate over analog techniques with the introduction of low-cost DSP chips and kits. Kits are already available at low cost. These have a processor board with input and output interfaces, and assembler and debugger software that enable code to be tested. Texas Instruments produces a kit called the DSK, which contains a TMS320C50 DSP device and an analog interface circuit. Analog Devices produces the EZ-KIT, which contains the AD2101 DSP and the EZ-KIT Lite, which contains the AD2115. Motorola has also joined in the low-cost introduction market with the DSP56002EVM, this contains a DSP56L002, which is a 24-bit fixed-point processor.

Many DSP specific programs such as FIR filters and Fast Fourier Transforms need multiplication and addition, and possibly subtraction, so before the budding DSP engineer can have programs running he should understand the basics of signal-processing mathematics.

Many devices (particularly low-cost ones) use fixed point numbers, and the mathematics requires some thought. Fixed point DSP devices that have 16-bit busses usually use the notation of Q15: the Q represents the sign and the 15 represents the number of binary characters after the decimal point; thus in this notation all numbers have a magnitude of less than one.

Binary and Hexadecimal

Each data sample, or filter coefficient, is a binary word of 16 or more bits. A long series of ones and zeroes is difficult to describe. In the computing world, multiples of 4 bits are grouped together. Each group is then converted into a hexadecimal number and thus represents a decimal number between 0 and 16. To signify a hexadecimal number a lower case h is appended. Decimal numbers 0 to 9 are the same in hexadecimal, but decimal numbers 10 to 16 are represented by A to F in hexadecimal. For the whole range, binary 0000 becomes 0h and binary 1111 becomes Fh. Many processors have a 16-bit bus; in 16-bit format 1100,1011,0110,1101 becomes CB6Dh.

How do we handle negative numbers? For these there is a different system where the most significant bit represents the sign. The sign bit is a 0 for positive numbers and a 1 for negative numbers. In 16-bit hexadecimal, 7FFFh is the highest positive number (which is 32767 in decimal). Numbers 8000h to FFFFh are negative, with a maximum value of −32768. It would be possible for us to simply change the most significant bit, depending on whether the number is positive or negative. Unfortunately this makes the mathematics difficult, so an alternative method is used where all the bits are inverted and then one is added. This system is known as two's complement. To see where the name comes from, and to explain further, I will have to use binary notation.

Two's Complement

The complement of a binary number is found by inverting each bit in turn. This is known as one's complement:

Binary Number	1001,0101,1101,1011
One's Complement	0110,1010,0010,0100

A two's-complement number is found by taking the one's complement and then adding a binary one to it. In simple terms this can be remembered as one plus one equals two. This is a bit corny but "one's-complement, plus one, equals two's-complement."

Binary Number	0111,0111,0001,1010	(771Ah)
One's Complement	1000,1000,1110,0101	(88E5h)
+1	0000,0000,0000,0001	
Two's Complement	1000,1000,1110,0110	(88E6h)

The two's complement number 88E6h is therefore the negative signed equivalent of 771Ah.

We can transform a hexadecimal number directly into a number with the same magnitude but with a negative sign. This can be done without having to convert it into binary first. Transformation is done by subtracting the hexadecimal number from FFFFh and then adding one:

	FFFF
Subtract Number	771A
Result	88E5
Add 0001h	88E6

The result is a number with a magnitude equal to 771Ah but with a negative sign. Alternatively, subtract the hexadecimal number from 1,0000h, or 2^n where n is the number of bits in the original number. In this case $n = 4$.

Strangely, these techniques also work when going from a negative number to find its positive magnitude. Logic indicates that a 1h should be subtracted instead of added in this case. However, by using the same numbers, it can be easily shown that an addition of 1h is still required. The inversion process works whether the number being inverted is positive or negative. The technique should perhaps be thought of as a sign-changing process.

	FFFF
Subtract Number	88E6
Result	7719
Add 0001h	771A

The result is the magnitude of the negative number. I will be using this technique in examples given later.

Now the fundamentals have been explained, we must examine some arithmetic.

Adding Two's Complement Numbers

Addition is the most basic of processes. Samples can be averaged by adding them together and then dividing by the number of samples. Filtering processes need addition (as well as multiplication).

Adding two's complement numbers is as simple as adding decimal numbers; the advantage is that the signs are automatically taken care of. A simple equation using two positive numbers will be given first.

Number	0111,0011,1011,1011	(73BBh or 29627)
Add	0011,1010,1001,1010	(3A9Ah or 15002)
Result	1010,1110,0101,0101	(AE55h or −51ABh or −20907)

So, what went wrong? The problem here was not taking into account the size of the numbers. The maximum number that can be used in 16-bit two's complement arithmetic is 32767, or $2^{15} - 1$. The addition of decimal 29627 and 15002 results in 44629, which is beyond the allowed range.

Adding a negative and a positive number together gives the correct result. This is shown in the following example.

Number		0111,0011,1011,1011	(73BBh or 29627)
Add		1011,1010,0001,1000	(BA18h or −45E8h or −17896)
Result	(1),	0010,1101,1101,0011	(2DD3h or 11731)

Notice that there has been a carry, which is ignored.

In DSPs that use Q15 format numbers, the accumulator has a Q30 structure. A Q15 number must be converted into a Q30 number before addition. This can be done by shifting the number left by 15 places in the 32-bit accumulator. This means that, after addition, the two most significant bits comprise the sign and carry information. The accumulator's result is then shifted left by one bit, so the carry falls off the end. The most significant bit in the accumulator is now the sign bit, and the upper half of the accumulator can be stored in a 16-bit register.

Subtracting Two's Complement Numbers

The principle of a carry still applies (where 1 is being subtracted from 0). Consider a straightforward equation with a smaller number subtracted from a larger one.

Number	0011,1110,1101,1001	(3ED9h or 16089)
Subtract	0010,0101,1101,0101	(25D5h or 9685)
Result	0001,1001,0000,0100	(1904h or 6404)

Now consider what happens when the number being subtracted from is negative.

Number	1011,1110,1111,1000	(BEF8h or −4108h or −16648)
Subtract	0010,0101,1101,0101	(25D5h or 9685)
Result	1001,1001,0010,0011	(9923h or −66DDh or −26333)

Two's complement numbers are well behaved in both addition and subtraction operations. The exception to this is when the number range is exceeded.

Multiplication

Consider a multiplication of two simple binary numbers.

Number #1	1011	(Bh or 11)
Number #2	0111	(7h or 7)
Product	1011	
	+ 1,0110	
	+ 10,1100	
Result	0100,1101	(4Dh or 77)

So binary multiplication gives us the correct result, but in two's complement notation there are some complications. Multiplying two positive numbers gives the correct result. However, the product of a positive and negative number, or the product of two negative numbers, gives the wrong answer.

As an example, I will multiply two positive numbers $3Fh \times 55h = 14EBh$ (or $63 \times 85 = 5355$). This is shown by a series of binary additions.

	0011,1111	(3Fh)
×	0101,0101	(55h)

By repeated additions we have:

	0011,1111	$(1 \times 3Fh)$
+	1111,1100	$(100h \times 3Fh)$
=	1,0011,1011	
+	11,1111,0000	$(1,0000h \times 3Fh)$
=	101,0010,1011	
+	1111,1100,0000	$(100,0000h \times 3Fh)$
=	1,0100,1110,1011	
=	14EBh	(This is correct)

Repeating the process with one negative number results in a wrong answer. For example, multiplying 99h × 3Dh gives an answer of 2475h, which is wrong. The correct answer should be 99h × 3Dh = E775h. This is explained by the following sequence of calculations. $-103 \times 61 = -6283$ or $-188Bh$. Converting $-188Bh$ into a two's complement form we have FFFFh − 188Bh + 1h = E775h, as required.

How are negative numbers handled? DSP devices perform two's complement multiplication by sign-extending negative numbers. Sign-extending means filling higher-order bits with logic 1 s. I will use the two numbers given in the previous example, 99h and 3Dh. Taking 99h and sign-extending it gives: 1111,1111,1001,1001 in binary. In hexadecimal this is FF99h. Multiplying FF99h × 3Dh results in an answer in the 32-bit accumulator that is 3CE775h. Since this number extends beyond the required 16-bit number format, the eight most significant bits (= 3C) fall off the end. The answer is then E775h, as expected.

If both numbers are negative, multiplying them together produces a positive answer. However, simply multiplying two negative two's-complement numbers together does not give the right answer. Neither does sign-extending both numbers before multiplying.

For example, suppose two simple negative numbers are multiplied: $-5 \times -4 = 20$. Let #1 = -5 and #2 = -4. The two's complement of these are #1′ and #2′, respectively. If one number #2′ is sign-extended and then $2^n \times$ #2 is added, the correct result is obtained. Similarly, #1′ could have been sign-extended and then $2^n \times$ #1 would need to be added.

Expressing these as 4-bit two's complement numbers with an 8-bit accumulator:

| #1′ | 1011 | (−5 or Bh) |
| #2′ | 1100 | (−4 or Ch) |

Sign-extend #2′ and multiply 1111,1100

$$\times \quad 1011$$
$$1111,1100$$
$$+ \ 1,1111,1000$$
$$= \ 10,1111,0100$$
$$+ \ 111,1110,0000$$
$$= \ 1010,1101,0100$$

The answer is AD4h, or to 8 bits it is D4h, which is wrong. The correct result is 14h. A correction factor must be added to the previous result in order to obtain the right answer. The correction factor is the positive value of #2 multiplied by 2^n. In this case #2 is 4h (not two's complement of #2) and $n = 4$ (the number of bits in #2). So the correction factor is 4h multiplied by 2^4 (or 10h), which is 40h.

Previous answer =	1010,1101,0100	(AD4h)
+	0100,0000	(40h)
=	1011,0001,0100	(B14h)

The last 8 bits are what was required (14h).

Multiplying two's-complement numbers is made simple using a DSP, because the internal processes do all this binary number juggling. Two Q15 numbers produce a Q30 result. To use this result, the number has to be converted back into Q15 format. The way this is done is to shift the number left by one bit; the upper 16 bits are then equivalent to a Q15 number, which can then be stored. Since this is a common requirement for DSP functions, many devices allow the upper and lower halves of the accumulator to be stored separately. This is illustrated in Figure 15.9.

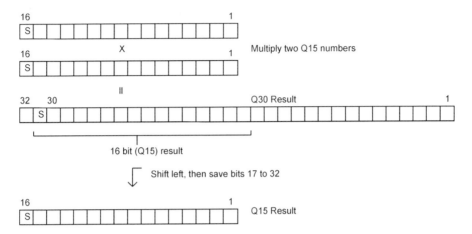

Figure 15.9

Multiplication of Q15 Numbers

Division

Division is usually accomplished by repeated subtraction. It is not required very often in signal processing, except during result scaling. Scaling is usually done by a left or right shift of the data, thus multiplying or dividing by a power of two.

Signal Handling

In a fixed-point processor the maximum value is 32767, or 7FFF in hexadecimal (normally written as 7FFFh). Negative values are given by a two's complement number where the most significant bit is a logic 1. For example, 8000h is −32768. Decimal numbers can be multiplied by 2 to the power of 15 to find their hexadecimal equivalent. A filter coefficient value of 0.2 will thus be $32767 \times 0.2 = 6553$ or 1999h.

There are two types of memory used in a DSP; program memory and data memory. The program memory address is abbreviated to pma and the data memory address is abbreviated to dma. The advantage of having separate memory is that separate busses are used within the processor to access the data. This enables multiplication operations to be carried out quickly by reading in two sets of data simultaneously. The disadvantage is that data in program space cannot be modified during the running of a program. Program space is suitable for storing fixed FIR filter coefficients, but not digitized signal data.

Two instructions perform the FIR filter routine (for a Texas Instruments TMS320C52). Program memory FF00h to FF10h has been allocated for storing filter coefficients. Data memory 300h to 311h has been reserved for signal data (input from an analog to digital converter).

```
RPTZ        #16
MACD        0FF00h,*–.
```

The RPTZ #16 zeroes the accumulator and then causes the instruction that follows to be repeated 17 times (do once, then repeat 16 times). MACD {pma, dma} is a multiply and data move instruction. Before this instruction is run an auxiliary register must be set as the data pointer; in this case we have used AR3, because we ran the instruction MAR *,AR3. The pma is incremented each time the instruction is repeated, so it starts at FF00h and ends up at FF10h. The *– that replaces the dma is an instruction to use the data pointed to by the auxiliary register, during the multiplication, and then decrement the address. Thus the initial address in AR3 has to be the highest used during multiplication (310h), which is then decremented step by step to 300h for the last multiplication.

The address FF00h was used as the pma and was located in memory block 0. The 17 FIR filter coefficients were stored here, starting at address FF00h. The dma was in block 1; note that in order to use block 1 as data memory, rather than program memory, the CNF bit must be set. Eighteen 16-bit words were reserved when the program was assembled (.space 120h), beginning at address 300h. Seventeen addresses were used to store a history of previous input data, and these were multiplied by the 17 coefficients.

In the program, each MACD instruction moved previous samples of input data to the next highest address. Seventeen previous input samples were needed to multiply by the filter coefficients. However, 18 spaces were reserved in the program for input data samples (in assembler this is denoted by .space 120h). This was because data in the highest address used by the instruction MACD (310h) was moved to the next highest address (311h). Data in address 311h did not take part in any further processes; this address is a waste bin!

After a repeated multiplication the accumulator and the product register must be added, to add the last result to the previous ones. Finally, the data must be shifted left by one bit to put the upper half of the accumulator in Q15 format, instead of Q30, and then stored. This final action is a single instruction. In the 'C50 DSP these instructions are:

```
APAC              ; add product register and accumulator
SACH    REG,1     ; shift left one bit and store result in register REG.
```

So, Why Use a Digital Filter?

Some advantages of a digital filter are: (1) reproducible response; (2) not temperature sensitive; (3) programmable. In a few cases the signal may be digitally processed in some way, then filtered, then digitally processed some more; digital filtering is the most obvious technique to use. Digital filters can have a sharp cutoff in the frequency domain, combined with linear phase in the time domain.

Some disadvantages of a digital filter are: (1) unable to pass power; (2) sampling effects; (3) requires a power supply; and (4) frequency range limitations.

The advantages are somewhat obvious: for example, logic levels are unaffected by temperature. The disadvantages are perhaps more subtle. Many passive filter designs are used to remove noise from power supplies—no one would consider using active analog or digital filters for this. Sampling effects include aliasing, where out-of-band signals are frequency shifted and appear in the passband. A digital filter, like an active analog filter, requires a power supply. In applications where low power consumption is important, passive filters are often used. The frequency range of the filter will depend upon the resolution needed to give us a high dynamic range, for example, whether we need 8-bit or 16-bit performance.

Passive filters are usually used at radio frequencies, although the need for them is decreasing. High-speed analog-to-digital converters are being produced with ever increasing sampling rates. These can take their input signals direct from the RF stage of a radio receiver, without the need for further filtering and demodulation stages.

Reference

1.   Rorabaugh, C. Britton. *Digital Filter Designer's Handbook*. New York: McGraw-Hill, 1993.

Exercises

15.1   What are the basic components of a digital filter?

15.2   Is the response of a digital filter defined in the time or frequency domain? How does this compare with the definition of an analog filter?

15.3   What is the fundamental difference in architecture between that of an IIR filter and an FIR filter?

CHAPTER 16

DIGITAL FIR FILTER DESIGN

Digital FIR filters were briefly introduced in Chapter 15. This chapter builds on that introduction. In particular, windows to shape the digital filter's frequency response are described in more detail, with equations for all the popular types.

You may recall from Chapter 15 that a digital filter works by processing a signal in the time domain. The time domain representation of a "brick wall" filter has a sinc(x) function. Multiplying a digitized signal by the sinc(x) function produces a filtering effect. Unfortunately, the sinc(x) function extends to infinity, so it is truncated (cut short) in a practical filter, and this limits the number of taps that are needed. To prevent a sudden change in the time-domain response, a window is used to gradually reduce the amplitude of the sinc(x) function at its limits. The window changes the amplitude of some of the filter tap coefficients and results in a nonperfect frequency domain response, but at least it is practical.

An FIR filter comprises an array of delay elements connected in series. A tap is taken after each element, and, at any sample instance, the value of the sample is multiplied by a filter coefficient. Thus a multiplier is needed for each delay element. Finally, the outputs of all the multipliers are added together to give the output.

The number of taps is given by N, but there are N-1 delay elements; the term N-1 is sometimes referred to as the filter order. It is common to use an odd number of taps, which results in an even number of delay elements. An example of a 7-tap FIR filter, which has an order of 6, is given in Figure 16.1.

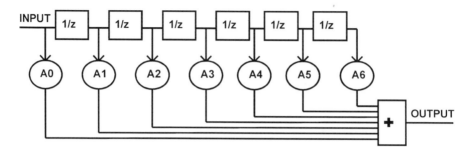

Figure 16.1

FIR Filter Design

Often, the filter coefficients are symmetrical. This allows us to design a hardware-reducing configuration where the delayed signal is fed back to halve the number of multipliers required. The circuit is folded around so that the first and last outputs from the delay line are added together and then multiplied by a common coefficient. Extra summing circuits are required, but the output stage adder has only half the number of inputs and therefore is simpler to implement. The folded FIR filter is illustrated in Figure 16.2.

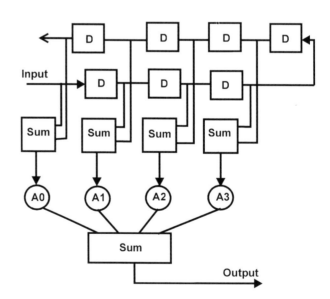

Figure 16.2

Reduced Digital Filter Circuit
(Folded FIR)

Suppose six delay elements are used in a folded FIR filter. The multiplying coefficient A4 has the same value as A2. Similarly, A5 has the same value as A1, and A6 has the same value as A0. This symmetry can be used to reduce the number of multiplier required: the signals from the output of delay elements 2 and 4 can be added before being multiplied by A2. The signals from the output of delay elements 1 and 5 can be added before being multiplied by A1. Finally, the signals at the input and from the output of delay element 6 can be added before being multiplied by A0. For the cost of three adders, multipliers A4, A5, and A6 can be removed.

If the folded FIR filter is implemented in a digital signal processor (DSP), it requires far less computational effort than the linear FIR filter. Summing circuits use little processor time, but multiplication requires a number of shift and add operations. Also, reading the filter coefficients from memory takes time. The processor is only required to read half the coefficients in a folded FIR filter.

In an FIR filter the delay to all signals is the same and does not depend upon the signal frequency, therefore the group delay is constant. This is important for filters handling impulsive signals because impulses contain a wide band of frequencies; if the group delay is not constant, so that some frequencies are delayed more than others, the impulse will have ringing superimposed on its waveform. This is an undesirable distortion of the signal. On the other hand, basic speech transmission is largely unaffected by group delay variations; for these applications IIR filters are more efficient.

The cutoff frequency (Fc) of an FIR filter is directly proportional to the data-sampling clock frequency. Using a single set of coefficients, the cutoff frequency can be doubled by doubling the sampling clock frequency. The normalized clock frequency for a digital filter is 1 Hz or 2π rad/s.

The sinc function passes through zero at multiples of 1/Fc, so a 0.25 Hz lowpass filter will have zero value coefficients at multiples of ±4 taps from the center value. For an odd-order filter these zero values will coincide exactly at the sample period, so the corresponding filter coefficients will be zero. If this particular filter were even-order there would not be any coefficients with a value of zero. This is because the center of the sinc function is midway between samples, and therefore the zeroes occur at points midway between filter taps. This is shown in Figure 16.3.

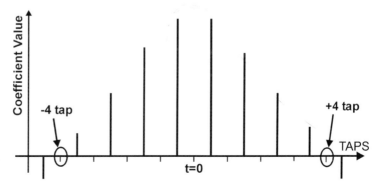

Figure 16.3

FIR Filter Coefficients
(Even N)

Frequency versus Time-Domain Responses

The following subsections provide coefficient values for lowpass, highpass, bandpass, and bandstop filters. In each case the central coefficient value h[0] is given separately and is derived using L'Hopital's rule (see Appendix). The coefficients h[n] apply to all nonzero values of n.

Denormalized Lowpass Response Coefficients

As discussed previously, in Chapter 15, the lowpass frequency domain response becomes a sinc(x) function in the time domain. Denormalization to give a particular lowpass response is quite simple. The normalized response has a sampling rate of 1 Hz (2π radians per second), so the cutoff frequency is relative to this (cutoff at ω_c) and the value of ω_c is given by the equation:

$$\omega_c = \frac{F_c}{F_s}$$

The relationship between sampling frequency and the filter cutoff frequency for lowpass filters is shown in Figure 16.4.

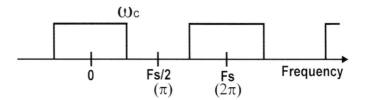

Figure 16.4

Sampled Lowpass
Frequency Response

For example, let $F_c = 3.4\,\text{kHz}$ and $F_s = 16\,\text{kHz}$. The value of $\omega_c = 3.4/16 = 0.2125$.

The value of the central coefficient is given by:

$$h[0] = \frac{\omega_c}{\pi}$$

The values of the other coefficients are given by:

$$h[n] = \frac{\sin(\omega_c n)}{\pi n}$$

Using the example value of $\omega_c = 0.2125$ in the above equations gives a set of coefficient values, which are: h[0] = 0.06764, h[1] = 0.067133, h[2] = 0.065623, ..., and so on.

Denormalized Highpass Response Coefficients

The highpass frequency domain response becomes a negative sinc(x) function in the time domain. Denormalization to give a particular highpass response is a similar process as the one just described for lowpass response denormalization. The normalized response has a sampling rate of 1 Hz (2π radians per second), so the cutoff frequency is relative to this (cutoff at ω_c); the value of ω_c is given by the equation:

$$\omega_c = \frac{F_c}{F_s}$$

The relationship between sampling frequency and the filter cutoff frequency for highpass filters is shown in Figure 16.5.

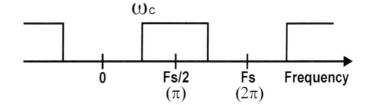

Figure 16.5

Sampled Highpass
Frequency Response

For example, let $F_c = 4\,\text{kHz}$ and $F_s = 8\,\text{kHz}$. The value of $\omega_c = 4/8 = 0.5$.

The value of the central coefficient is given by:

$$h[0] = 1 - \frac{\omega_c}{\pi}$$

The values of the other coefficients are given by:

$$h[n] = \frac{\sin(\omega_c n)}{\pi n}$$

Using the example value of $\omega_c = 0.5$ in the above equations gives a set of coefficient values, which are: $h[0] = 0.840845$, $h[1] = -0.152606$, $h[2] = -0.133924$, ..., and so on.

Denormalized Bandpass Response Coefficients

The bandpass frequency domain response becomes a modified sinc(x) function in the time domain. Denormalization to give a particular bandpass response requires the lower and upper passband limits (cutoff frequencies) to be specified. The normalized response has a sampling rate of 1 Hz (2π radians per second), so the cutoff frequencies are relative to this (cutoff points at ω_{C1} and ω_{C2}). Cutoff point ω_{C1} is the relative frequency of the lower passband edge. Cutoff point ω_{C2} is the relative frequency of the upper passband edge.

The relationship between sampling frequency and the filter cutoff frequency for bandpass filters is shown in Figure 16.6.

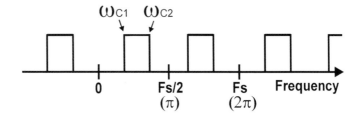

Figure 16.6

Sampled Bandpass
Frequency Response

The values of ω_{C1} and ω_{C2} are given by the equations:

$$\omega_{C1} = \frac{F_{C1}}{F_S} \qquad \omega_{C2} = \frac{F_{C2}}{F_S}$$

For example, let $F_{C1} = 2\,\text{kHz}$, $F_{C2} = 6\,\text{kHz}$, and $F_S = 16\,\text{kHz}$. The value of $\omega_{C1} = 2/16 = 0.125$ and the value of $\omega_{C2} = 6/16 = 0.375$.

The value of the central coefficient is given by:

$$h[0] = \frac{\omega_{C2} - \omega_{C1}}{\pi}$$

The value of the other coefficients is given by:

$$h[n] = \frac{\sin(\omega_{C2}n) - \sin(\omega_{C1}n)}{\pi n}$$

Using the example value of $\omega_{C1} = 0.125$ and $\omega_{C2} = 0.375$ in the above equations gives a set of coefficient values, which are: $h[0] = 0.079577$, $h[1] = 0.076903$, $h[2] = 0.0691106$, ..., and so on.

Denormalized Bandstop Response Coefficients

The bandstop frequency domain response becomes a modified $\text{sinc}(x)$ function in the time domain. Just as the equations for the lowpass coefficients were modified to give highpass coefficients, the bandpass coefficients are modified to give bandstop coefficients. Denormalization to give a particular bandstop response requires the lower and upper passband limits (cutoff frequencies) to be specified. The normalized response has a sampling rate of 1 Hz (2π radians per second), so the cutoff frequencies are relative to this (cutoff points at ω_{C1} and ω_{C2}). Cutoff point ω_{C1} is the relative frequency of the lower stopband edge. Cutoff point ω_{C2} is the relative frequency of the upper stopband edge.

The relationship between sampling frequency and the filter cutoff frequency for bandstop filters is shown in Figure 16.7.

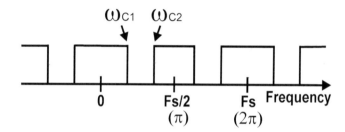

Figure 16.7

Sampled Bandstop
Frequency Response

The values of ω_{C1} and ω_{C2} are given by the equations:

$$\omega_{C1} = \frac{F_{C1}}{F_S} \qquad \omega_{C2} = \frac{F_{C2}}{F_S}$$

For example, let $F_{C1} = 2\,\text{kHz}$, $F_{C2} = 6\,\text{kHz}$, and $F_S = 16\,\text{kHz}$. The value of $\omega_{C1} = 2/16 = 0.125$ and the value of $\omega_{C2} = 6/16 = 0.375$.

The value of the central coefficient is given by:

$$h[0] = 1 - \frac{\omega_{C2} - \omega_{C1}}{\pi}$$

The values of the other coefficients are given by:

$$h[n] = \frac{\sin(\omega_{C1}n) - \sin(\omega_{C2}n)}{\pi n}$$

Note that, relative to the bandpass equations, the two sin functions are reversed. Using the example value of $\omega_{C1} = 0.125$ and $\omega_{C2} = 0.375$ in the above equations gives a set of coefficient values, which are: $h[0] = 1 - 0.079577 = 0.920423$, $h[1] = -0.076903$, $h[2] = -0.0691106$, ..., and so on.

The sinc function envelope to give the lowpass response, and the variants for other responses, extends to plus and minus infinity—a little impractical! Truncating the envelope, by limiting its extent to a certain time limit, causes ripple in the frequency response passband and stopband, and limits the achievable stopband attenuation. Truncation can be applied gradually using specially designed window functions; these reduce the ripple effects and improve the stopband attenuation. Windows are applied by multiplying the window coefficients by the sinc(x) coefficients.

Windows

There are two types of FIR filter design methods. The first uses the relationship between the time and frequency domains and is known as the Fourier Transform method. The Fourier Transform of the "brick wall" frequency response gives the $\sin(x)/x$ time domain response. The second FIR filter design method uses a mathematical process known as the Remez exchange algorithm, which is described later in this chapter. Basically, the filter coefficients are found by varying them until the desired frequency response is produced.

Fourier Method of FIR Filter Design

Windows are designed to truncate the sinc(x) function to a certain number of taps. The simplest is a rectangular window; each coefficient, $h(n)$, has a value of 1 for N taps. The next simplest is the triangular window that has a maximum value for the center value, but tapers down to zero at either side. Windows more complicated than those just described use cosine functions to shape the final frequency domain response. Cosine functions give either greater stopband attenuation or a steeper skirt, as required.

Before describing the window functions, a brief note about the terminology is required. The midway point in an odd number of N taps is referred to as the zero time sample. Signals are then considered to exist in the filter between $-(N-1)/2$ and $+(N-1)/2$ sample periods, relative to the zero time sample. The reason for this is that the ideal frequency domain "brick wall" becomes a sinc(x) function in the time domain, centered on zero and with a response from minus infinity to plus infinity. The FIR filter approximates to a sinc(x) function but with a truncated time domain, so the central sinc(x) coefficient in the filter is still referred to as "zero time." In practice, all signals are delayed by $(N-1)/2$, so that the "zero time" occurs at the $(N-1)/2$ sample time; that is, with a sample every 1 ms through a 21-tap filter, the "zero time" is at 10 ms.

A delay has to be introduced in order to make the filter realizable, since negative time is not allowed! Most windows are symmetrical, thus all that is necessary is to arrange the equations so those tap coefficients up to the "zero time" sample are calculated. For an odd number of taps, these coefficients are $h(0)$ to $h([N-1]/2)$, where $h(0)$ is the first sample and $h(N-1)$ is the last. The remaining tap coefficients can be equated to corresponding values either side of the "zero time" sample, thus: $h(0) = h(N-1)$; $h(1) = h(N-2)$; $h(2) = h(N-3)$; and so forth.

Filters with an odd number of taps have the "zero time" coefficient occurring at $h([N-1]/2)$. On either side of this tap we have $h([N-1]/2-1) = h([N-1]/2 + 1)$. For example, if $N = 21$, the "zero time" coefficient is $h(10)$. On either side of this, the coefficient of $h(9)$ equals that of $h(11)$, the coefficient of $h(8)$ equals $h(12)$, and so on.

For filter with an even number of taps, there is no tap at $h([N-1]/2)$, nor is there a "zero time" coefficient. However, there are taps symmetrically on either side of this point, at $h([N-2]/2)$ and $h([N-2]/2 + 1)$. For example, if $N = 20$, the coefficient of $h(9)$ equals that of $h(10)$. "Zero time" is midway between $h(9)$ and $h(10)$.

Window Types

1 Rectangular Window

The rectangular window has a value of unity over the whole of its length. The $\mathrm{sinc}(x)$ function is used for the filter coefficients, but outside the window they are set to zero.

$$h(n) = 1.0, \quad n = -(N-1)/2, \ldots, -1, 0, 1, 2, \ldots (N-1)/2.$$

Using the rectangular window, the first side-lobe stopband attenuation is limited to 13.2 dB, increasing by 6 dB per octave at higher frequencies.

2 Triangular (Bartlett) Window

The triangular window has coefficients that decrease linearly on either side of the zero time value. The first side-lobe stopband attenuation is limited to 27 dB for this window, which increases by 12 dB per octave for higher frequencies. One way of calculating the tap coefficients is to simply scale the values so that they end up at zero:

$$h(n) = 1.0 - |n|/(N-1)/2,$$
$$n = -(N-1)/2, -(N-3)/2, \ldots -1, 0, 1, \ldots (N-3)/2, (N-1)/2.$$

The value of $h(n)$ falls by 0.1 per tap, either side of the zero time coefficient (which has a value of 1). At the window edge, when $n = -(N-1)/2$ and $(N-1)/2$, the window coefficient is equal to zero. This is a waste of computing power since the first and last filter taps are having no effect. In other words, the filter length has effectively been reduced by a factor of 2. For example, a 21-tap filter will only have 19 nonzero coefficients.

A better algorithm assumes that the window is 2 taps longer than the number of taps actually available. The zero-valued coefficients are then placed outside the array of tap multipliers; that is, they are not used. Thus all taps have non-zero value multipliers and contribute to the filter.

$$h(n) = 1.0 - |n|/(N+1)/2,$$
$$n = -(N-1)/2, -(N-3)/2, \ldots -1, 0, 1, \ldots (N-3)/2, (N-1)/2.$$

Now, when $n = -(N-1)/2$ and $(N-1)/2$, the window coefficient is equal to:

$$h(n) = 1.0 - (N-1)/(N+1), \quad \text{when } n = -(N-1)/2 \text{ and } (N-1)/2$$

In a 21-tap filter the final tap's coefficient is $h(n) = 1.0 - 20/22 = 1/11$. Thus, using the revised equation, the coefficients reduce by 1/11 per tap on either side of the zero time coefficient.

3 Von Hann (Raised Cosine) Window

The Von Hann window is sometimes known as the Raised Cosine window because its values are calculated from a cosine raised to the power two. It is derived from a simple expression:

$$h(n) = \cos^2\left[\frac{n\pi}{N}\right], \text{ alternatively this is given as:}$$

$$h(n) = 0.5 + 0.5\cos\left[\frac{2\pi.n}{N}\right],$$
$$\text{where } n = -(N-1)/2, \ldots -1, 0, 1, \ldots (N-3)/2, (N-1)/2.$$

When time-shifted, so that the window edges occur at $n = 0$ and $n = N - 1$, the second half of the equation changes sign:

$$h(n) = 0.5 - 0.5\cos\left[\frac{2\pi.(n+1)}{N+1}\right], \quad \text{where } n = 0, 1, \ldots (N-3)/2, (N-1)/2.$$

Notice that it is necessary to add 1 to the value of n in the numerator, so that the end values of the window are not zero. The denominator also has to become $N + 1$, so that $h(n) = 1$ when $n = (N - 1)/2$ (the zero time value).

With this window, the first side-lobe stopband attenuation is 35 dB, increasing by 18 dB per octave at higher frequencies.

4 Hamming Window

Modifying the Von Hann window improves the stopband attenuation, giving up to 43 dB for the first side-lobe. At higher frequencies the attenuation increases by 6 dB per octave.

$$h(n) = 0.54 + 0.46\cos\left[\frac{2\pi \cdot n}{N}\right],$$
where $n = -(N-1)/2, \ldots -1, 0, 1, \ldots, (N-1)/2$.

Note: More accurate values for the constants in this equation are 0.54347826 and 0.45652174, to eight decimal places.

When time-shifted, so that the window edge begins at $n = 0$, the second half of the equation changes sign. It is not necessary to increment n and N (as we did with the Von Hann window) because the window edges are not zero-valued.

$$h(n) = 0.54 - 0.46\cos\left[\frac{2\pi \cdot n}{N}\right], \quad \text{where } n = 0, 1, \ldots, (N-1)/2.$$

5 Blackman and Exact Blackman Windows

The Blackman window requires an additional element in the cosine series:

$$h(n) = 0.42 + 0.5\cos\left[\frac{2\pi \cdot n}{N}\right] + 0.08\cos\left[\frac{2\pi \cdot n}{N}\right],$$
where $n = -(N-1)/2, \ldots -1, 0, 1, \ldots, (N-1)/2$.

Using these values, the first side-lobe is attenuated by 59 dB relative to the main lobe. Higher frequencies are attenuated by 18 dB per octave.

When time-shifting the window, so that the coefficients start at $n = 0$, the central part of the equation changes sign. Also the values of n and N are incremented by 1 to prevent zero values of coefficients at the window edges and to produce a window coefficient of 1 at the center tap:

$$h(n) = 0.42 - 0.5\cos\left[\frac{2\pi \cdot (n+1)}{N+1}\right] + 0.08\cos\left[\frac{4\pi \cdot (n+1)}{N+1}\right],$$
where $n = 1, 0, \ldots, (N-1)/2$.

The Exact Blackman uses the same basic formula as the Blackman window, but with exact values (to eight decimal places) for the multiplying coefficients. A coefficient value of 0.42659071 for the Exact Blackman response replaces the first coefficient value of 0.42 used in the Blackman window. Similarly, 0.49656062 replaces the second coefficient value of 0.5 that was used in Blackman window, and 0.07684867 replaces the last coefficient value of 0.08 that was used by Blackman. Using the Exact Blackman coefficients given, the first side-lobe is attenuated by 69 dB. However, higher-frequency side-lobes have an amplitude above this level, so a stopband attenuation of about 65 dB is achieved at the seventh side-lobe. The attenuation then increases at higher frequencies.

Exact Blackman coefficients do not produce zero-valued coefficients at the window edges. It is not necessary to add 1 to the value of n and N. The value of the window edge coefficient is very small, though: $h(0) = 0.00687876$.

6 Blackman-Harris Window

Harris improved the stopband attenuation of the Blackman window by adjusting the values slightly for the three-term cosine series. The first side-lobe attenuation produced by the three-term series coefficients is 61 dB. At higher frequencies the attenuation is greater.

Harris also produced a four-term series that gave even better stopband attenuation. The four-term series has a first side-lobe attenuation of 74 dB.

(a) Three-term Blackman-Harris coefficients:

$$h(n) = 0.44959 + 0.49364 \cos\left[\frac{2\pi.n}{N}\right] + 0.05677 \cos\left[\frac{2\pi.2n}{N}\right],$$

where $n = -(N-1)/2, \ldots, -1, 0, 1, \ldots, (N-1)/2$.

When time-shifted this equation becomes:

$$h(n) = 0.44959 - 0.49364 \cos\left[\frac{2\pi.n}{N}\right] + 0.05677 \cos\left[\frac{4\pi.n}{N}\right],$$

where $n = 0, 1, \ldots, (N-1)/2$.

(b) Four-term Blackman-Harris coefficients:

$$h(n) = 0.40217 + 0.49703 \cos\left[\frac{2\pi.n}{N}\right] + 0.09892 \cos\left[\frac{2\pi.2n}{N}\right]$$
$$+ 0.00183 \cos\left[\frac{2\pi.3n}{N}\right]$$

where $n = -(N-1)/2, \ldots -1, 0, 1, \ldots, (N-1)/2$.

When time-shifted, this equation becomes:

$$h(n) = 0.40217 - 0.49703 \cos\left[\frac{2\pi.n}{N}\right] + 0.09892 \cos\left[\frac{2\pi.2n}{N}\right]$$
$$- 0.00183 \cos\left[\frac{2\pi.3n}{N}\right]$$

where $n = 0, 1, \ldots, (N-1)/2$.

7 Harris-Nutall Window

Nutall improved the Blackman-Harris coefficients to produce greater stoband attenuation. The three-term series produced a first side-lobe attenuation of 67 dB. The four-term series produced a magnificent 94 dB first side-lobe attenuation.

(a) Three-term Harris-Nutall coefficients:

$$h(n) = 0.42323 + 0.49755\cos\left[\frac{2\pi.n}{N}\right] + 0.07922\cos\left[\frac{2\pi.2n}{N}\right].$$

where $n = -(N-1)/2, \ldots, -1, 0, 1, \ldots, (N-1)/2$.

When time-shifted, this equation becomes:

$$h(n) = 0.42323 - 0.49755\cos\left[\frac{2\pi.n}{N}\right] + 0.07922\cos\left[\frac{2\pi.2n}{N}\right].$$

where $n = 0, 1, \ldots, (N-1)/2$.

(b) Four-term Harris-Nutall coefficients:

$$h(n) = 0.35875 + 0.48829\cos\left[\frac{2\pi.n}{N}\right] + 0.14128\cos\left[\frac{2\pi.2n}{N}\right]$$
$$+ 0.01168\cos\left[\frac{2\pi.3n}{N}\right]$$

where $n = -(N-1)/2, \ldots -1, 0, 1, \ldots, (N-1)/2$.

When time-shifted, the second and fourth terms change sign:

$$h(n) = 0.35875 - 0.48829\cos\left[\frac{2\pi.n}{N}\right] + 0.14128\cos\left[\frac{2\pi.2n}{N}\right]$$
$$- 0.01168\cos\left[\frac{2\pi.3n}{N}\right]$$

where $n = 0, 1, \ldots, (N-1)/2$.

8 Kaiser-Bessel Window

This window is generally known as a Kaiser window. It is not a fixed window; instead, a formula is given in which a factor α can be varied to give different levels of stopband attenuation. The factor α should be between 0 and 4.

The value of equation constants for $h(n)$ are $a(0) = \dfrac{H_1(0)}{c}$ and

$a(m) = \dfrac{2H_1(m)}{c}$, where $m = 1, 2, 3$, and $H_1(m) = \dfrac{\sinh\left(\pi\sqrt{\alpha^2 - m^2}\right)}{\pi\sqrt{\alpha^2 - m^2}}$

$c = H(0) + 2.H(1) + 2.H(2) + 2.H(3)$.

The Kaiser-Bessel window, with $\alpha = 3.0$, produces a first side-lobe attenuation of 70 dB. The coefficients are very similar to those of the four-term Blackman-Harris window:

$$h(n) = 0.40243 + 0.49804\cos\left[\frac{2\pi.n}{N}\right] + 0.09831\cos\left[\frac{2\pi.2n}{N}\right]$$
$$+ 0.00122\cos\left[\frac{2\pi.3n}{N}\right]$$

where $n = -(N-1)/2, \ldots, -1, 0, 1, \ldots, (N-1)/2$.

Once again, when time-shifted, the second and fourth terms change sign:

$$h(n) = 0.40243 - 0.49804\cos\left[\frac{2\pi.n}{N}\right] + 0.09831\cos\left[\frac{2\pi.2n}{N}\right]$$
$$- 0.00122\cos\left[\frac{2\pi.3n}{N}\right]$$

where $n = 0, 1, \ldots, (N-1)/2$.

Summary of Fixed FIR Windows

Table 16.1 gives details of the fixed window filters discussed in this chapter.

WINDOW	(−3 dB) Bandwidth	Attenuation
Rectangular	0.89	−13.2
Bartlett	1.28	−27
Hamming	1.3	−43
Von Hann	1.54	−35
Blackman	1.52	−51
Exact Blackman	1.42	−69
3-term Blackman-Harris	1.56	−61
4-term Blackman-Harris	1.74	−74
3-term Harris-Nutall	1.66	−67
4-term Harris-Nutall	1.9	−92

Table 16.1

Window Bandwidth and Stopband Attenuation

Number of Taps Needed by Fixed Window Functions

1   Find the steepness of the slope between passband and stopband.

The number of taps needed depends on the steepness of the slope between the passband and the stopband. In lowpass and highpass filter designs this is the difference between the passband and the stopband frequencies. In bandpass and bandstop filter designs there are two slopes, one on either side of the passband or stopband. In these designs the smaller of the two values (the steepest slope) should be chosen.

2   Find the filter ratio.

Using the value of the slope obtained by the method outlined above, a ratio can be obtained. This ratio is dependent on whether the response is lowpass, highpass, bandpass, or bandstop. The ratio required for all filters to determine the number of taps is:

ratio = (clock frequency − slope)/slope.

3   Decide on the window.

The number of taps required also depends upon the window function used. A rectangular window requires the least number of taps. In order of increasing number of taps required we have: Von Hann (or Hanning), Hamming, Bartlett (triangular window), and finally Blackman.

4   Calculate the number of taps.

The number of taps required for a rectangular window is:

$N = 1 + (\text{integer})\, 0.95 \times \text{ratio}$.

For example, if the clock frequency is 1 kHz and the filter cutoff frequency is 80 Hz, the ratio = (1000 − 80)/80 = 11.5, then 0.95 × 11.5 = 10.925. The ratio equals (integer) 10

N = 1 + 10 = 11.

In Table 16.2 are empirical formulae for the number of taps required for some basic types of fixed window.

Window Function	Number of Taps, N
Rectangular	1 + (int) 0.95 × ratio
Bartlett (Triangular)	1 + (int) 4.15 × ratio
Von Hann	1 + (int) 3.3 × ratio
Hamming	1 + (int) 3.44 × ratio
Blackman	1 + (int) 6.0 × ratio
Exact Blackman	1 + (int) 6.8 × ratio
3-term Blackman Harris	1 + (int) 6.0 × ratio
4-term Blackman Harris	1 + (int) 5.8 × ratio
3-term Harris Nutall	1 + (int) 6.8 × ratio
4-term Harris Nutall	1 + (int) 6.8 × ratio

Table 16.2

Empirical Formulae for Number of Taps

FIR Filter Design Using the Remez Exchange Algorithm

The second FIR filter design method uses a mathematical process called the Remez exchange algorithm. This sounds complicated, and indeed the algorithm itself is, but the functionality of the algorithm is quite simple in principle. Sample values of the desired frequency response that is selected by the designer are used as a model. The Remez exchange algorithm then tries to generate a set of filter coefficients that will produce the same response as the model. The algorithm is a curve-fitting method that minimizes the error between the model and the filter. It is equi-ripple, in that the final response has equal errors above and below the desired response. The equi-ripple method sometimes fails to find a suitable solution, but it is still useful.

One of the first considerations when designing an FIR filter is the number of taps required to achieve the desired performance. Providing more taps than necessary raises the cost by having to use a higher processor speed or additional processors. The desired performance will not be met if insufficient taps are provided. The rules for determining the number of taps are somewhat empirical but valuable nevertheless.

Number of Taps Needed by Variable Window Functions

Variable windows have coefficient values that are dependent upon the attenuation required. Also, the number of taps required varies with the desired level of attenuation. Kaiser and Dolph-Chebyshev windows are good examples of variable windows.

The number of taps required for a Kaiser window is given by $N = 2M + 1$, where

$$M = \frac{(A_S - 7.95).\pi}{14.36.(\omega_S - \omega_P)}$$

The term A_S is the stopband attenuation in decibels, ω_S is the stopband frequency, and ω_P is the passband frequency. Hence,

$$N = 1 + (\text{int})\frac{(A_S - 7.95)}{14.36.(f_S - f_P)/f_{clock}}$$

f_S is the stopband frequency, f_P is the passband frequency, and f_{clock} is the clock frequency, all in Hertz.

Remember that the equation for M assumes that the sample clock is 1 Hz, or 2π rad/s, so to convert to Hertz the passband and stopband frequencies have to be multiplied by 2π. However, the π term introduced cancels the one in the numerator. The term 2 introduced into the denominator can be cancelled by multiplying M by 2, giving an equation for $2M$; this is exactly what is needed to find N, because $N = 2M + 1$.

FIR Filter Coefficient Calculation

An example of how FIR filter coefficients are calculated is shown by the following exercise. Find the coefficients for a bandpass filter (cutoff at ω_{C1} and ω_{C2}). The sinc(x) function for a bandpass design is given by the equations:

$$h[0] = \frac{\omega_{C2} - \omega_{C1}}{\pi} \qquad h[n] = \frac{\sin(\omega_{C2}n) - \sin(\omega_{C1}n)}{\pi n}$$

These must be multiplied by a Window function in order to obtain the coefficient values. Using the Hann Window:

$$h(n) = 0.5 + 0.5\cos\left[\frac{2\pi.n}{N}\right],$$
where $n = -(N-1)/2, \ldots -1, 0, 1, \ldots (N-3)/2, (N-1)/2.$

For the center tap of the Vonn Hann Window, where $n = 0$, $h[0] = 1$.

$$h(0) = \frac{\omega_{C2} - \omega_{C1}}{\pi} \qquad h[n] = \frac{\sin(\omega_{C2}n) - \sin(\omega_{C1}n)}{\pi n} \cdot \left\{0.5 + 0.5\cos\left[\frac{2\pi n}{N}\right]\right\}$$

When time-shifted, a filter with N taps will have window edges at $n = 0$ and $n = N - 1$. These modified values can be used in the equation. Note that with some mathematical manipulation, the second half of the equation changes sign:

$$h[n] = \frac{\sin(\omega_{C2}n) - \sin(\omega_{C1}n)}{\pi n} \cdot \left\{0.5 - 0.5\cos\left[\frac{2\pi(n+1)}{N+1}\right]\right\}$$
where $n = 0, 1$, through to $(N-3)/2$ and then $(N+1)/2$ through to N.

Consider the midvalue coefficient for a bandpass sinc(x) function:

$$h[0] = \frac{\omega_{C2} - \omega_{C1}}{\pi}$$

The Vonn Hann Window has value $h[0] = 1$, so the sinc(x) value is unchanged after multiplying by the Window value. After time-shifting, where $n = (N-1)/2$, this becomes:

$$h[(N-1)/2] = \frac{\omega_{C2} - \omega_{C1}}{\pi}$$

A similar process can be applied to other Window functions and other frequency responses. That is, use the appropriate sinc(x) function and multiply this by a Window function for each value of n. The Window will be a function of N, the total number of coefficients (taps) required. Time-shifting must then be applied to make the first tap coefficient become $h[0]$ and the last tap coefficient become $h[N-1]$.

A Data-Sampling Rate-Changer

Apart from filtering, it is also possible to use FIR filters to perform a change of the data-sampling rate. Suppose a system is receiving signals from two sources that have different sampling rates and that the system clock is operating at the higher rate. Provided that one of these sampling rates is an integer multiple of the other, it is possible to convert the signal that has the slower sampling rate into one with a higher sampling rate.

An example of where differing sampling rates could occur is a system that is receiving digitized telecommunication signals sampled at 8 kHz and 16 kHz. The 8 kHz sampled signals could be digitized speech in the analog band of 300 Hz to 3.4 kHz. The 16 kHz sampled signals could be wider bandwidth speech (up to 7 kHz). The ratio of the higher sampling rate to the lower sampling rate is two in this case.

Data from the speech channel sampled at 8 kHz is input to the system. Since the system is processing data samples at twice the rate that they are being received, intermediate samples are set to zero. Thus the data sequence is: D1, 0, D2, 0, D3, 0, D4, 0, D5, and so on. This process is equivalent to mixing in radio systems and results in aliases of the original spectrum being produced. Filtering is needed to remove these aliases, which can be within the frequency range of the wider bandwidth speech.

A suitable filter is known as an interpolator. The interpolator is a lowpass filter that has a passband cutoff frequency equal to the highest frequency of the sampled signals. The stopband of the filter must be below the alias frequency produced by the system over-sampling. The zero-valued samples are replaced by an average of the samples on either side. The exact value of the replacement depends upon the frequency of the signals being processed.

References

1.   Thede, Les. *Analog and Digital Filter Design Using C*. New Jersey: Prentice-Hall, 1996.

2.   Sanjit J. Mitra and James F. Kaiser. *Handbook for Digital Signal Processing*. New York: John Wiley & Sons, 1993.

CHAPTER 17

IIR FILTER DESIGN

Infinite impulse response (IIR) filters are more efficient than FIR filters because, for a given frequency response, they require fewer delay elements, adders, and multipliers. The disadvantage of IIR filters is their nonlinear phase response (nonconstant group delay). Group delay has been discussed previously in Chapters 2 and 9 in relation to analog filters: a nonconstant group delay means that not all frequencies experience the same delay. Thus, impulses containing components with a wide range of frequencies will be distorted when passed through an IIR filter.

Most IIR filters are designed using an analog filter model. Analog filter models are the familiar Butterworth, Chebyshev, Cauer (Elliptic), Inverse Chebyshev, and Bessel types. Generally speaking, Bessel models are not converted into digital filters. You may remember from Chapters 2 and 9 that the advantage of a Bessel response in an active or passive linear filter is the constant group delay, at the expense of a poor skirt response (the filter attenuation increases very slowly). FIR filters can produce a constant group delay with far superior skirt response, so they are used where group delay is important.

The linear frequency response formulae $H(\omega)$ can be converted into the digital equivalent using Impulse Invariant, Step Invariant, or Bilinear Transformation. Only the bilinear transform provides a general-purpose conversion function that can be used for lowpass, highpass, bandpass, and bandstop responses. The impulse invariant and step invariant conversion functions are quite difficult to apply and can only be used for lowpass filters (and bandpass with great care); these conversion functions cannot be used with highpass or bandstop responses. For these reasons, only bilinear transforms are considered in this chapter.

The basic IIR filter is based on the biquadratic (biquad) structure, which is shown in Figure 17.1. The delay elements are denoted as $1/z$ in this diagram. The $1/z$ term is sometimes written as z^{-1}, especially in transfer function equations.

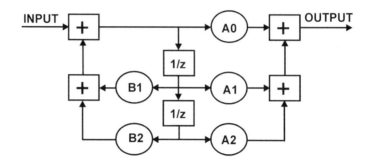

Figure 17.1

Biquad Structure

During the study of analog filters in Chapters 4 to 7, it was shown that an analog biquad filter could perform lowpass, highpass, bandpass, and bandstop functions; this is also true for digital biquad filters. The digital biquad uses four adders, two delays, and four multipliers. The multiplier coefficients are $A0$, $A1$, $A2$, $B1$, and $B2$. These coefficients are calculated during the filter design process. The transfer function of the biquad structure is:

$$H(z) = \frac{Y(z)}{X(z)} = \frac{A0 + A1.z^{-1} + A2.z^{-2}}{1 - B1.z^{-1} - B2.z^{-2}}$$

The feed-forward element $A0$ gives the DC gain and is often unity. There is no feedback element $B0$, which is replaced by a unity-valued element because the signal path through this element is forward, not backward. Note: in some textbooks, the terms AN and BN are interchanged.

High-order filters are designed by cascading biquad stages. Each biquad gives a second-order response, so a fourth-order filter uses two biquad stages in series. Consider the case where an odd-order lowpass filter is required. This requires one or more second-order stages, followed by a first-order stage. A first-order stage is simply a delay and feedback coefficient, as shown in Figure 17.2.

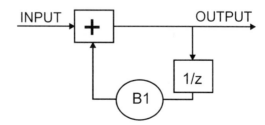

Figure 17.2

First-Order Filter

The first-order section is the same as a second-order section with coefficients $A1$, $A2$, and $B2$ set to zero.

Bilinear Transformation

The bilinear transform is used to convert the analog frequency response into a digital domain response. The advantage of the bilinear transform is that any response, be it lowpass, highpass, bandpass, or bandstop, can be converted. The digital domain is also known as the Z-domain.

The transformation from the analog S-plane into the digital Z-plane is quite simple to visualize. The S-plane frequency (jω) axis is wrapped around onto itself into the Z-plane to form a circle. One side of the circle is the zero frequency point, which is the origin on the S-plane diagram. The other side of the circle is where the +infinity and −infinity points meet. This is shown in Figure 17.3.

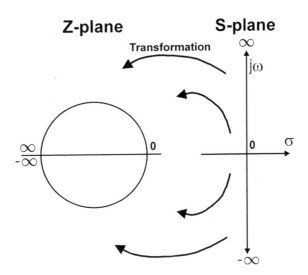

Figure 17.3

S-Plane to Z-Plane Transformation

In the S-plane, a zero on the jω axis becomes a zero on the edge of the unit circle in the Z-plane. Poles in the S-plane should be located to the left of the jω axis for stability; these are then transformed to be inside the unit circle of the Z-plane. Poles in the S-plane to the right of the jω axis indicate instability in an analog filter. In the Z-plane these poles move outside the unit circle and also indicate instability.

The transformation of a first-order analog filter S-plane diagram into a digital Z-plane diagram will be illustrated. In the S-plane, the pole is close to the origin on the negative real axis. After transformation, this pole will appear inside the unit circle of the Z-plane, to the left of the zero frequency point. This is shown in Figure 17.4.

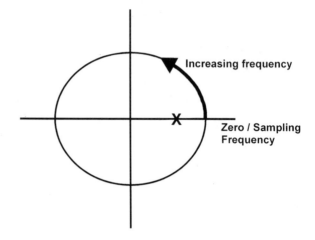

Figure 17.4

First-Order Z-Plane

When a digital filter's poles and zeroes are plotted onto the Z-plane, and the frequency response is calculated, the response repeats itself at multiples of the sampling frequency. Consider the first-order lowpass filter described in Figure 17.4, which has a pole on the real axis, close to +1. The frequency response can be found by moving a reference point around the edge of the unit circle and measuring from this point to the position of the pole. Starting at +1 on the circle, signals are close to zero frequency, and the filter will have an output level determined by the inverse distance from the pole.

As the signal frequency increases, the reference point moves around the unit circle toward the −1 point (± infinite frequency). The distance from the pole to the reference point is at a maximum and, therefore, the output signal amplitude is at a minimum. By moving the reference point further around the unit circle, it begins to approach the point where it started. During this half of the circle the (negative) frequency decreases and approaches zero once again. Thus, the distance from the pole decreases and the amplitude of the signal increases. This pattern repeats itself, as may be seen in Figure 17.5.

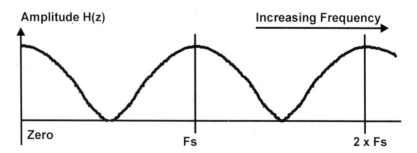

Figure 17.5

First-Order Frequency Response

The bilinear transform is a simple mathematical process. Starting with an analog frequency response, $H(s)$, bilinear transformation to produce $H(z)$ is carried out by substitution of s.

$$s = \left[\frac{z-1}{z+1}\right]$$

To see how this works, let's use a second-order analog (Butterworth) transfer function:

$$H(s) = \frac{1}{s^2 + \sqrt{2}s + 1}$$

Substituting for s gives:

$$H(z) = \frac{1}{\left\{\left[\dfrac{z-1}{z+1}\right]\right\}^2 + \sqrt{2}.\left\{\left[\dfrac{z-1}{z+1}\right]\right\} + 1}$$

This can be simplified by multiplying everything by the highest power denominator, which is $(z + 1)$ squared, or $(z^2 + 2z + 1)$.

The equation then becomes:

$$H(z) = \frac{(z^2 + 2z + 1)}{(z^2 - 2z + 1) + \sqrt{2}.(z^2 - 1) + (z^2 + 2z + 1)}$$

Now z^{-1} is a single clock cycle delay, which can be achieved easily in digital systems. The equation can be restated in terms of delays by multiplying top and bottom by z^{-2}, giving:

$$H(z) = \frac{(1 + 2z^{-1} + z^{-2})}{(1 - 2z^{-1} + z^{-2}) + \sqrt{2}.(1 - z^{-2}) + (1 + 2z^{-1} + z^{-2})}$$

Collecting terms on z^{-1}, z^{-2}, and so on, to give us coefficients for each delay term, this becomes:

$$H(z) = \frac{(1 + 2z^{-1} + z^{-2})}{3.4142 + 0.585786z^{-2}}$$

This equation can be compared to the equation for the biquad that follows:

$$H(z) = \frac{Y(z)}{X(z)} = \frac{A0 + A1.z^{-1} + A2.z^{-2}}{1 - B1.z^{-1} - B2.z^{-2}}$$

The first term in the denominator is required to be 1.0 instead of 3.4142, so all terms in the equation must be divided by 3.4142. Also the last term in the denominator should be subtracted, so B2 must be a negative value. Carrying out these changes gives:

$$H(z) = \frac{0.292894 + 0.5857887.z^{-1} + 0.292894.z^{-2}}{1 - (-0.585786.z^{-2})}$$

Now this means that the coefficient values are $A0 = 0.292894$, $A1 = 0.5857887$, $A2 = 0.292894$, $B1 = 0$, and $B2 = -0.585786$.

Pre-Warping

Unfortunately, the simple bilinear transform approach is an approximation and will not produce the exact frequency response required. If an analog S-plane transfer function is converted into a Z-plane transfer function, as previously shown, the frequency response will be distorted. The relationship between analog and digital responses is given by:

$$\omega_{ana\log} = \tan\left(\frac{\omega_{digital}}{2}\right)$$

If the analog frequency response is distorted prior to applying the bilinear transform, the desired final response can be obtained. This distortion is called pre-warping. To pre-warp an analog response, the following equation should be used:

$$\omega_{ana\log} = \tan\left(\frac{\omega_C}{2}\right)$$

The desired filter cutoff frequency ω_C should be used to give a new analog cutoff frequency $\omega_{ana\log}$. This should be used in the S-plane transfer function before applying the bilinear transform. Thus the cutoff frequency of the normalized lowpass response will be slightly modified. The term ω_C represents the normalized frequency of $2\pi(Fc/Fs)$.

Denormalization

Suppose that the desired response is a cutoff frequency of 3.4 kHz and the sampling clock is 8 kHz. Then $\omega_C = 3.4/8 = 0.425$. When pre-warped this becomes $\tan(2\pi\ 0.215757) = \tan(1.335176878) = 4.1652998$. In the analog transfer function, s can be replaced by $s/4.1652998$ ($= 0.2400788\,s$) before the bilinear transform is applied. If a highpass filter having the same cutoff frequency were required, s would have to be replaced by $4.1652998/s$.

Lowpass Filter Design

Design a second-order IIR lowpass filter with a passband of 3.4 kHz and a sampling clock of 8 kHz using the analog (Butterworth) transfer function:

$$H(s) = \frac{1}{s^2 + \sqrt{2}s + 1}$$

There are several ways to produce this design. Two methods will be described. The first will follow the procedure outlined previously above: pre-warp the analog equation and then use the bilinear transform to produce the coefficients. The second method is more complex but produces the design in a single step.

Design Method 1

The analog frequency response must be pre-warped using the following equation:

$$\omega_{analog} = \tan\left(\frac{\omega_C}{2}\right)$$

The desired filter cutoff frequency ω_C should be used to give a new analog cutoff frequency ω_{analog}. The term ω_C represents the normalized frequency of $2\pi(Fc/Fs)$, thus $\omega_C = 6.8\pi/8 = 2.6703538$. When pre-warped, this becomes 4.1652998. In the analog transfer function, s can be replaced by $s/4.1652998$, or $0.2400788\,s$, giving:

$$H(s) = \frac{1}{0.0576378s^2 + 0.3395227s + 1}$$

The bilinear transform can now be carried out by substitution of s.

$$s = \left[\frac{z-1}{z+1}\right]$$

Substituting for s gives:

$$H(z) = \frac{1}{0.0576378.\left\{\left[\dfrac{z-1}{z+1}\right]\right\}^2 + 0.3395227.\left\{\left[\dfrac{z-1}{z+1}\right]\right\} + 1}$$

This can be simplified by multiplying everything by the highest power denominator, which is $(z + 1)$ squared, or $(z^2 + 2z + 1)$.

The equation then becomes:

$$H(z) = \frac{(z^2 + 2z + 1)}{0.0576378.(z^2 - 2z + 1) + 0.3395227.(z^2 - 1) + (z^2 + 2z + 1)}$$

Now z^{-1} is a single clock cycle delay, which can be achieved easily in digital systems. The equation can be restated in terms of delays by multiplying top and bottom by z^{-2}, giving:

$$H(z) = \frac{(1 + 2z^{-1} + z^{-2})}{0.0576378.(1 - 2z^{-1} + z^{-2}) + 0.3395227.(1 - z^{-2}) + (1 + 2z^{-1} + z^{-2})}$$

Collecting terms on z^{-1}, z^{-2}, and so forth, to give us coefficients for each delay term, this becomes:

$$H(z) = \frac{(1 + 2z^{-1} + z^{-2})}{1.3971605 + 1.8847244z^{-1} + 0.7181151z^{-2}}$$

This equation can be compared to the equation for the biquad that follows:

$$H(z) = \frac{Y(z)}{X(z)} = \frac{A0 + A1.z^{-1} + A2.z^{-2}}{1 - B1.z^{-1} - B2.z^{-2}}$$

The first term in the denominator is required to be 1.0 instead of 1.3971605, so all terms in the equation must be divided by 1.3971605. Also the last term in the denominator should be subtracted, so $B2$ must be a negative value. Carrying out these changes gives:

$$H(z) = \frac{0.7157374 + 1.4314748.z^{-1} + 0.7157374.z^{-2}}{1 - (-1.3489677.z^{-1}) - (-0.5139818.z^{-2})}$$

Now this means that the coefficient values are $A0 = 0.7157374$, $A1 = 1.4314748$, $A2 = 0.7157374$, $B1 = -1.3489677$, and $B2 = -0.5139818$. This completes the design process using method 1.

Design Method 2

In the second lowpass filter method, s is replaced by:

$$s = \cot(\omega_D/2).\left[\frac{z-1}{z+1}\right]$$

Now by substituting into the original Butterworth transfer function

$$H(s) = \frac{1}{s^2 + \sqrt{2}s + 1}$$

Using $\omega_D = 2\pi(3.4/8) = 2.6703538$, the function $\cot(\omega_D/2) = \cot(1.33517688) = 1/\tan(1.33517688) = 0.2400788$:

$$s = \cot(\omega_D/2).\left[\frac{z-1}{z+1}\right]$$

$$H(z) = \frac{1}{0.0576378.\left[\dfrac{z-1}{z+1}\right]^2 + 0.3395226.\left[\dfrac{z-1}{z+1}\right] + 1}$$

Comparing to design method 1 and 1, it can be seen that the equation for $H(z)$ using design method 2 is identical (within the error limits of my calculator). The only difference is that the second method is a single step.

Highpass Frequency Scaling

Design a second-order IIR highpass filter with the same characteristics as the lowpass design produced in the previous section. The filter should have a passband edge at 3.4 kHz, a sampling clock of 8 kHz, and should be based on the lowpass analog (Butterworth) transfer function:

$$H(s) = \frac{1}{s^2 + \sqrt{2}s + 1}$$

As in the case of lowpass design, there are several ways to produce a highpass IIR filter design. First I will follow the step-by-step procedure outlined previously: pre-warp the analog equation and then use the bilinear transform to produce the coefficients. The second method is more complex, but produces the design in a single step.

Design Method 1

The analog frequency response must be pre-warped using the following equation:

$$s = \cot(\omega_D/2).\left[\frac{z-1}{z+1}\right]$$

The desired filter cutoff frequency ω_C should be used to give a new analog cutoff frequency $\omega_{ana\,log}$. The term ω_C represents the normalized frequency of $2\pi(Fc/Fs)$, thus $\omega_C = 2\pi(3.4/8) = 2.6703538$. When pre-warped, this becomes $\tan(1.335177) = 4.1652998$. In the analog transfer function, s can be replaced by $4.1652998/s$, which is the inverse of the lowpass case and gives:

$$H(s) = \frac{1}{17.349722/s^2 + 5.8906235/s + 1}$$

The bilinear transform can now be carried out by substitution of $1/s$.

$$s = \left[\frac{z-1}{z+1}\right] \qquad \therefore 1/s = \left[\frac{z+1}{z-1}\right]$$

Substituting for $1/s$ gives:

$$H(z) = \frac{1}{17.349722.\left\{\left[\dfrac{z+1}{z-1}\right]\right\}^2 + 5.8906235.\left\{\left[\dfrac{z+1}{z-1}\right]\right\} + 1}$$

This can be simplified by multiplying everything by the highest power denominator, which is $(z - 1)$ squared, or $(z^2 - 2z + 1)$.

The equation then becomes:

$$H(z) = \frac{(z^2 - 2z + 1)}{17.349722.(z^2 + 2z + 1) + 5.8906235.(z^2 - 1) + (z^2 - 2z + 1)}$$

Now z^{-1} is a single clock cycle delay, which can be achieved easily in digital systems. The equation can be restated in terms of delays by multiplying top and bottom by z^{-2}, giving:

$$H(z) = \frac{(1 - 2z^{-1} + z^{-2})}{17.349722.(1 + 2z^{-1} + z^{-2}) + 5.8906235.(1 - z^{-2}) + (1 - 2z^{-1} + z^{-2})}$$

Collecting terms on z^{-1}, z^{-2}, and so on, to give us coefficients for each delay term, this becomes:

$$H(z) = \frac{(1 - 2z^{-1} + z^{-2})}{24.2403455 + 32.699444z^{-1} + 12.4590985z^{-2}}$$

This equation can be compared to the equation for the biquad that follows:

$$H(z) = \frac{Y(z)}{X(z)} = \frac{A0 + A1.z^{-1} + A2.z^{-2}}{1 - B1.z^{-1} - B2.z^{-2}}$$

The first term in the denominator is required to be 1.0 instead of 24.2403455, so all terms in the equation must be divided by 24.2403455. The $A1$ term in the numerator for the biquad is positive, so the coefficient $A1$ will be negative. Also the $B1$ and $B2$ terms in the denominator should both be negative, so coefficients $B1$ and $B2$ have negative values. Carrying out these changes gives:

$$H(z) = \frac{0.041253537 + (-0.082507074).z^{-1} + 0.041253537.z^{-2}}{1 - (-1.348967736) - (-0.513981886.z^{-2})}$$

Now this means that the coefficient values are $A0 = 0.041253537$, $A1 = -0.082507074$, $A2 = 0.041253537$, $B1 = -1.348967736$, and $B2 = -0.513981886$. This completes the design process using method 1.

Design Method 2

In the second highpass filter method, s is replaced by:

$$s = \tan(\omega_D/2).\left[\frac{z+1}{z-1}\right]$$

Now by substituting into the original Butterworth transfer function

$$H(s) = \frac{1}{s^2 + \sqrt{2}s + 1}$$

Using $\omega_D = 2\pi(3.4/8) = 2.670353756$, the function $\tan(\omega_D/2) = \tan(1.335176878) = 4.165299774$. It is now possible to substitute for s:

$$s = \tan(\omega_D/2).\left[\frac{z+1}{z-1}\right]$$

$$H(z) = \frac{1}{17.34972221.\left[\frac{z+1}{z-1}\right]^2 + 5.890623432.\left[\frac{z+1}{z-1}\right] + 1}$$

By comparing design method 1 and 2, it can be seen that the results using design method 2 are identical (within calculator error limits). The only difference is that the second method is a single step.

Bandpass Frequency Scaling

Bandpass frequency scaling is more complex than either lowpass or highpass transformation and scaling. Lowpass to bandpass transformation and frequency scaling requires three steps: find the values of α and β, and then use these to convert the analog transfer function $H(s)$ into the digital transfer function $H(z)$.

Find the value of α, using the upper and lower passband frequencies (F_U and F_L) and the sampling clock frequency Fs:

$$s = \tan(\omega_D/2).\left[\frac{z+1}{z-1}\right]$$

Similarly, find the value of β

$$\beta = \cot[2\pi(F_U - F_L)/2\,Fs]$$

This can also be written as:

$$\beta = 1/\tan[2\pi(F_U - F_L)/2Fs]$$

From this, the equation for $H(s)$ can be transformed into an equation for $H(z)$ by substituting for s using the values of α and β previously obtained:

$$s = \beta.\left(\frac{1-2\alpha z^{-1}+z^{-2}}{1-z^{-2}}\right)$$

For example, let $F_U = 2\,\text{kHz}$, $F_L = 1\,\text{kHz}$, and $F_S = 8\,\text{kHz}$.

The values of α and β are $\alpha = 0.414213561$ and $\beta = 2.414213562$.

Hence, in this case, the analog transfer function $H(s)$, s is replaced by:

$$s = 2.414213562.\left(\frac{1-0.828427122z^{-1}+z^{-2}}{1-z^{-2}}\right)$$

Consider a second-order Butterworth filter, with transfer function:

$$H(s) = \frac{1}{s^2 + \sqrt{2}s + 1}$$

This has a s^2 term, and when the replacement for s is substituted, factors of up to z^{-4} are produced. Therefore a simple digital biquad stage is not sufficient for a second-order bandpass filter: two biquad stages will be necessary.

Bandstop Frequency Scaling

Bandstop frequency scaling is as complex as bandpass transformation and scaling. Lowpass to bandstop transformation and frequency scaling requires three steps: find the values of α and β, and then use in the frequency transformation equation.

The first step is to find the value of α:

$$\alpha = \frac{\cos[2\pi(F_U + F_L)/2Fs]}{\cos[2\pi(F_U - F_L)/2Fs]}$$

Next, find the value of β:

$$\beta = \tan[2\pi(F_U - F_L)/2Fs]$$

Now we can find the equation for $H(z)$ by replacing s in the equation for $H(s)$:

For example, suppose $F_U = 3.0\,\text{kHz}$, $F_L = 0.5\,\text{kHz}$, and $F_S = 8\,\text{kHz}$.

$$s = \beta \cdot \left(\frac{1 - 2\alpha z^{-1} + z^{-2}}{1 - z^{-2}} \right)$$

First find the values of α and β. $\alpha = 0.351153302$ and $\beta = 1.496605763$.

Hence, in this case, the substitution for s becomes:

$$s = 1.496605763 \left(\frac{1 - z^{-2}}{1 - 0.702306604 z^{-1} + z^{-2}} \right)$$

Consider a second-order Butterworth filter, with transfer function:

$$H(s) = \frac{1}{s^2 + \sqrt{2}s + 1}$$

This has a s^2 term, and when the replacement for s is substituted factors of up to z^{-4} are produced. Therefore, as in the bandpass filter case, a simple digital biquad stage is not sufficient for a second-order bandstop filter: two biquad stages will be necessary.

IIR Filter Stability

Stability is guaranteed in FIR filters because they have no feedback path. This is not the case with IIR filters. Using a linear prototype, which is inherently stable, will produce a stable IIR equivalent when processed by bilinear transform. However, if the filter coefficients are rounded up, or down, it is possible to introduce instability. Rounding of coefficient values occurs when a fixed-point DSP device is being used.

An 8-bit data word produces a coefficient range of $\pm$ 128 levels, a step increment of 0.0078125. A 16-bit data word produces a coefficient range of $\pm$ 32768 levels; this time the step increment is reduced to about 0.0000305176. As the data word length increases, the risk of instability reduces because rounding up or down to a step increment level will be little different from the original value. Floating-point DSP devices have the advantage of being able to use more precise coefficients and are therefore less likely to introduce instability.

Reference

1. Cunningham, E. P. *Digital Filtering: An Introduction.* Boston: John Wiley & Sons, 1992.

APPENDIX

DESIGN EQUATIONS

Bessel Transfer Function

The Bessel response is produced by a transfer function that is derived from Bessel polynomials, and using the Bessel transfer function produced the graph in Figure 2.6.

As previously stated, the Bessel response is produced from a time-delay function. The time-delay for all filter orders is normalized to one second, which results in a frequency response that is dependent on the order, n. The transfer function for a pure delay is given by:

$$H(s) = e^{-sT}, \text{ and where normalization gives } T = 1, \text{ and } H(s) = e^{-s}$$

$$H(s) = e^{-s} = \frac{1}{\sin h(s) + \cos h(s)}$$

Hyperbolic sine and cosine functions can be expressed as a series, with the sine functions having even powers of s and the cosine function having odd powers of s. The transfer function, $H(s)$ then becomes a simple polynomial.

$$H(s) = \frac{a_0}{B_n(s)}$$

$B_n(s)$ is the Bessel polynomial, $B_n(s) = \sum_{i=0}^{n} a_i s^i$

This looks complex but consider values of $B_n(s)$ for orders up to three:

$$B_0(s) = 1$$
$$B_1(s) = s + 1$$
$$B_2(s) = s^2 + 3s + 3$$
$$B_3(s) = s^3 + 6s^2 + 15s + 15$$

The general expression is given by:

$$B_n(s) = (2n-1)B_{n-1}(s) + s^2 B_{n-2}(s)$$

This expression can be used to find $B_3(s)$ by letting $n = 3$ in the expression. The expression will now be broken down into manageable pieces:

$(2n-1) = 5$

$B_{n-1}(s) = B_2(s) = s^2 + 3s + 3$

Therefore, $(2n-1)B_{n-1}(s) = 5(s^2 + 3s + 3) = 5s^2 + 15s + 15$

$B_{n-2}(s) = B_1(s) = s + 1$

Therefore, $s^2 B_{n-2}(s) = s^2(s+1) = s^3 + s^2$.

Finally, $B_n(s) = (s^3 + s^2) + (5s^2 + 15s + 15) = s^3 + 6s^2 + 15s + 15$

Considering the original expression, $a_3 = 1$, $a_2 = 6$, $a_1 = 15$ and $a_0 = 15$.

These coefficients can be found using the equation: $a_i = \dfrac{(2n-1)!}{2^{(n-i)}i!(n-i)!}$, for $i = 0,1,2\ldots n$

Where has this got us? Well, we have found the transfer function, which is:

$$H(s) = \frac{15}{s^3 + 6s^2 + 15s + 15}$$

In terms of frequency, substituting $s = j\omega$, we have:

$$H(\omega) = \frac{15}{-j\omega^3 - 6\omega^2 + 15j\omega + 15}$$

The transfer function is used to calculate the attenuation and phase of the Bessel response at any frequency. The denominator must be broken down into odd and even parts. The real parts are the even powers of ω and their coefficients. Imaginary parts are the odd powers of ω and their coefficients. The attenuation is the zero power coefficient divided by the magnitude of real and imaginary parts.

In decibels, the attenuation is: $-20.LOG_{10}\left(\dfrac{a_0}{\sqrt{real^2 + imaginary^2}}\right) dB$.

The phase is given by the expression:

$$\phi = \tan^{-1}(\text{Imaginary } H(j\omega)/\text{Real } H(j\omega))$$

and group delay is given by:

$$T_d = \frac{d\phi}{d\omega}$$

Unfortunately the attenuation formula is for a Bessel response with a one-second delay. We must scale it using Table 2.1. Let's show how this works for $n = 3$.

real part $= 15 - 6\omega^2$; imaginary part $= 15j\omega - j\omega^3$; $a_0 = 15$.

In the case of the third-order response, the scaling factor is 1.75. This factor must be multiplied by the frequency ratio. For instance, to find the attenuation at 5 rad/s with a normalized response (1 rad/s cutoff), the frequency ratio is 5 and, therefore, in both real and imaginary parts, $\omega = 8.75$.

$$-20 . LOG_{10}\left(\frac{a_0}{\sqrt{real^2 + imaginary^2}}\right)$$
$$= -20 . LOG(15/\sqrt{197469 + 641876}) = 35.7 \, dB$$

Compare this with the graph in Figure 2.6 showing attenuation versus frequency and order. Polynomials for the Bessel response up to third order, Table A.1 gives some of the Bessel polynomial coefficients for higher orders.

The phase of a third-order response can be analyzed in a similar way to show the near constant group delay.

$$\phi = \tan^{-1}(\text{Imaginary } H(j\omega)/\text{Real } H(j\omega))$$

real part $= 15 - 6\omega^2$; imaginary part $= 15j\omega - j\omega^3$; $a_0 = 15$.

$$\phi = \tan^{-1}(j\omega.(15 - \omega^2)/(15 - 6\omega^2))$$
$$= \tan^{-1}\left[\frac{j\omega(15 - \omega^2)}{(15 - 6\omega^2)}\right]$$

If $\omega = 0.25$, $\phi = \tan^{-1}[j0.25.(14.9375)/(14.625)] = 0.249999 \, rad$

If $\omega = 0.5$, $\phi = \tan^{-1}[j0.5.(14.75)/(13.5)] = 0.499995 \, rad$

If $\omega = 0.75$, $\phi = \tan^{-1}[j0.75.(14.4375)/(11.625)] = 0.749922 \, rad$

If $\omega = 1$, $\phi = \tan^{-1}[j1.(14)/(9)] = 0.999459 \, rad$

If $\omega = 1.5$, $\phi = \tan^{-1}[j1.5.(12.75)/(1.5)] = 1.492525 \, rad$

Since the magnitude of the phase is almost the same as the frequency, the rate of change is almost constant (i.e., $= 1 \, rad/rad/s = 1 \, s$). As the frequency increases to $\omega = 1.5$ the error is only 0.007475, or 0.5%. In the normalized response, a scaling factor of 1.75 must be used. Rather than scale ω before the phase is calculated, calculate the phase for a value of ω, then scale the frequency. Essentially, the phase shifts indicated above occur at a lower frequency in the case of the normalized 3 dB cutoff response. A 0.249999 rad phase shift will occur at $\omega = 0.25/1.75 = 0.143$ rad/s.

The reader may like to work out the amplitude and phase of higher-order Bessel responses and for this will need to work out Bessel polynomials. Bessel polynomials for orders up to three have already been given, in this section and 4th, 5th and 6th-order values are listed in Table A.1. The rate of increase in the coefficient values with order can be seen from this limited list. Seventh-order polynomials begin with $a_0 = 135{,}135$, $a_1 = 135{,}135$, $a_2 = 62{,}370$, and so on. In all cases the highest-order coefficient is one.

Order	a_0	a_1	a_2	a_3	a_4	a_5	a_6
4	105	105	45	10	1		
5	945	945	420	105	15	1	
6	10,395	10,395	4,725	1,260	210	21	1

Table A.1

Bessel Polynomial Coefficients

Butterworth Filter Attenuation

The attenuation curves in the graph in Figure 2.10 were plotted using the following equation:

$$A(dB) = 10.\log[1 + \omega^{2n}]$$

The group delay of the Butterworth response rises as the cutoff frequency is approached, but this rise is smooth and can be compensated for by adding all-pass filter stages.

Butterworth Transfer Function

The Butterworth transfer function is very simple. It is merely:

$$H(\omega) = \frac{1}{\sqrt{1 + \omega^{2n}}}, \text{ where } n \text{ is the order.}$$

It follows from this that at $\omega = 1$, $H(\omega) = 1/\sqrt{2}$, or 0.7071. This is −3 dB relative to the zero frequency point, no matter what value of filter order is considered, since $1^n = 1$. For this reason the −3 dB cutoff frequency is considered the natural passband edge. Attenuation for any filter order at other frequencies can be found by substituting different values of ω and n.

For example, at $\omega = 3$ and $n = 5$. Attenuation = $10.\log(1/(1 + 3^{10})) = 47.7$ dB, which agrees with the graph in Figure 2.10. Another way of looking at this is to find the filter order that will satisfy the attenuation requirements.

$$n = \frac{\ln A}{\ln R}$$
$$A = \sqrt{\frac{10^{0.1\,Ks} - 1}{10^{0.1\,Kp} - 1}}$$
$$R = \omega_s / \omega_p$$

In the case of the normalized 3 dB filter, $K_p = 3$ dB so $10^{0.1 Kp} = 2$ and $A = \sqrt{(10^{0.1 Ks} - 1)}$. The passband frequency is $\omega_p = 1$ and so $R = \omega_s$.

For example, to find the filter order required for a normalized filter with 65 dB attenuation at $\omega = 2.5$:

$$A = 1778 \quad \text{and} \quad R = 2.5.$$
$$n = \frac{\ln 1778}{\ln 2.5} = 8.1669$$

An eighth-order filter is just outside the attenuation limit, so a ninth-order design is necessary.

Butterworth Phase

To find the phase response of Butterworth filters it is necessary to expand the transfer function, to give a denominator with a polynomial.

$$H(s) = \frac{1}{a_0 + a_1 s + a_2 s^2 + a_3 s^3 + \ldots}$$

Fortunately there is an iterative equation that can be used to find the polynomial coefficients, a_k. The initial coefficient, $a_0 = 1$.

$$a_k = \frac{\cos[(k-1)\pi/2n]}{\sin(k\pi/2n)}.a_{k-1} \quad k = 1, 2, 3, \ldots n$$

The coefficients are symmetric, so it is only necessary to calculate half the values.

$a_0 = a_n = 1$

$a_1 = a_{n-1}$, etc. To simplify matters, some values are listed in Table A.2.

Order, n	a_1	a_2	a_3	a_4	a_5
2	1.41421				
3	2.00000				
4	2.61313	3.41421			
5	3.23607	5.23607			
6	3.86370	7.46410	9.14162		
7	4.49396	10.09784	14.59179		
8	5.12583	13.13707	21.84615	25.68836	
9	5.75877	16.58172	31.16344	41.98639	
10	6.39245	20.43173	42.80206	64.88240	74.23343

Table A.2

Butterworth Denominator Coefficients

The phase can be calculated by substituting $j\omega = s$. Consider the phase of a fourth-order Butterworth response.

$$H(s) = 1/(1 + 2.61313s + 3.41421s^2 + 2.61313s^3 + s^4)$$
$$H(\omega) = 1/(1 + j2.61313\omega - 3.41421\omega^2 - j2.61313\omega^3 + \omega^4)$$

Real parts of the denominator are: $1 - 3.41421\omega^2 + \omega^4$.

Imaginary parts are: $j2.61313\omega - j2.61313\omega^3$.

The phase is given by $-\tan^{-1}(\text{real/imaginary})$

At $\omega = 1$, $n = 4$. $-\tan^{-1}(-1.41421/0) = -\tan^{-1}(\infty) = -\pi/2$ or $-90°$.

At $\omega = 0.5$, $\phi = -\tan^{-1}[(1.0625 - 0.85355)/(1.306565 - 0.32664)]$

$\phi = -\tan^{-1}[0.21323] = 0.21$ rad/s or about $-12°$.

Nonstandard Butterworth Passband

It is possible to scale the response to have other attenuation levels at $\omega = 1$. For an attenuation of Kp:

$$\omega_{Kp} = (10^{0.1Kp} - 1)^{1/(2n)}$$

For example, suppose you want to know the frequency at which a 3 dB normalized fifth-order filter has 1 dB attenuation.

$$\omega_{Kp} = (10^{0.1} - 1)^{0.1} = 0.8736$$

In order for the filter to have 1 dB attenuation at $\omega = 1$, pole positions or component values must be scaled to give the filter a higher 3 dB attenuation frequency. The scaling factor will be

$$K_P = \frac{1}{\omega_{Kp}}, \quad \text{which is approximately } 1.1447.$$

Normalized Component Values for Butterworth Filter with RL ≫ RS or RL ≪ RS

For zero or infinite impedance load, the following equations give the element values. Values are given as C_n and are the nth component. The components are alternating inductors and capacitors; the sequence depends on the load, as just described.

$$a_R = \sin\left[\frac{(2R-1)\pi}{2n}\right] \quad R = 1, 2, 3, \ldots n$$

$$b_R = \cos^2\left(\frac{R\pi}{2n}\right) \quad R = 1, 2, 3, \ldots n$$

$$C_1 = a_1$$

$$C_R = \frac{a_R a_{R-1}}{b_R C_{R-1}} \quad R = 2, 3, 4, \ldots n$$

These equations produce component values in an order that assumes they are normalized against source impedance, rather than the load ($Rs = 1$).

Normalized Component Values for Butterworth Filter: Source and Load Impedances within a Factor of Ten

The following equations are used to find the element values.

$$\text{Termination factor, } T = \frac{4R_L}{(R_L + 1)^2}$$

$$\delta = (1-T)^{1/(2n)}$$

$$a_R = \sin\left[\frac{(2R-1)\pi}{2n}\right] \qquad R = 1, 2, 3, \dots n$$

$$b_R = 1 + \delta^2 - 2\delta \cos\left(\frac{R\pi}{n}\right) \qquad R = 1, 2, 3, \dots n$$

$$C_1 = \frac{2a_1}{1-\delta}$$

$$C_R = \frac{4a_R a_{R-1}}{b_{R-1} C_{R-1}} \qquad R = 2, 3, 4, \dots n$$

Chebyshev Filter Response

Attenuation of Chebyshev filters is more difficult to calculate than for Butterworth filters. The following expression is used.

$$A = 10.\log(1 + \varepsilon^2 Cn^2(\Omega))\ dB$$

$\varepsilon = \sqrt{10^{0.1Ap} - 1}$, where Ap is the passband ripple in decibels (e.g., 0.1 dB)

$Cn(\Omega)$ is the Chebyshev polynomial and can be found from the equation:

$$C_{n+1}(\Omega) = 2\omega C_n(\Omega) - C_{n-1}(\Omega)$$

$C_0(\Omega) = 1$ and $C_1(\Omega) = \omega$, hence a table can be built up:

$$C_2(\Omega) = 2\omega^2 - 1$$
$$C_3(\Omega) = 4\omega^3 - 3\omega$$
$$C_4(\Omega) = 8\omega^4 - 8\omega^2 + 1$$

The Chebyshev polynomial can be reproduced using this iterative process, but there exists an alternative—an entirely equivalent solution:

Up to the ripple bandwidth, $Cn(\Omega) = \cos(n.\cos^{-1}\Omega)$.

Beyond the ripple bandwidth, $Cn(\Omega) = \cos h(n.\cos h^{-1}\Omega)$.

The ratio of cutoff frequency to stopband edge is represented by the symbol Ω. When a 3 dB cutoff frequency is required, Ω must be multiplied by a function to give the correct results.

$$\Omega(3\ dB) = \Omega.\cos h\left[\frac{1}{n}.\cos h^{-1}\left(\frac{1}{\varepsilon}\right)\right]$$

In other words, $Cn(\Omega_{3\,dB}) = \cos h\left(n.\cos h^{-1}\left(\Omega.\cos h\left(\frac{1}{n}.\cos h^{-1}\frac{1}{\varepsilon}\right)\right)\right).$

For example, consider a fifth-order 0.1 dB ripple Chebyshev filter. What attenuation does it produce at twice the 3 dB cutoff frequency?

$$\varepsilon = \sqrt{10^{0.1Ap} - 1} = 0.15262$$

$$\Omega(3\,dB) = \Omega.\cos h\left[\frac{1}{n}.\cos h^{-1}\left(\frac{1}{\varepsilon}\right)\right] = 2.\cos h(0.513415) = 2.26944$$

$$Cn(\Omega) = \cos h(n.\cos h^{-1}\Omega) = 740.77$$

$$A = 10.\log(1 + \varepsilon^2 Cn^2(\Omega))\,dB = 10.\log(12783) = 41\,dB.$$

The required filter order is given by:

$$n = \frac{\cos h^{-1}(Cn)}{\cos h^{-1}(\Omega)}$$

where $Cn = \sqrt{(10^{0.1As} - 1)}$, As = desired stopband attenuation and Ω is the ratio of stopband to passband frequency.

Equations to Find Chebyshev Element Values

Chebyshev with Zero or Infinite Impedance Load

For a Chebyshev filter with zero or infinite impedance load, the following equations give the element values. Values are given as C_n that are the nth component. These are alternating inductors or capacitors, and the sequence depends on the load, as just described. The passband ripple, in decibels, is Ap.

$$\varepsilon = \sqrt{10^{0.1Ap} - 1}$$

$$\beta = 2\sin h^{-1}\left(\frac{1}{\varepsilon}\right)$$

$$\gamma = \sin h\left(\frac{\beta}{2n}\right)$$

$$a_R = \sin\left[\frac{(2R-1)\pi}{2n}\right] \qquad R = 1, 2, 3, \ldots n$$

$$b_R = \left[\gamma^2 + \sin^2\left(\frac{R\pi}{2n}\right)\right].\cos^2\left(\frac{R\pi}{2n}\right) \qquad R = 1, 2, 3, \ldots n$$

Now, the component values are given by C_R

$$C_1 = \frac{a_1}{\gamma}$$

$$C_R = \frac{a_R a_{R-1}}{b_{R-1} C_{R-1}} \quad R = 2, 3, 4, \ldots n$$

Chebyshev Filter with Source and Load Impedances within a Factor of Ten

If the load is not much greater than or less than the source, either shunt C or series L can be used as the first component. The last component will depend on whether the filter has an odd or even order. The following equations are used to find the element values.

$$\text{Termination factor}, T = \frac{4R_L}{(R_L + 1)^2} \quad (\text{odd order})$$

$$T = \frac{4R_L 10^{0.1Ap}}{(R_L + 1)^2} \quad (\text{even order})$$

$$a_R = \sin\left[\frac{(2R-1)\pi}{2n}\right] \qquad\qquad R = 1, 2, 3, \ldots n$$

$$\delta = \sin h\left[\left(\frac{1}{n}\right)\sin h^{-1}\left(\frac{\sqrt{1-T}}{\varepsilon}\right)\right]$$

$$\gamma = \sin h\left(\frac{\beta}{2n}\right)$$

$$b_R = \gamma^2 + \delta^2 - 2\gamma\delta \cos\left(\frac{R\pi}{2n}\right) + \sin^2\left(\frac{R\pi}{2n}\right) \quad R = 1, 2, 3, \ldots n$$

$$C_1 = \frac{2a_1}{\gamma - \delta}$$

$$C_R = \frac{4a_R a_{R-1}}{b_{R-1} C_{R-1}} \qquad\qquad R = 2, 3, 4, \ldots n$$

Component values must be normalized for a 3 dB cutoff frequency. This is done by increasing all values, by multiplying those obtained in the formulae by $\cos h = \left[\frac{1}{n} . \cos h^{-1}\left(\frac{1}{\varepsilon}\right)\right]$ for odd order. For example, for a fifth-order 0.1 dB Chebyshev filter, component values must be multiplied by 1.13472.

The values obtained for even-order filters should be increased in value by

$$\cos h\left[\frac{1}{n}.\cos h^{-1}\left(\frac{\sqrt{2.10^{0.1Ap}}-1}{\varepsilon}\right)\right].$$

Load Impedance for Even-Order Chebyshev Filters

For even-order Chebyshev filters, equal source and load impedance is not possible. It must have a normalized load of greater than unity if the first component is a series inductor (the last component is therefore a shunt capacitor). The minimum value of the normalized load is $R_L \geq \coth^2\left(\frac{\beta}{4}\right)$. Conversely, if the first component is a shunt capacitor, the last component is a series inductor, and the load must be less than unity. The maximum value of the normalized load is $R_L \leq \tan h^2\left(\frac{\beta}{4}\right)$.

Inverse Chebyshev Filter Equations

To find the filter order required for Inverse Chebyshev filters, use the following equation.

$$n = \frac{\cos h^{-1}(Cn)}{\cos h^{-1}(\Omega)}$$

Where $Cn = \sqrt{10^{0.1A} - 1}$, with A being the desired attenuation, and Ω is the ratio of stopband to passband.

For example, if the 3 dB point is at 10 Hz and the stopband begins at 15 Hz, $\Omega = 1.5$. If 20 dB of stopband attenuation is required, $Cn = 9.95$; $n = 2.988/0.9624 = 3.1$; the filter order must be 4 or more.

Inverse Chebyshev filters have to be adjusted to give a 3 dB passband edge at $\omega = 1$ rad/s. Consider a third-order lowpass filter, as shown in Figure A.1, having equal source and load impedance and a stopband beginning at $\omega = 1$ rad/s:

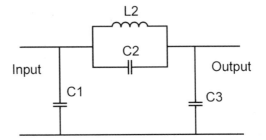

Figure A.1

Third-Order Inverse Chebyshev
Filter

If Ks is the stopband attenuation in decibels, then the following equations apply.

$$\varepsilon = \sqrt{1/(10^{0.1Ks} - 1)}$$
$$\eta = \sin h(1/n . \sin h^{-1}(1/\varepsilon))$$

The values of capacitors $C1$ and $C3$ are equal. Their value is given by the equation:

$$C_r = 2.\eta.\sin[(2r-1).\pi/(2.n)]$$

Where $r = 1$ or 3, and $n = 3$.

Inductor $L1$ is given by the same equation

$$L_r = 2.\eta.\sin[(2r-1).\pi/(2.n)]$$

Where $r = 2$ and $n =$ filter order, in this case $n = 3$. η is the value obtained to find $C1$ and $C3$.

Capacitor $C2$ forms a parallel tuned circuit with inductor $L2$. The frequency of resonance determines the null position. For a third-order design this frequency will be $\beta = 1.15470$, for a stopband starting at $\omega = 1$ rad/s. For a tuned circuit, resonance occurs at $1/\sqrt{LC}$. So the values of $C2$ is given by: $C2 = 1/\beta^2 L$.

The zero frequency locations for any order Inverse Chebyshev are given by the equations below. Zero locations are given as β_K, since $Z_K = \alpha_K + \beta_K$ and the real part $\alpha_K = 0$. Applying the equations produces both positive and negative frequencies, but only the positive frequencies are used. The proof for finding the equations is given in Huelsman (See Chapter 2, Reference 2).

The zero frequency is now higher than the value given by β, because it is relative to the 3 dB cutoff frequency. In fact, the new zero frequency is β/ω_{3dB}. Previously given in this section was the formula for calculating the tuned circuit capacitor: $C2 = 1/\beta^2L$. But the new zero is at β/ω_{3dB} and the new inductor value is $L\omega_{3dB}$.

So $C2 = \omega_{3dB}^2/\beta^2L\omega_{3dB}$. This simplifies to $C2 = \omega_{3dB}/\beta^2L$. In other words, $C2$ is also multiplied by ω_{3dB}.

$$U_k = \frac{(2K-1)\pi}{2n}$$

$$\beta_k = \frac{1}{\cos U_k}$$

$$K = 1, 2, \ldots, n$$

An equation exists to find the 3 dB point for any Inverse Chebyshev filter:

$$\omega_{3dB} = \frac{1}{\cos h\left[\dfrac{\cos h^{-1}\left(\sqrt{10^{0.1Ks}-1}\right)}{n}\right]}$$

The 3 dB cutoff frequency depends on both the stopband attenuation Ks and the filter order n. Component values, or pole positions in the case of active filters, can be scaled to give a 3 dB cutoff point.

Elliptic or Cauer Filter Equations

The filter order required for a given passband ripple, stopband attenuation, and ratio of passband to stopband frequency can be calculated by use of complicated algorithms. These algorithms use something called an elliptic integral. Elliptic integrals are easier than they look, and examples will be given to show this. To find the filter order required we must note the passband ripple and frequency as well as the stopband ripple and frequency.

ripple = maximum passband ripple (in dB).

amin = minimum stopband ripple (in dB).

pass = passband frequency (cutoff frequency).

stop = stopband frequency (where the stopband has been reached).

The ratio of stopband ÷ passband is given as k. Another variable, L, is dependent upon the passband ripple and the stopband attenuation.

$$L = \sqrt{\frac{10^{Amin/10}-1}{10^{Ripple/10}-1}}$$

Now, the required order is equal to n. This is given by an equation that uses the two variables k and L:

$$n = \frac{K(1/k).K(\sqrt{1-1/L^2})}{K(1/L).K(\sqrt{1-1/k^2})}$$

Function $K(x)$ is an elliptic integral of x, so four elliptic integrals are needed to find the filter order. The elliptic integral itself takes the value to be integrated as the starting point. It is a recursive equation. Here are the equations.

$$x_0 = x$$
$$x'_m = \sqrt{1-x_m^2}$$
$$x_{m+1} = \frac{1-x'_m}{1+x'_m}$$
$$K(x) = \frac{\pi}{2} \prod_{m=0}^{\infty} (1+x_{m+1})$$

These equations look horrific, so an example of the elliptic integral algorithm's use is now given.

If $k = 2$, $K(1/k) = K(0.5)$ and the value to be integrated is $X_0 = 0.5$. $X'_0 = \sqrt{(1-0.5^2)} = 0.8660$. This is then used to find X_1.

$$X_1 = (1-0.866)/(1+0.866) = 0.0718$$
$$X'_1 = \sqrt{(1-0.0718^2)} = 0.9974. \text{ This is used to find } X_2.$$
$$X_2 = (1-0.9974)/(1+0.9974) = 1.3017 \times 10^{-3}$$
$$X'_2 = \sqrt{(1-0.0013017^2)} = 0.999999152. \text{ This is used to find } X_3.$$
$$X_3 = 4.2361 \times 10^{-7}$$

As X becomes small its effect diminishes, and when it is less than 10^{-7} it can be ignored. Therefore the infinite limit to the product of $(1 + X)$ is actually truncated after a few iterations. In this case we can see that X_4 will be very small.

$$K(0.5) = \pi/2 (1+0.0718).(1+0.0013017).(1+4.2361 \times 10^{-7}) = 1.68577.$$

Noise Bandwidth

Knowledge of a filter's noise bandwidth can help in system design and testing. Suppose you want to find the noise figure of an amplifier. You first filter the amplifier's output and measure the RMS (root mean square) output voltage. You then divide by the amplifier's gain, the filter bandwidth,

Boltzmann's constant, and the temperature. The result of these calculations is the noise figure.

$$F = \frac{N_0}{GkTB} \quad \text{when } T = 290\,K, kT = 4.14 \times 10^{-21}\,W.$$

Obviously, unless the bandwidth B is known, it is not possible to know what the noise figure is either. The noise bandwidth of a filter is given by:

$$B_n = \frac{1}{g}.\int_0^\infty |H(f)|^2 .df$$

Butterworth Noise Bandwidth

$$B_n = \int_0^\infty \frac{1}{1+\omega^{2n}}.d\omega$$

Note that the $1/g$ term has been removed; this is because Butterworth filters have a smooth response with a gain of one. Note also that g is the power gain. According to Carlson[1] this equation can be simplified to:

$$B_n = \frac{\pi.B}{2n.\sin(\pi./2n)}$$

The noise bandwidth of a first-order filter is $\pi/2$, or 1.570796 times the 3 dB bandwidth.

Table A.3 gives normalized Butterworth filter noise bandwidths.

Filter Order	Bandwidth
1	1.570796
2	1.110721
3	1.047198
4	1.026172
5	1.016641
6	1.011515
7	1.008442
8	1.006455
9	1.005095
10	1.004124

Table A.3

Noise Bandwidth of Butterworth Filters

As the filter order increases, the noise equivalent bandwidth approaches the 3 dB bandwidth. To find the actual bandwidth, simply multiply the figure given by the 3 dB bandwidth of the filter that you wish to assess.

Chebyshev Noise Bandwidth

The noise bandwidth of Chebyshev filters having a 3 dB cutoff point is given by:

$$\omega_n = \frac{\pi.\cosh\left[\frac{1}{n}\ln\left(\frac{1+\sqrt{1+\varepsilon^2}}{\varepsilon}\right)\right]}{\omega_{3dB}.\sqrt{1+\varepsilon^2}.2n.\sin\left(\frac{\pi}{2n}\right)}$$

Applying this formula gives a noise bandwidth of less than the 3 dB bandwidth for high-order filters. This can be explained by remembering that Chebyshev filters have ripple in the passband. This means that, on average, the gain in the passband will be less than one. Also, Chebyshev filters have a steeper skirt response; the attenuation beyond the filter cutoff rises sharply with frequency. Therefore the noise equivalent bandwidth of Chebyshev filters can be expected to be less than a Butterworth filter of the same order.

In fact the amplitude of the voltage ripple is:

$$R = \frac{1}{\sqrt{1+\varepsilon^2}}$$

In terms of power this becomes:

$$g = \frac{1}{1+\varepsilon^2}$$

The term ε is equal to $\sqrt{10^{\frac{RdB}{10}}-1}$

The average power gain within the ripple part of the passband can therefore be approximated by assuming that the ripple is symmetrical. In that case the average power gain is halfway between the gain at the peaks and the gain in the troughs, that is, between one and g. The equation for this is $\frac{1+g}{2}$. Table A.4 shows the passband power gain for Chebyshev filters that have a 3 dB cutoff point.

Ripple, dB	ε	power gain, g	average power gain
0.01	0.048	0.997701	0.998855
0.1	0.153	0.977126	0.988563
0.25	0.243	0.944243	0.972122
0.5	0.349	0.891423	0.945716
1.0	0.509	0.794230	0.897115

Table A.4

Chebyshev Filter Passband
Power Gain (with 3 dB Cutoff)

When the Butterworth filters were considered, in the previous subsection a tenth-order filter had a noise bandwidth that was slightly greater than unity. Now Chebyshev filters have a faster rate of attenuation outside the passband, so their noise bandwidth should be very close to, but slightly higher than, the average power gain. To check this, Table A.5 (with a correction for 3 dB cutoff) has been calculated using *MATHCAD*.

order\ripple	0.01 dB	0.1 dB	0.25 dB	0.5 dB	1.0 dB
2	1.110051	1.103508	1.091806	1.071364	1.029180
3	1.046332	1.038008	1.023595	0.999233	0.950851
4	1.025208	1.016098	1.000631	0.974868	0.924478
5	1.015619	1.006114	0.990130	0.963717	0.912432
6	1.010458	1.000722	0.984456	0.957688	0.905929
7	1.007363	0.997482	0.981043	0.954057	0.902022
8	1.005359	0.995384	0.978832	0.951716	0.899491
9	1.003989	0.993948	0.977318	0.950106	0.897758
10	1.003010	0.992921	0.976236	0.948956	0.896519

Table A.5

Noise Bandwidth of Chebyshev Filters (Cutoff at 3 dB Point)

Comparing the average gain with the tenth-order noise bandwidths, it can be seen that the noise bandwidth almost reduces to the 3 dB passband, taking into account the average power gain. The exception to this is the 1.0 dB ripple filter that seems to have a noise bandwidth of less than an equivalent "brick wall." This shows that the passband power approximation was close to the actual figure, but slightly too high. Guessing, the passband power should be about 1% lower.

Table A.6 gives Chebyshev filter noise bandwidth, without correcting for the 3 dB cutoff point. The bandwidth values given are for Chebyshev filters that have a cutoff point equal to the ripple value.

order\ripple	0.01 dB	0.1 dB	0.25 dB	0.5 dB	1.0 dB
2	3.667181	2.144358	1.744861	1.488922	1.253156
3	1.964153	1.441787	1.282450	1.166589	1.041057
4	1.503881	1.232628	1.140487	1.065631	0.973477
5	1.311385	1.141656	1.077978	1.020826	0.943285
6	1.211956	1.093719	1.044843	0.996982	0.927166
7	1.153700	1.065312	1.025137	0.982764	0.917541
8	1.116561	1.047071	1.012457	0.973615	0.911331
9	1.091400	1.034655	1.003813	0.967367	0.907091
10	1.073553	1.025817	0.997654	0.962912	0.904066

Table A.6

Chebyshev Filter Noise Bandwidth (Cutoff at Ripple)

Complications arise if a noise figure measurement or a signal-to-noise calculation is required when Chebyshev filters are being used. Assuming a 1 dB ripple Chebyshev filter is used, the signal could vary in amplitude by 1 dB as the signal frequency is changed. So, at what frequency is signal-to-noise measured? Do you take the average of the minimum and maximum values? I will leave those thoughts with you, Dear Reader!

Pole and Zero Location Equations

Butterworth Pole Locations

As briefly described above, the poles of the Butterworth response all lie on the unit circle; because of this they are the easiest to find out of all the filter designs. The following formula gives the normalized pole positions for a Butterworth response with a 3 dB cutoff point at $\omega = 1$:

$$-\sin\frac{(2K-1)\pi}{2n} + j\cos\frac{(2K-1)\pi}{2n}$$

for $K = 1, 2, \ldots, n$ and where n is the required filter order.

For example, find the poles of a normalized fifth-order Butterworth response; $n = 5$.

To find the first pole, let $K = 1$. This is at $-\sin(\pi/10) + j\cos(\pi/10)$, or $-0.309 + j0.9511$.

The second pole uses $K = 2$. This is at $-\sin(3\pi/10) + j\cos(3\pi/10)$, or $-0.809 + j0.5878$.

The third pole, with $K = 3$, is at $-\sin(5\pi/10) + j\cos(5\pi/10)$, or -1.0

The fourth and fifth poles are at the complex conjugate positions relative to the first and second poles. The complex conjugate has the same real part, but the imaginary part has the opposite sign. That is $-0.309 - j0.9511$ and $-0.809 - j0.5878$, which are the negative frequency complements to the first and second poles. Note that the third pole is real, is on the $-\sigma$ axis, and has a magnitude of one. All odd-order Butterworth responses have a pole in this position.

If a cutoff point other than the 3 dB attenuation frequency is required there is a simple formula that can be used to scale the pole positions given:

$$\omega_{KP} = (10^{0.1KP} - 1)^{\frac{1}{2N}}$$

KP is the desired cutoff point attenuation.

For example, say we want a fifth-order response with a 1 dB cutoff, then:

$$\omega_{KP} = (10^{0.1} - 1)^{0.1}$$
$$= 0.2589^{0.1}$$
$$= 0.8736$$

The new cutoff point occurs at a lower frequency (as you would expect, since the attenuation is lower), so the pole magnitudes have to be divided by ω_{KP}, which increases their value and moves them away from the origin of the S-plane. The real pole moves from -1.0 to -1.14467. Both real and imaginary parts of the other poles are increased in magnitude by 1.14467. In fact, all poles will lie on a circle that is 1.14467 in diameter. Now the 1 dB point is at $\omega = 1$ and the 3 dB point is at $\omega = 1.14467$.

Do not worry about scaling the normalized Butterworth response unless you have some particular reason for using anything other than a 3 dB cutoff point.

Pole locations for the Butterworth response are given by the formula:

$$-\sin\frac{(2K-1)\pi}{2n} + j\cos\frac{(2K-1)\pi}{2n}$$

for $K = 1, 2, \ldots, n$ and where n is the required filter order.

Chebyshev Pole Locations

Considering the comment made earlier about the relationship between Butterworth and Chebyshev response pole locations, close correspondences between their pole-locating formulae are expected. In fact, pole locations for the Chebyshev response are given by:

$$-\sin\frac{(2K-1)\pi}{2n}.\sinh n + j\cos\frac{(2K-1)\pi}{2n}.\cosh \nu$$

Where $\nu = \frac{1}{n}.\sinh^{-1}\frac{1}{\varepsilon}$

The filter order is given by n, and ε depends on the passband ripple.

$\varepsilon = \sqrt{10^{0.1R}-1}$ where R is the ripple in decibels (dB).

Correction is necessary if a 3 dB cutoff point is required. The correction factor is:

$$C_{3dB} = \cosh\left(\frac{1}{n}.\cosh^{-1}\frac{1}{\varepsilon}\right)$$

The 3 dB point occurs at a higher frequency than the natural cutoff point at the ripple attenuation. To have a 3 dB cutoff point, the magnitude of each pole location must be reduced by C_{3dB}.

For example, suppose we have a third-order Chebyshev response with 0.15 dB ripple in the passband.

$$R = 0.15, \text{ so } \varepsilon = \sqrt{(10^{0.015}-1)}$$
$$\varepsilon = \sqrt{0.03514} = 0.1875$$
$$1/\varepsilon = 5.3344$$
$$\nu = \frac{1}{n}.\sinh^{-1}\frac{1}{\varepsilon} = \frac{1}{3}.\sinh^{-1}5.3344$$
$$\nu = 2.376/3 = 0.792$$
$$\sinh\nu = 0.8774$$
$$\cosh\nu = 1.3304$$

The real pole given by the equation for the Butterworth response is $S = -1.0$. This moves towards the imaginary axis to give a Chebyshev pole location at $S = -0.8774$, since the real part of the Butterworth pole location has to be multiplied by $\sinh\nu$.

The real part of the two imaginary poles for the Butterworth response is $\sin(\pi/6)$ or $S = -0.5$, so multiplying this by $\sinh\nu$ to give the Chebyshev response moves it to $S = -0.4387$ in the horizontal direction. The imaginary part of these poles for the Butterworth response is at jcos(π/6) or $S = \pm j0.866$, so multiplying this by $\cosh v$ to give the Chebyshev response moves it to $S = \pm j1.1522$ in the vertical direction.

The pole locations for a third-order Chebyshev response with 0.15 dB ripple in the passband are:

$$-0.8774$$

$$-0.4387 \pm j1.1522.$$

The Chebyshev pole locations produce a normalized frequency response with attenuation equal to the ripple (0.15 dB) at $\omega = 1$. The 3 dB point will have a frequency greater than $\omega = 1$. The magnitude of the pole locations given for the Chebyshev response must now be reduced to correct for the 3 dB cutoff point; they must each be divided by C_{3dB}. Dividing by a constant factor (that is greater than one) makes the pole positions move towards the origin of the S-plane.

$$C_{3dB} = \cosh\left(\frac{1}{n}.\cos h^{-1}\frac{1}{\varepsilon}\right) = \cosh\left(\frac{1}{3}.\cos h^{-1}5.3344\right)$$

$$C_{3dB} = \cosh 0.78614 = 1.32525$$

In other words, the 0.15 dB point occurs at $\omega = 1$, and the 3 dB point occurs at $\omega = 1.32525$. So dividing the pole locations by C_{3dB} gives:

$$-0.6621$$

$$-0.3310 \pm j0.8694.$$

All three poles are now within the unit circle, and the 3 dB point occurs at $\omega = 1$.

Inverse Chebyshev Pole and Zero Locations

As suggested by their name, Inverse Chebyshev filters are derived from Chebyshev filters. The pole positions are the inverse of those given for Chebyshev filters. The frequency response of Chebyshev filters was described in Chapter 2. There are ripples in the passband with a smoothly decaying response in the stopband. Inverting the pole positions produces a filter with a smooth passband. The zeroes produce ripple in the stopband.

Pole locations for the Chebyshev response have been described earlier in the previous subsection and are given by:

$$-\sin\frac{(2K-1)\pi}{2n}.\sinh v + j\cos\frac{(2K-1)\pi}{2n}.\cosh v$$

Where $n = \frac{1}{n}.\sinh^{-1}\frac{1}{\varepsilon}$ and $K = 1, 2, \ldots, n$

The filter order is given by n, and ε depends on the passband ripple.

$$\varepsilon = \sqrt{10^{0.1R} - 1} \quad \text{where } R \text{ is the ripple in decibels (dB)}.$$

Pole locations for the Inverse Chebyshev response are based on the Chebyshev response. There is no passband ripple, but the value for ε can be found from the stopband attenuation:

$$\varepsilon = \frac{1}{\sqrt{10^{0.1A} - 1}} \quad \text{where } A \text{ is the stopband attenuation (dB)}.$$

For example, if $A = 20\,\text{dB}$:

$$\varepsilon = \frac{1}{\sqrt{10^2 - 1}} = \frac{1}{\sqrt{99}} = 0.1005$$

Using this and the Chebyshev pole-locating formulae, the pole locations can be found. These can be expressed in the form:

$$-\sigma_i \pm j\omega_i \quad \text{where } i \text{ is an integer from 1 to } n/2.$$

These can now be transformed to give Inverse Chebyshev poles using the following equations:

$$Pole_i = \frac{-\sigma_i}{\sigma_i^2 + \omega_i^2} \pm j\frac{\omega_i}{\sigma_i^2 + \omega_i^2}$$

This gives us the natural pole locations where the stopband equals 1 rad/s. A more practical normalized response with a 3 dB passband cutoff point can be obtained by modifying these values. The 3 dB frequency is given by:

$$\omega_{3dB} = \frac{1}{\cosh\left(\frac{1}{n}\cosh^{-1}(C_n)\right)}$$

$$C_n = \sqrt{10^{0.1A} - 1}$$

For example, if $A = 20\,\text{dB}$, $C_n = \sqrt{99} = 9.499$ and $n = 3$.

$$\omega_{3dB} = 1/\cosh(0.99605) = 0.65\,\text{rad/s}.$$

To make $\omega_{3dB} = 1$ rad/s, the pole locations must be scaled by 1/0.65 or 1.53846. The following formula incorporates this scaling factor.

$$Pole_i = \frac{-\sigma_i . \cos h\left(\frac{1}{n}.\cos h^{-1}C_n\right)}{\sigma_i^2 + \omega_i^2} \pm j\frac{\omega_i . \cos h\left(\frac{1}{n}.\cos h^{-1}C_n\right)}{\sigma_i^2 + \omega_i^2}$$

Tables 3.15, 3.17, and 3.19, in Chapter 3, show the pole locations for the Inverse Chebyshev response having a normalized 3 dB passband cutoff and 20 dB, 30 dB, and 40 dB stopband attenuation.

To find the filter order required, use the following equation:

$$n = \frac{\cos h^{-1}(C_n)}{\cos h^{-1}(\Omega)}, \text{where } \Omega \text{ is the ratio of stopband to passband.}$$

For example, if the 3 dB point is at 10 Hz and the stopband begins at 15 Hz, $\Omega = 1.5$. If 20 dB of stopband attenuation is required, $Cn = 9.95$. This gives $n = 2.988/0.9624 = 3.1$; the filter order must be four or more.

Inverse Chebyshev Zeroes

The zero frequency locations for any order of Inverse Chebyshev filter were given by equations in Chapter 2 and are repeated below. Zero locations are given as β_K since $Z_K = \alpha_K + \beta_K$ and the real part $\alpha_K = 0$. Applying the equations produces both positive and negative frequencies, but only the positive frequencies are used. The proof for finding the equations is given in Huelsman[2].

$$U_k = \frac{(2K-1)\pi}{2n}$$

$$\beta_k = \frac{1}{\cos U_k}$$

$$k = 1, 2, \ldots, n$$

Inverse Chebyshev zero locations found using these equations should be used with pole locations for the natural (normalized to stopband) response. The Inverse Chebyshev response can be normalized to have a 3 dB passband attenuation. The zero locations for this response can be found by modifying these values. Previously, I showed that the poles moved away from the origin by a frequency-scaling factor:

$$\cos h\left(\frac{1}{n}.\cos h^{-1}(C_n)\right)$$

Cauer Pole and Zero Locations

Formulae to find the normalized pole and zero locations of Cauer filters will now be presented. Inputs to the equations are stopband frequency ωs (assuming that the passband equals unity), passband ripple Ap, stopband attenuation As.

We first need to find the order of the filter. The method shown here is an alternative to that shown in Chapter 2. This method avoids the need for elliptic integrals; it uses an approximation to it instead.

$$k = 1/\omega s$$

$$u = \frac{1 - \sqrt[4]{1-k^2}}{2\left(1 + \sqrt[4]{1-k^2}\right)}$$

$$q = u + 2u^5 + 15u^9 + 150u^{13}$$

$$D = \frac{10^{0.1As} - 1}{10^{0.1Ap} - 1}$$

$$n \geq \frac{\log_{10} 16D}{\log_{10}(1/q)}$$

Now that we have the filter order required we can find the factors in the transfer function, using the filter order n.

$$V = \frac{1}{2n} . \ln\left[\frac{10^{0.05Ap} + 1}{10^{0.05Ap} - 1}\right]$$

The real pole $P(0)$ for odd-order filters can now be found. This pole is required to calculate the values of the complex poles. Even-order filters only have complex poles, so the real pole should not be used directly to find component values.

$$P(0) = \left| \frac{\sqrt[4]{q} . \sum_{m=0}^{\infty} (-1)^m q^{m(m+1)} \sin h[(2m+1)V]}{0.5 + \sum_{m=1}^{\infty} (-1)^m q^{m^2} \cos h(2mV)} \right|$$

$$W = \sqrt{\left(1 + \frac{P(0)^2}{k}\right)\left(1 + k . P(0)^2\right)}$$

Now comes several recursive equations. The limit is $i = r$, where $r = n/2$ for even-order filters and $r = (n - 1)/2$ for odd-order filters. For $i = 1, 2, 3, \ldots r$ compute X_i.

$$X_i = \left| \frac{2 . \sqrt[4]{q} . \sum_{m=0}^{\infty} (-1)^m \, q^{m(m+1)} \sin[(2m+1)\mu\pi/n]}{1 + 2 . \sum_{m=1}^{\infty} (-1)^m \, q^{m^2} \cos(2m\mu\pi/n)} \right|$$

$\mu = i$ for odd-order filters, $\mu = i - 0.5$ for even-order filters. For $i = 1, 2, 3, \ldots r$ compute Y_i.

$$Y_i = \sqrt{\left(1 - \frac{X_i^2}{k}\right) . (1 - kX_i^2)}$$

Now we can find the transfer function coefficients, a_i, b_i, and c_i. From these we can find the pole and zero locations.

$$a_i = \frac{1}{X_i^2}$$

$$b_i = \frac{2 . P(0) . Y_i}{1 + P(0)^2 . X_i^2}$$

$$c_i = \frac{(P(0) . Y_i)^2 + (X_i . W)^2}{(1 + P(0)^2 . X_i^2)^2}$$

The zeroes are at $S_i = \pm j\sqrt{a_i}$.

The real pole is at $P(0)$.

The remaining poles are at $P(i) = \dfrac{-b_i \pm \sqrt{b_i^2 - 4 . c_i}}{2}$, for $i = 1, 2, \ldots, r$.

Using these pole and zero locations we find that the filter's passband is less than $\omega = 1$, the normalized frequency. The reason is that the poles are placed symmetrically about the geometric mean frequency, compared to the zeroes. Poles are at frequencies lower than the geometric mean; zeroes are at frequencies above the geometric mean. Calculations are simplified if frequency scaling is applied after the pole and zero locations are found. Frequency scaling corrects for this response, and the passband cutoff frequency increases to $\omega = 1$. All pole and zero locations must be multiplied by $\sqrt{\omega s}$.

The zeroes are at $S_i = \pm j\sqrt{\omega s . a_i}$.

The real pole, for odd-order filters, is at $P(0) . \sqrt{\omega s}$.

The remaining poles are at $P(i) = \sqrt{\omega s} . \dfrac{-b_i \pm \sqrt{b_i^2 - 4 . c_i}}{2}$, for $i = 1, 2, \ldots, r$.

Some insight into the development of the design equations can be found if the circuit of a second-order Sallen and Key filter section is analyzed. The transfer function is given by the following equations.

$$T(s) = \frac{K/R_1R_2C_1C_2}{s^2 + (1/R_1C_1 + 1/R_2C_1 + 1/R_2C_2 - K/R_2C_2)s + 1/R_1R_2C_1C_2}$$

This is simplified if $K = 1$.

$$T(s) = \frac{1/R_1R_2C_1C_2}{s^2 + (1/R_1C_1 + 1/R_2C_1)s + 1/R_1R_2C_1C_2}$$

The general equation for a second-order transfer function is given by the following equation.

$$T(s) = \frac{\omega_n^2}{s^2 + (\omega_n/Q)s + \omega_n^2}$$

Therefore the two functions can be equated, and we have:

$$\omega_n^2 = \frac{1}{R_1R_2C_1C_2}$$

$$\frac{\omega_n}{Q} = \frac{1}{R_1C_1} + \frac{1}{R_2C_1}$$

If both resistors are equal to one,

$$\omega_n^2 = \frac{1}{C_1C_2} \quad \text{and} \quad \frac{\omega_n}{Q} = \frac{1}{C_1} + \frac{1}{C_1} = \frac{2}{C_1}$$

$$\text{Therefore } C_1 = \frac{2Q}{\omega_n} = \frac{\sqrt{\sigma^2 + \omega^2}}{\sigma\sqrt{\sigma^2 + \omega^2}} = \frac{1}{\sigma}$$

and replacing C_1 in the equation for ω_n^2 we get $\omega_n^2 = \frac{\sigma}{C_2}$ or $C_2 = \frac{\sigma}{\omega_n^2}$

but $\omega_n^2 = \omega^2 + \sigma^2$, so $C_2 = \frac{\sigma}{\sigma^2 + \omega^2}$

These equations were simplified by letting the two resistors have equal values. If you try making the capacitor values equal instead, the equations are harder to simplify. Finding resistor values to meet the specification is more difficult. In fact, the resistor values relate to the pole locations by parallel and series combination. I will not give the details here, but try it for yourself if you want to.

Scaling Pole and Zero Locations

Important factors that are related to the pole locations are ω_n and Q. The origin to pole distance is equal to ω_n. The Q is given by the distance from the pole to the origin, divided by twice the real coordinate.

Pole to origin is $\sqrt{\omega^2 + \sigma^2} = \omega_n$. Real part of pole coordinate $= \sigma$

$$Q = \frac{\sqrt{\omega^2 + \sigma^2}}{2\sigma}$$

Another way of expressing these is:

$$\omega_n = 2Q\sigma \quad \text{and} \quad 2Q = \sqrt{\left(\frac{\omega}{\sigma}\right)^2 + 1}.$$

Notice that these equations show that Q depends on the ratio of ω/σ so, as the pole-zero diagram is scaled for a higher cutoff frequency, the value of Q remains unchanged. The natural frequency ω_n is dependent upon σ, and this changes in proportion to the scaling of the diagram. Zero locations are scaled in a similar way, moving away from the origin ($s = 0$) and along the imaginary axis.

Digital Filter Equations

Finding FIR Filter Zero Coefficient Using L'Hopital's Rule

The maximum coefficient value is at $n = 0$, but this cannot be calculated because we would be dividing by zero. The value $h[0]$ is calculated by differentiating the numerator and denominator separately, and then letting $n = 0$. This is known as L'Hopital's rule, named after a French mathematician.

Let us look at a sinc function where the first zero coefficient is at $n = 5$. The sampled $\mathrm{sinc}(x)$ function has values given by $h[n]$.

$$h[n] = \frac{1}{n\pi}.\sin\left(\frac{n\pi}{5}\right)$$

The value of $h[0]$ for this equation can be found using L'Hopital's rule:

$$h[0] = \frac{1}{\dfrac{d}{dn}\{n\pi\}}.\frac{d}{dn}\left\{\sin\left(\frac{n\pi}{5}\right)\right\}$$

$$h[0] = \frac{1}{\pi}.\frac{\pi}{5}\cos\left(\frac{n\pi}{5}\right) = \frac{1}{5} = 0.2$$

The next value, $h[1]$, is simply $h[1] = \frac{1}{\pi}.\cos\left(\frac{\pi}{5}\right) = 0.1871$. Likewise, by substituting values for n, other values are calculated.

Appendix References

1. Carlson, A. B. *Communication Systems*, Singapore, McGraw-Hill, 1986.

2. Huelsman, L. P. *Active and Passive Analog Filter Design*, New York, McGraw-Hill, 1993.

BIBLIOGRAPHY

I have used the following books and other references over the past few years. Readers may find the list useful in building up their own knowledge of filters.

Abrie, Pieter L. D. *The Design of Impedance Matching Networks*. Norwood, MA: Artech House, 1985.

Adel S. Sedra and Peter Brackett. *Filter Theory and Design: Active and Passive*. New York and London: Pitman, 1979.

Antoniou, Andreas. *Digital Filters: Analysis, Design, and Applications*. New York: McGraw-Hill, 1993.

Arthur B. Williams and Fred Taylor. *Electronic Filter Designer's Handbook*, 2nd Ed.ew York: McGraw-Hill, 1988.

Carlson, A. B. *Communication Systems*. New York: McGraw-Hill, 1986.

Chen, W. K. *Passive and Active Filters: Theory and Implementations*. New York: John Wiley & Sons, 1986.

Chen, W. K. *Broadband Matching Theory and Implementations*. New Jersey and London: World Scientific, 1988.

Chen, W. K. *The Circuits and Filters Handbook*. Boca Raton, FL: CRC Press/IEEE, 1995.

Cuthbert, Thomas R. *Circuit Design Using Personal Computers*. New York: John Wiley & Sons, 1983.

Daryanani, Gobind. *Principles of Active Network Synthesis and Design*. New York: John Wiley & Sons, 1976.

Ellis, Michael G. *Electronic Filter Analysis and Synthesis*. Norwood, MA: Artech House, 1994.

G. C. Temes and J. W. LaPatra. *Circuit Synthesis and Design*. New York: McGraw-Hill, 1977.

Guillemin, *Synthesis of Passive Networks*. New York: John Wiley & Sons, 1957.

Helszajn, Joseph. *Synthesis of Lumped Element, Distributed and Planar Filters*. New York: McGraw-Hill, 1990.

Herman J. Blinchikoff and Anatol I. Zverev. *Filtering in the Time and Frequency Domains*. ew York: John Wiley & Sons, 1976.

Huelsman, L. P. *Active and Passive Analog Filter Design*. New York: McGraw-Hill, 1993.

Humphreys, Ernst A. *The Analysis, Design, and Synthesis of Electrical Filters*. New Jersey: Prentice Hall, 1977.

Maddock, R. J. *Poles and Zeros in Electrical and Control Engineering*. Austin, TX: Holt, Rinehart, and Winston, 1982.

Middlehurst, Wiley. *Practical Filter Design*. New Jersey: Prentice Hall, 1993.

Niewiadomski, Stefen. *Filter Handbook: A Practical Design Guide*. Boca Raton, FL: CRC Press, 1989.

R. Schaumann, M. S. Ghausi, and K. R. Laker. *Design of Analog Filters*. New Jersey: Prentice Hall, 1990.

Rhodes, J. D. *Theory of Electrical Filters*. New York: John Wiley & Sons, 1976.

Rorabaugh, C. Britton. *Digital Filter Designer's Handbook*. New York: Tab Books/McGraw-Hill, 1993.

Sanjit K. Mitra and Carl F. Kurth. *Miniaturized and Integrated Filters*. New York: John Wiley & Sons, 1989.

Stephenson, R. C. *Active Filter Design Handbook*. New York: John Wiley & Sons, 1985.

Su, K. L. *Analog Filters*. London: Chapman and Hall, 1996.

Terrell, Trevor J. *Introduction to Digital Filters*. New York: MacMillan Press, 1980.

Tomlinson, G. H. *Electrical Networks and Filters: Theory and Design*. New York: Prentice Hall, 1991.

Van Valkenburg, Mac E. *Analog Filter Design*. London: Holt Saunders, 1982.

Vlach, Jiri. *Computerized Approximation and Synthesis of Networks*. New York: John Wiley & Sons, 1969.

Zverev, A. I. *Handbook of Filter Synthesis*. New York: John Wiley & Sons, 1967.

ANSWERS

Chapter 1

1.1 The ratio of output power to input power is $0.3/6.0 = 0.05$. The "gain" is $10.\log(0.05) = -13\,dB$, therefore the attenuation, or signal loss, is $+13\,dB$. Relative to the volt, an input voltage of $2\,V = 20.\log(2) = +6\,dBV$. With the attenuation being $13\,dB$, the output voltage (in dBs) will be $6 - 13 = -7\,dBV$. The actual voltage is $10^{(-7/20)} = 10^{(-0.35)} = 0.4467\,V$.

1.2 $24\,dB$. The filter gives a $12\,dB$ per octave attenuation rate; $2\,MHz$ is an octave above $1\,MHz$ and $4\,MHz$ is an octave above $2\,MHz$. Two octaves $\times$ $12\,dB = 24\,dB$.

1.3 A $10\,mW$ input signal has a level (in dBs) of $10\,dBm$. At $2\,MHz$ the attenuation is $12\,dB$, so the output level is $-2\,dBm$. This is $10^{(-2/10)} = 0.63\,mW$. At $4\,MHz$ the attenuation is $24\,dB$, so the output level is $-14\,dBm$. This is $10^{(-14/10)} = 0.04\,mW$.

1.4 At the $-3\,dB$ point the voltage across the output will be 0.7071 ($1/\text{root}$ of 2) times V_{in}. Therefore the voltage across the capacitor will be $7.071\,V$. The current through the capacitor also flows through the resistor, and since they have equal impedance at the $-3\,dB$ point, the voltage across the resistor is also $7.071\,V$. The peak in resistor voltage is $90°$ ahead of the peak in capacitor voltage. (a) $7.071\,V$, (b) $7.071\,V$.

Chapter 2

2.1 Lowpass and bandstop.

2.2 The passband describes a range of frequencies that allow signals to pass with little or no attenuation. The stopband describes a range of frequencies that attenuate signals by at least the design specification limit. The skirt is the range of frequencies between the passband and the stopband, where attenuation will be more than $3\,dB$ but less than the stopband level.

2.3   Chebyshev and Cauer (elliptic).

2.4   Inverse Chebyshev and Cauer (elliptic).

2.5   Cauer filters have ripple in both passband and stopband. They are used because they have a very steep skirt (almost a "brick wall" response).

2.6   Bessel filters have a constant delay in the passband. Unfortunately, they have a very shallow skirt response.

2.7   Component values are normalized so that one set of data (usually written in a table) can be applied to any cutoff frequency or load impedance by simply scaling the values.

Chapter 3

3.1   An output step followed by a smooth exponential decay.

3.2 $-0.3 - j0.67$.

3.3   Imaginary axis.

3.4   A null in the stopband; otherwise known as stopband ripple. The presence of two zeroes implies a Cauer (elliptic) or Inverse Chebyshev response.

3.5   Butterworth poles are located on the unit circle. Each pole is equidistant from the origin, and they have equal angular distance between each other.

3.6   Chebyshev poles are located on an ellipse. The Butterworth filter pole locations are shifted towards the imaginary axis (to the right) and away from the real axis (up or down). The amount of pole movement is mathematically derived.

Chapter 4

4.1   An inductor value has to be increased in proportion to the load value, so to denormalize for impedance we get $0.8212H \times 50 = 41.06H$. To scale for cutoff frequency of 20 kHz, remember that we want the inductor to have an impedance equivalent to a $41.06H$ inductor at 1 radian per second (which is 41.06 ohms at 1 rad/s). This means that we have to divide by $2\pi F$ rad/s, where F is the cutoff frequency. In this case $2\pi F = 125,664$ rad/s. In summary, $L = 0.8212 \times R/2\pi F$. This gives a denormalized value of $327\mu H$, which has an impedance of 41.06 ohms at 20 kHz.

4.2   The impedance of a capacitor is inversely proportional to its value. Thus for a 600-ohm load, denormalization requires the value to be reduced: $0.5532/600 = 922\mu F$. To scale for a frequency of 100 kHz, we want to find a capacitor value that has the impedance of the $922\mu F$ capacitor at 1 rad/s, which is 1 rad/s/$922\mu F = 1085\,\Omega$. The impedance of a capacitor reduces with frequency, and, in order to maintain the $1085\,\Omega$ impedance as the cutoff frequency is increased, the capacitance value must be reduced in proportion. The denormalized capacitor is thus $0.5532/(2\pi FR) = 922\mu F/628,318 = 1467pF$ or $1.467nF$.

4.3   They move away from the origin along a line that passes through the original pole position.

4.4 $C1 = 1/\sigma = 1/0.7071 = 1.4142F$. $C2 = \sigma/(\sigma^2 + \omega^2) = 0.7071/(0.5 + 0.5) = 0.7071F$.

4.5   To denormalize $C1$ and $C2$, divide the values found in Exercise 4.4 by impedance and frequency. $C1 = 1.4142/2\pi FR = 1.4142/62,831,853 = 22.508nF$. $C2 = 0.7071/2\pi FR = 0.7071/62,831,853 = 11.254nF$.

Chapter 5

5.1   An inductor value has to be increased in proportion to the load value, so to denormalize for impedance we get $0.6834H \times 100 = 68.34H$. To scale for cutoff frequency of 12 kHz we have to divide by $2\pi F$ rad/s, where F is the cutoff frequency. In this case $2\pi F = 75,398$ rad/s. $L = 0.6834 \times R/2\pi F$. This gives a denormalized value of $906\mu H$.

5.2   The impedance of a capacitor is inversely proportional to its value. For a 75-ohm load, denormalization requires the value to be reduced: $0.7490/75 = 9.9867mF$. To scale for a frequency of 10 kHz, the capacitance value must be reduced in proportion to frequency. The denormalized capacitor is $0.7490/(2\pi FR) = 9.9867mF/62,832 = 158.94nF$.

5.3 $R1 = \sigma_{(LP)} = 0.6205\,\Omega$. $R2 = (\sigma^2_{(LP)} + \omega^2_{(LP)})/\sigma_{(LP)} = (0.6205^2 + 0.9075^2)/0.6205 = 1.2085/0.6205 = 1.9477\,\Omega$.

5.4   If $C1 = C2 = 1nF$ and cutoff frequency $= 15$ kHz, $R1 = 0.6205/2\pi FC = 0.6205/94.2478 \times 10^{-6} = 6.584k\Omega$. $R2 = 1.9477/2\pi FC = 20.666k\Omega$.

Chapter 6

6.1   Denormalize the lowpass design to have a cutoff frequency equal to the required bandwidth. Resonate each series arm with a series connected capacitor. Resonate each shunt capacitor arm with a parallel inductor. For both series and parallel tuned circuits, the resonant frequency is equal to the bandpass center frequency. Finally, denormalize for load impedance.

6.2 Each series arm will have two parallel LC circuits connected in series. One LC circuit has high impedance above the passband, and the other has high impedance below the passband. Each series arm thus gives two notches in the frequency response: one above and one below the filter's passband.

6.3 $R3 = 10/(\pi\, 80\,.\,10^3\,.\,220\,.\,10^{-12}) = 180.86\,k\Omega$
$R1 = R3/40 = 4.521\,k\Omega$
$R2 = R3/(400 - 40) = R3/360 = 502\,\Omega$

6.4 $R = R3 = R4 = 1/(2\,.\,\pi\, 35\,.\,10^3\; 4.7\,.\,10^{-9}) = 967.5\,\Omega$
$R1' = QR = 50 \times 967.5 = 48.375\,k\Omega$
$R1 = 2 \times 48.275\,k/1.5 = 64.5\,k\Omega$
$R2 = 2 \times 48.275\,k/0.5 = 193.5\,k\Omega$

Chapter 7

7.1 Start with lowpass prototype, and convert to highpass prototype using reciprocal values (i.e., a lowpass prototype inductor with a value of 2.0 becomes a highpass prototype capacitor with a value of 0.5). Frequency scale the highpass prototype to have a cutoff frequency equal to the bandstop filter's stopband. The highpass design must now be translated into a bandstop design by resonating each component at the stopband center frequency. Series capacitors require a parallel connected resonant inductor. Shunt inductors should have a series-connected resonant capacitor. Finally, scale the components for the correct load impedance.

7.2 Lowpass prototype $C1 = C3 = 1.0$ and $L2 = 2.0$.

$$C(\text{shunt}) = \frac{(F_U - F_L).1}{2\pi F_U F_L R}$$

$Fu = 1.05\,MHz$ and $F_L = 0.95\,MHz$

$$C(\text{shunt}) = \frac{100k}{2\pi 9.975 10^{11} 50} = 319\,pF$$

$$L(\text{shunt}) = \frac{50}{2\pi 100k} = 79.58\,\mu H$$

$$C(\text{series}) = \frac{1}{2\pi 100k 50\ 2} = 15.9\,nF$$

$$L(\text{series}) = \frac{100k 50\ 2}{2\pi 9.975 10^{11}} = 1.595\,\mu H$$

7.3 $\quad R1 = R4 = \dfrac{40}{2\pi 50k\,47010^{-12}} = 270.902\ k\Omega$

$\quad\quad R2 = R3 = \dfrac{270.9k}{40} = 6.772\ k\Omega$

$\quad\quad R5 = \dfrac{2.510^{13}}{40(2.510^{9} - 2.40110^{9})} = 6.313\ k\Omega$

$\quad\quad R6 = \dfrac{2.510^{9}.10^{4}}{2.40110^{9}} = 10.412\ k\Omega$

Chapter 8

8.1   Delta, 6 dB.

8.2   3 dB, because half the power goes into each load (−3 dB = half power). The splitter circuit absorbs no power when the impedance of each load is equal.

8.3   Because the passband of one filter section coincides with the stopband of the other. In the passband, a filter presents the source with its load impedance. The other filter section is connected in parallel with this and must therefore present high impedance to avoid impedance mismatch of the source.

8.4   At the −3 dB point only half of the available power from the source enters the filter. When both lowpass and highpass filters are connected in parallel, half the power enters the lowpass filter and half enters the highpass filter. All the power is thus absorbed and no reflections occur—the source is matched to the load. At frequencies below the −3 dB point, the lowpass filter absorbs a greater proportion of power and the highpass filter absorbs less. Similarly, at frequencies above the −3 dB point, the highpass filter absorbs more power and the lowpass filter absorbs less. Thus the source power is absorbed at all frequencies.

Chapter 9

9.1   It is the unequal delay of some frequencies relative to others.

9.2   Square wave signals contain a fundamental frequency and all its odd harmonics. Group delay causes some harmonic signals to be delayed relative to the fundamental, so the waveform is distorted. The rise time of the wave is slowed and the peak level contains ripple.

Chapter 10

10.1 Because surface-mount components have no wire leads, any series inductance is minimal, and therefore the self-resonant frequency is high.

10.2 The ferrite core concentrates the magnetic flux to create a high inductance. The pot-core design ensures that the flux is contained within the center of the component, which allows designers to place other inductors adjacent on the circuit board with little likelihood of altering the inductance or unwanted coupling. Air-gaps are not easily saturated, thus allowing the inductor to carry DC or high power levels without affecting the inductance.

10.3 Surface-mount resistors are ideal for radio frequency work because of their low inductance. Carbon composition resistors are also used because of their low inductance, although they can be noisy. Wire-wound resistors have high inductance and are avoided in high-frequency circuits.

10.4 Using

$$Fc = \frac{Fu}{N^2}, \quad Fc\, N^2 = Fu$$

Where $Fc = 20\,\text{kHz}$ and $N = 6$

$Fu = 20\,\text{kHz} \times 36 = 720\,\text{kHz}$.

Op-amps with a gain-bandwidth of 720 kHz or more are easily available.

Chapter 12

12.1 $L > \lambda/20$.

12.2 $\dfrac{w}{h} = \dfrac{8}{e^A - 2e^{-A}}$

$$A = \frac{50\sqrt{11.4}}{119.9} + \frac{3.7}{11.4}\left(0.46 + \frac{0.22}{4.7}\right)$$

$A = 1.408 + 0.16449 = 1.57249$

$$\frac{w}{h} = \frac{8}{4.818641 - 0.415055} = 1.8167$$

$h = 1.6\,\text{mm}$, hence $w = 1.8167 \times 1.6\,\text{mm} = 2.9\,\text{mm}$.

Chapter 15

15.1   Delay, sum and multiply.

15.2   An analog filter has a response defined in the frequency domain. A digital filter has a response defined in the time domain, which is an Inverse Fourier transform of the required frequency response.

15.3   An IIR filter has a feedback path from its output back to its input. This type of filter requires fewer components to produce a particular frequency response, compared to a FIR filter, but it produces a nonlinear phase response, which can be a disadvantage.

INDEX